Conditioning, Spawning and Rearing of Fish. Emphasis on Marine Clownfish

Author

Frank H. Hoff Jr.

Edited by

Martin Moe
Joe Lichtenbert
Don Johnson
Jeff Neslen

Published By

Florida Aqua Farms Inc.

33418 Old Saint Joe Rd.
Dade City, FL 33525

Third Printing

ISBN: 978-0-9662960-1-3

1st Printing 1996
2nd Printing 2001
3rd Printing 2009

FORWARD

There is AQUACULTURE, which is the culture of commercially valuable, edible invertebrates and finfish in freshwater or saltwater environments -- and there is the culture of ornamental or "aquarium" invertebrates and finfish, also in freshwater and saltwater environments. One may think that there is little difference between AQUACULTURE and the culture of ornamental fish, and technically, there is not a great deal of difference. Technically, the differences are primarily the scale of operations and the size of the cultured organisms at the point of sale. But perceptually, in the public eye, there is a great difference. AQUACULTURE is an agricultural industry, a major contributor to the economy, a developing source of protein to supplement overexploited natural fish populations, a research subject of private, governmental and university studies, a field with great economic potential for third world and industrialized countries, and one worthy of significant research grants. On the other hand, the culture of ornamental fish is seen as a hobby, a frivolous pursuit, a pastime that plucks the flowers from oceans, reefs, lakes, ponds and streams and encourages the incarceration of pretty little fish born to live free. It is only a small, dark corner somewhere in the giant recesses of the local super store.

This is a very inaccurate perception of the aquarium industry, of course. The hobby of keeping aquatic organisms in home aquariums feeds the mind and spirit of people all over the world. It encourages study and respect of natural environments, and it is far more worthwhile and educational to maintain an aquarium than to watch sitcoms on TV or see the good guys destroy the bad guys in a Hollywood movie. Culture of ornamental aquarium fish and invertebrates is big business but, not as big as even bad films. Yet, according to Pet Dealer Magazine, just in the United States alone, recent annual sales of only marine aquarium livestock are about 123 million dollars and the sale of freshwater livestock is many times that figure. Most of the freshwater fish in the aquarium trade are carefully cultured in ponds and tanks. All but a very few tropical marine fish, however, are taken from coral reefs all over the world. It is not easy to culture marine tropical fish. They have very small eggs, planktonic larvae that require microscopic live foods and nutritional and environmental requirements that are keyed to the vast planktonic pastures of the open ocean. If it were easy to grow these fish, technically and economically, there would be farms and hatcheries all over the world, as there are for freshwater fish.

Frank Hoff and I believe that successful, economically sustainable culture of marine tropical fish will eventually happen. We have spent most of our professional lives working toward this end, sometimes competitively and sometimes in concert. We know the bitterness of failure and we also know the sweetness of success. We know it can be done and that it is so close that one can smell it in the air. Just a few more technical breakthroughs, a few more well sited and managed operations, more diversification of products and higher prices for tank raised fish, will initiate the dawn of a new era. Dreamers and pioneers may retire but they never say uncle.

Frank has distilled his considerable expertise over the last 30 years of marine fish culture into the pages of this book. Much of his lifeblood is recorded here. He has done this no small task not only to record the work he and his cohorts have accomplished, but also to help bring the culture of marine ornamental fish into the fold of AQUACULTURE. Ornamental fish culture is now gaining respect and significance in aquaculture science and Frank's manual will do much to narrow the gaps between scientist, fish farmer, entrepreneur, and hobbyist. Anyone even remotely interested in the culture of tropical fish, whether fresh or salt water species, will find a wealth of information within these pages.

Martin A. Moe, Jr.
1996

Table Of Contents

BACKGROUND, BIOLOGICAL & SYSTEM INFORMATION **1-51**

Introduction **1-6**

USA Roots of Marine Tropical Fish Culture 2

Who And What Was Instant Ocean Hatcheries 3

Reason For This Manual 6

Taxonomy and Distribution **7-8**

Taxonomy 7

Distribution 7

Common And Scientific Names 8

General Biology and Physiology **9-16**

External Anatomy 9

Internal Anatomy 12

Clownfish Biology 12

Pairing 14

Spawning And Hatching 14

Sex Changes 15

Cross Breeding 15

Feeding Habits 16

Gonadal Cycling and Egg Development .... **17 -23**

Oocyte Sampling 17

Anesthetizing Fish 18

Gonadal Cycling 19

Egg Development 20

Species Selection **24-27**

Criteria for Species Selection 24

Sourcing Literature And Unpublished Data 26

System Design **28-51**

System Verses Singular Tanks 28

Types Of Systems 28

System Arrangement At IOH 30

Key Points For Hatchery And System Designs 33

Estimating Rate of Water Exchange 34

Particle Accumulation Verses Filtration 36

Mechanical Filters 37

Foam Skimmers 38

Principles of Biological Filtration 39

Types of Biological Filters 39

Artificial Biological Media 40

Wet/Dry filters 41

Upwellers or Fluidized Beds 44

Reservoirs 45

Ultraviolet Sterilization 46

Ozone 47

Aeration 49

Carbon Filtration 51

BROODSTOCK CARE **52-106**

Broodstock Housing **52-55**

Tank Selection 52

Tank Filtration 54

Spawning Habitat 54

Pairing Broodstock **56-60**

Sexual Maturity 56

Sex Reversal And Pairing 56

Pairing By Coloration 57

Pairing By Size 58

Pairing By Behavior 59

Determining If You Have A Real Pair 60

Conditioning For Spawning **61-71**

Inadvertent Conditioning 61

Natural Conditioning 61

Controlled Conditioning 61

Clownfish Conditioning 65

Conditioning Temperature 66

Conditioning Light 66

Water Quality Conditions 67

Conditioning Habitat 67

Habitat For Clownfish 68

Conditioning Diets 69

Gelatin Diets 70

Procedure For Making Gelatin Diets 70

Live Feeds 71

Broodstock Maintenance **72-74**

Disruption 72

Feeding Schedule 72

Tank Maintenance 72

Broodstock Technician Checklist 73

Nest Consumption 74

Record And Data Keeping **75-81**

Historical Records 75

Broodstock Records 75

Broodstock Tank Identification Tags 77

Pair Record Form 78

Pair Inventory Form 79

Monthly Pair Summary Form 80

Spawning And Culling of Broodstock **82-92**

Maturation time 82

Spawning activity 82

Nest Care 83

Nest location 83

Nest Coloration, Fertilization, and Hatch 84

Nest Changes Prior To Hatch 87

Egg Development 87

Egg (oocyte) Development 88

Egg and Nest Size 89

Nest Configuration 89

Spawning Duration ... 89
Spawning Versus Season ... 92
Criteria For Culling Pairs ... 92
Hatching Procedures ... 93-100
Hatching Time ... 93
Theory Of Hatching Nests ... 93
Hatching In The Broodstock Tank ... 94
Hatching Nests Remotely ... 95
Hatching Method Used At IOH ... 96
Broodstock Diseases ... 100-106
Cause of Diseases ... 100
Detection of Pending Diseases ... 100
Common Seasonal Diseases ... 101
Less Common Diseases ... 102
"Popeye" And Mechanical Injuries ... 103
Gas Bubble Disease ... 104
Constipation ... 104
Blindness ... 104
Disease Prevention ... 105
Egg Quality Versus Treatments ... 106

LARVAL REARING & CARE ... 107-142
Larval Rearing ... 107-109
Introduction ... 107
Adequate Environments ... 107
Foods Versus Larvae ... 108
Maintenance Procedures ... 108
Record Keeping ... 109
Larval Rearing Tank ... 110-118
Selecting Larval Rearing Tanks ... 110
Tank Coloration ... 111
Lighting ... 112
Commercial Tank Configuration ... 113
Other Commercial Tank Considerations ... 115
Hobby Level Tank Configuration ... 116
Larval Foods And Feeding ... 119-131
Larvae Prior to Metamorphosis ... 119
Metamorphosis ... 120
Pigmentation Development in *A. ocellaris* ... 121
Larval Development In Relation To Food ... 122
Feeding Strategy And Regime ... 123
Integrated Diets ... 126
Problems In Feeding Live Foods ... 127
Food Quality Problems ... 128
Food Size ... 129
Feeding Live Foods ... 130
Larval Care And Maintenance ... 132-138
Water Exchanges And Flow Rates ... 132
Daily Larval Maintenance ... 132
Disease, Deformities, and Treatments ... 133

Use Of Iodine ... 135
Water Quality ... 137
Transferring Post Larvae ... 138
Larval Records And Data ... 139-142
Daily Larval Log ... 139
Monthly Post Larval Report ... 140
Analysis of Yearly Post-larval Production ... 140
Post-larval Survival on a Seasonal Basis ... 141
Typical Daily Mortality of Larvae ... 141
Percentage Nest Size, Hatch and Survival of 30 Day Larvae ... 141
JUVENILE REARING & CARE ... 143-182
Juvenile Rearing ... 143-144
Introduction ... 143
Juvenile Rearing Facilities ... 145-147
Growout Tanks ... 145
Filtration at IOH ... 146
Current Filtration Suggestions ... 146
Juvenile Rearing Procedure ... 148-153
Criteria For Post-Larval Transfer ... 148
Transfer Method ... 148
Stocking Density ... 149
Tank Identification ... 150
Feeding ... 150
Maintenance ... 151
Criteria For Sale And Culling ... 152
Nutrition And Diseases ... 154-168
Pathway Detection ... 154
Foods ... 156
Feeding Strategy ... 156
Pigmentation ... 158
Disease And Health Status ... 162
Nutrition And Internal Disorders ... 163
Food Storage ... 164
Vitamin Deficiencies ... 165
Essential Fatty Acids ... 166
Common Juvenile Diseases ... 169-178
Treatment Agents ... 169
Calculating Treatments ... 169
Oodinium ocellatum ... 172
Treatment for *Oodinium* ... 173
Cryptocaryon irratans ... 174
Treatment For *Cryptocaryon* ... 175
Bacterial Infections ... 175
Treatment For Bacteria ... 175
Fish Tuberculosis (*Mycobacterium*) ... 176
Treatment For *Mycobacterium* ... 176
General Alternative Treatments ... 177
Probiotics ... 178
Water Quality Stress ... 179-181
Shipping ... 182

FUTURE FISH CULTURE ENDEAVORS **183-199**

Future of Marine Ornamental Fish Culture **183-198**

Pet Sales 183

Past And Current Endeavors 184

Survival And Sales At IOH 189

Final Analysis of Operation Procedures At IOH 190

Future Considerations 192

Devotion 193

Location 193

Pilot Scale Farms 194

Facility Design 195

Budget 198

REFERENCES **199-207**

CONVERSION TABLES **208-210**

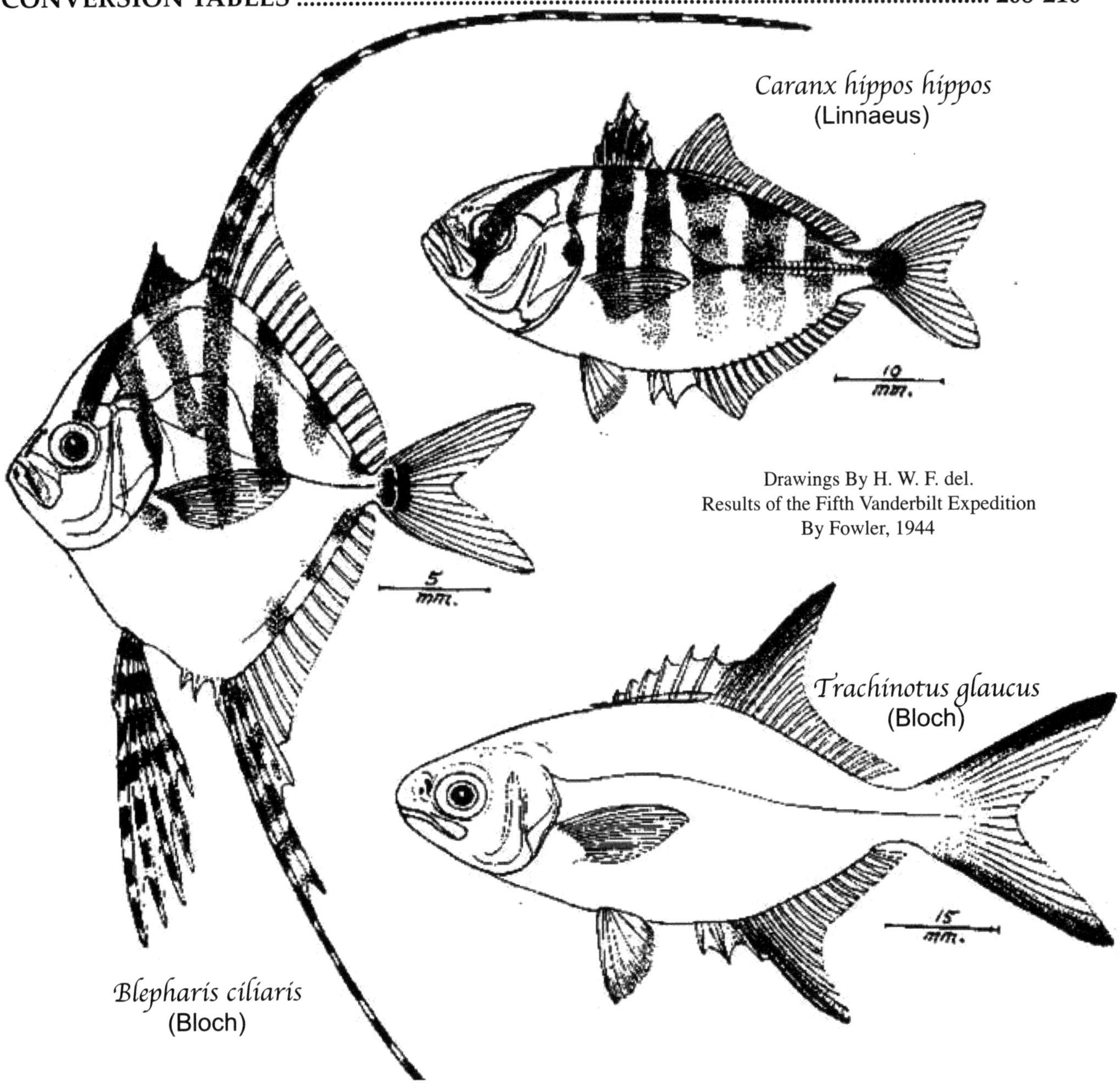

Drawings By H. W. F. del.
Results of the Fifth Vanderbilt Expedition
By Fowler, 1944

List of Figures

1 - Types of scales found on modern and primitive fish ..... Page 9
2 - Gill rakers in relation to feeding habits ..... Page 10
3 - Mouth location in relation of feeding habits ..... Page 10
4 - Representative digestive tracks from fish ..... Page 10
5 - External body parts and features ..... Page 11
6 - Internal body parts and features ..... Page 11
7 - Hormonal stimulation pathway controlled by environmental stimuli, etc. ..... Page 17
8 - Oocyte sampling device, Hoff 1972 ..... Page 17
9 - Egg sample preparation for microscopic viewing ..... Page 18
10 - Wet table for measuring and in-vivo oocyte sampling of large fish ..... Page 18
11 - Egg samples are obtained by insertion of sampling device into the genital, etc. ..... Page 19
12 - Gonadal conditions and states associated with gonadal cycles, etc. ..... Page 20
13 - Oocyte growth rates per hour from an intermediate state to a spawning state, etc. ..... Page 22
14 - Influence of oocyte diameter at initial injection on percent of fertilized eggs ..... Page 22
15 - Alternating reservoir open-water exchange system ..... Page 29
16 - Open water system utilizing well points and conditioning reservoirs ..... Page 29
17 - Arrangement of adult and juvenile tanks at IOH ..... Page 31
18 - Recirculating system suitable for marine and freshwater tropical fish culture ..... Page 32
19 - Typical components used in recirculating production systems etc. ..... Page 33
20 - Typical sand filter ..... Page 37
21 - Typical DE (diatomaceous earth) filter ..... Page 37
22 - Counterflow foam skimmer using venturies ..... Page 38
23 - Common acrylic wet/dry filter frequently used in the aquarium industry ..... Page 41
24 - Wet/dry degaser biofilter made of stackable styrofoam shipping boxes etc. ..... Page 42
25 - Large recirculating system using two pumps and large trickle filter, etc. ..... Page 43
26 - Typical design of a fluidized bed filter ..... Page 44
27 - Downflow bubble contactor after Colt & Watten, 1988 ..... Page 49
28 - U-tube oxygen concentrator ..... Page 49
29 - Carbon filter using a standard five-gallon bucket ..... Page 50
30 - Construction of four tanks from two sheets of marine grade 3/4" thick plywood. ..... Page 51
31 - External overflow arrangements which allow less inside obstructions, etc. ..... Page 52
32 - Unique, versatile design of all glass tanks used at C-Quest, etc. ..... Page 53
33 - Broodstock tank arrangement for benthic or substrate spawning fish ..... Page 54
34 - Pelagic egg collector mounted on the output line from the broodstock tank ..... Page 54
35 - Hanging acclimation basket ..... Page 58
36 - Out of phase light and temperature step conditioning regime, etc. ..... Page 61
37 - Constant photo period and temperature regime used on T. carolinus etc. ..... Page 62
38 - Temperature fluctuations to induce spawning in redfish ..... Page 63
39 - Adult identification tag system used at IOH etc. ..... Page 77
40 - External indications of oocyte development and spawning in clownfish ..... Page 81
41 - Egg size, *A. ocellaris*, Berschy 1979a ..... Page 86
42 - Side overflow collecting device (Lichtenbert 1992) ..... Page 94
43 - Remote flow through hatchers used at IOH ..... Page 94
44 - Setting up for hatch using aeration as a means of creating a current ..... Page 95
45 - Air lift water hatcher pumps used at IOH for providing circulation to nests ..... Page 96
46 - Jointed stand pipe used in larval rearing tanks for quick uniform water, etc. ..... Page 97
47 - Internal powered hatcher suspended inside the rearing tank ..... Page 98
48 - Larval rearing recirculating system used to culture clownfish at IOH ..... Page 113
49 - Details of an all glass larval rearing tank used at Instant Ocean Hatcheries ..... Page 113
50 - Optional larval rearing containers used in commercial facilities ..... Page 115
51 - Floating larval rearing tank designed by Florida Aqua Farms Inc. ..... Page 116
52 - Small self contained larval rearing system developed by Allen Bertschy. ..... Page 116
53 - Chronology of metamorphic pigmentation stages for *A. ocellaris,* etc. ..... Page 120
54 - Population curves based on good feeding and poor feeding procedures ..... Page 122

Figures Continued

55 - Feeding regime of several marine fish and shrimp utilizing live and dry feeds .... Page 124
56 - Method for transferring postlarvae into juvenile growout tanks .... Page 147
57 - Tank information tag used on juvenile growout tanks at IOH .... Page 148
58 - Comparison of ammonia and nitrite levels in eight juvenile systems etc. .... Page 149
59 - Nutritional disorders and disease pathway considerations in adults etc. .... Page 153
60 - Ideal feeding balance we want to achieve .... Page 154
59 - Nutritional disorders and disease pathway considerations in adults etc. .... Page 153
60 - Ideal feeding balance we want to achieve .... Page 154
61- Contributing external and internal stress, nutritional and pathogenic, etc. .... Page 178
62 - Sloping floors and drains .... Page 193
63 - Tropical fish hatchery designed based on C-Quest in Puerto Rico .... Page 193
64 - Detail "A" and "B" of fish hatchery design based on C-Quest, etc. .... Page 194
65 - Detail "C" culture room facilities based on C-Quest in Puerto Rico .... Page 194

List of Tables

1..... General distribution of clownfish species based in part from Allen 1972 .... Page 08
2..... Criteria for species selection for commercial culture .... Page 24
3..... Criteria for species selection of marine tropical fish based on previous etc. .... Page 25
4..... Approximate carrying capacity utilizing a fluidized column .... Page 45
5..... Killing dose of UV in microwatt/sec/cm squared .... Page 46
6..... Reported toxic ozone levels on fish and humans .... Page 47
7..... Oxygen transfer using atmospheric air stones with medium air bubbles .... Page 50
8..... Distinguishing characteristics for pairing clownfish .... Page 57
9..... Ingredients for conditioning and growout diet for clownfish .... Page 70
10... Broodstock technician duties .... Page 73
11... Historical records data sheet .... Page 76
12... Example of pair record form .... Page 79
13... Example of pair inventory record form .... Page 80
14... Example of the monthly pair summary form .... Page 81
15... Effect of pigmentation with regards to hatch and survival of post larvae .... Page 85
16... Color changes in developing nest prior to hatch .... Page 86
17... High, low and average nest size of captive clownfish. Based on 2 years data .... Page 87
18... Spawning records of clownfish pairs at IOH .... Page 89
19... Comparison of gross egg production at IOH over a three year period .... Page 90
20... Seasonal comparison of nests produced (NP) verses nest that were lost .... Page 91
21... Comparison between spawning pairs of clownfish verses non spawning pairs .... Page 91
22... Criteria for culling adult pairs .... Page 92
23... Hatch based on 26°C to 28°C .... Page 93
24... Initial indicators of possible disease or stress .... Page 101
25... Larval rearing aquaria used by various investigators .... Page 109
26... Tank coloration verses survival and size .... Page 111
27... Daily need of microalgae and rotifers to raise 10,000 to 250,000 larvae .... Page 121
28... Clownfish larval feeding regime modified slightly from original IOH data .... Page 124
29... Daily larval feeding schedule .... Page 124
30... Feeding regime of several marine fish and shrimp utilizing live & dry feeds .... Page 125
31... Water exchange in larval rearing tanks .... Page 131
32... Effect of high and low nitrate on A. ocellaris larva .... Page 135
33... Water quality in the larval rearing room at IOH .... Page 136
34... Larval rearing daily data sheet .... Page 138
35... Example of the monthly post larval report used at IOH .... Page 139
35... Example of the monthly post larval report used at IOH .... Page 139
36... Postlarval production at IOH over four years .... Page 139
37... Average post larval survival on a seasonal basis .... Page 140

Tables Continued

38... Typical daily larval mortality pattern based on the average of 20 hatches .... Page 140
39... Percentage hatch & survival of post larvae (20-30 day) of various species...................... Page 140
40... Feeding schedule for juvenile clownfish used at IOH... Page 149
41... Typical culling data sheet used at IOH... Page 151
42... Monthly juvenile growout data sheet which was completed after culling.......................... Page 152
43... Astaxanthin content found in natural food items .... Page 157
44... Xanthophyll content in plants.. Page 157
45... Pigmentation analysis of feeds being fed to clownfish compared to wild etc...................... Page 158
46... Symptoms that indicate pending water quality, disease and/or nutrient etc...................... Page 159
47... Points to check and measure to take when a disease is suspected................................ Page 160
48... Symptoms normally associated with nutritional disorders.. Page 164
49... Clinical signs and pathological effects of some vitamin deficiencies .... Page 166
50... Fatty acid composition (weight per cent) of samples of food items etc.......................... Page 165
51... Effects of antibacterial and therapeutic agents on recirculating systems .... Page 166
52... Weight of chemical needed to a unit volume of water to yield 1 ppm .... Page 168
53... Conversion factors % to ppm .... Page 168
54... Possible management options based on water quality and feed etc. .... Page 178
55... Possible options in managing a recirculating tank system based on etc .... Page 179
56... Comparison of facilities and production at IOH, Sea World and Aqualife etc. .... Page 182
57... Major differences between IOH, Sea World and Aqualife Research .... Page 184
58... Comparative histological study of clownfish from IOH, Sea World & Aqualife..................... Page 185
59... Post larval production data at IOH from 1981-1982.. Page 185
60... Juvenile percent survival vs percent marketable at IOH .... Page 186
61... Main problems at IOH and suggestions for future endeavors....................................... Page 186
62... Summation of production and sales .... Page 187
63... Final analysis of operation procedures at IOH... Page 188
64... Average percent of monthly costs for running IOH from 1980-81 .... Page 195

List of Photos

1..... Our first broodstock conditioning tank and system.. Page 2
2..... Author with our first larval rearing recirculating aquarium system in 1969 .... Page 2
3..... Resting state eggs .... Page 20
4..... Pre-spawning state eggs .... Page 20
5..... Intermediate state eggs .... Page 21
6..... Spawning state eggs .... Page 21
7..... Exterior view of tanks at IOH showing exterior stand pipe & elevated boardwalk.................. Page 32
8..... Interior view of IOH upper level showing rows of tanks etc. .... Page 32
9..... Interior view of IOH showing lower level with pumps on boardwalks etc........................... Page 32
10... Polyester felt bags used as primary particle filters .... Page 37
11... Rotating biological contactor RBC (Marineland) .... Page 39
12... Biological substrates... Page 40
13... Oocyte development of *A. ocellaris* .... Page 88
14... Oocyte development of *C. melanus* .... Page 88
15... Overflow and exchange standpipe used at IOH .... Page 99
16... Maroon clownfish with severe lymphocystis .... Page 103
17... Remove fish and physically scrape off nodules .... Page 103
18... Treating sores with iodine solution .... Page 102
19... Larval rearing tanks at IOH.. Page 114
19a. Example of gas bubble disease on larvae .... Page 122
19b. Rotifers consumed by a clownfish... Page 130
20... Deformities due to either genetic or environmental conditions at metamorphosis.................. Page 134
21... Juvenile *A. ocellaris* gowth stages, 30 day old post larvae .... Page 142
22... Juvenile *A. ocellaris* gowth stages, 60 day old juvenile... Page 142
23... Juvenile *A. ocellaris* gowth stages, 90 day old juvenile... Page 142
24... Suspended feeding method used at IOH .... Page 156
25... Cryptocaryan on gill filaments and caudal fin rays .... Page 172
25a. Tail and fin rot bacterial infection. .... Page 175

Go Green! Encourage And Support Farm Raised Aquatic Life

INTRODUCTION

Research biology can be challenging and rewarding. Being first to discover or do something is extremely fulfilling and rewarding. Since 1966, I have been actively involved in the research of reproductive biology and aquaculture of marine fish, macroalgae, microalgae and zooplankton. I have been fortunate to be regarded as one of the few pioneers with extensive expertise in conditioning, spawning, and rearing of marine fish. Yet, being a pioneer is often a lonely situation since there are few with whom you can compare or share your thoughts and work. Usually pioneer work does not result in great monetary returns but does instill a rewarding sense of accomplishment, personal satisfaction and opens new doors and directions.

Biology is actually accrued knowledge based on observations, experimentation, analysis, and theory. In the field of biology, you tend to build upon past experience towards an ultimate goal. Schooling helps you organize facts and understand principles but does not make you a biologist. With or without degrees, it takes time. When I received my bachelor's degree in zoology I walked out thinking, "I am a marine biologist". After three years of working in the field I came to realize, "I finally was a marine biologist".

When working with animals, it is essential to think like them. You must gain insight into their needs, their environment, their desires and their goals in life. To most of us, fish appear to have very simplistic, basic needs. Yet, there are many subtle internal and external physiological factors that determine the success or failure of fish populations. Success in breeding fish is based on how these internal and external factors occur and how long they last. A simple fish is not really so simple.

This manual is dedicated to the avid aquarist and beginning aquaculturist. It is especially dedicated to the many biologists and technicians who have worked with me over the years including: Martin Moe, Joe Mountain, Randy Hagood, Charles Dugan, Dan Roberts, Scott Willis, Robert Presley, Tom Frakes, Terry Pulver, Ken Halscott, Walter Havens, Chris Turk, King Wen Sun, Donna Hastie, Judy Johnson, Marsha Madill, Jack Madill, Vonnie De Rosure, Linn Corella, Rose Torintino, Sharon Autman and William Hoff. Lastly, this manual will climax my efforts to reach my goal of sharing my observations and knowledge of conditioning, spawning, and rearing fish.

A very special thanks is extended to my tolerant wife Nancy Jo and to my friends and reviewers, Martin Moe, Joe Lichtenbert, and Don Johnson, who pushed, prodded, and threatened me to complete this book. My hope and desire is that this manual will restimulate the field of aquaculture of marine tropical fish at all levels from advanced aquarists to advanced aquaculturists.

I attempted to make this manual an "ABC cookbook" in marine tropical fish culture with emphasis on marine tropical fish culture, specifically clownfish. Keep in mind that what is discussed can apply to many fish in either marine, brackish, or freshwater environments.

My desire is for the reader to understand as much as possible. Therefore, I have purposely avoided extensive scientific verbiage yet provide professional rather then popular amateur discussions. The information is broad-based so it can be easily applied to other than clownfish.

With the closing of Instant Ocean Hatcheries, one began to question whether tank raised marine tropical fish will ever become a standard, reliable commodity. First Sea World's facility, founded by Chris Turk, followed by Instant Ocean Hatcheries (IOH) founded by Frank Hoff, which ceased operations about 11 years ago. All facilities including Aqualife Research founded by Martin Moe, have ceased operations. Survival and demise of these facilities was not based on a lack of technology to culture clownfish but on investor inability or desire to sustain further losses (see last chapter for further information). Fortunately the work of these pioneers was not in vane. Cquest and Ocean Reefs and Aquariums (ORA) and many small garage operations exist. ORA is now considered the largest producer on cultured marine tropical fish and corals in the world.

Data presented in this manual was based on an a accumulated 17 years of personal research on the spawning and rearing marine fish. Emphasis on clownfish culture stems from 10 years of research (1974 to 1984) initially conducted at Neptune's Nursery in St. Petersburg, Florida and later at Instant Ocean Hatcheries located in Dade City, just north of Tampa, Florida. Both facilities were unique in that they utilized artificial sea salts (Instant Ocean™, Aquarium Systems) in closed recirculating systems. By today's aquaculture standards, IOH can be considered an exceptionally cheap investment into virtually unexplored fields. When IOH closed in 1984, it could be efficiently run with just 3 1/2 people. We had accomplished more than anyone else has ever done. Yet, we recognized specific problem areas that still needed attention. Some required considerable research and attention while most needed time to iron out the bugs. Yet, none of the problems were significant enough to deter production or a degree of sales (see the last chapter).

USA Roots Of Marine Tropical Fish Culture

The roots of commercial ornamental marine tropical fish culture in the United States actually began within the Bayboro Harbor area in St. Petersburg. Florida in 1972. During the following decade three facilities: Instant Ocean Hatcheries , Aqualife Research, and Sea World Research evolved and were totally dedicated to the culture of marine tropicals with emphasis on clownfish. Ironically, all three founders were marine biologists and were intertwined early in their lives within the research activities being conducted at the Florida Marine Research Laboratory located at Bayboro Harbor in St. Petersburg, Florida.

Photo 1 - Our first broodstock conditioning tank and system

My first job as a marine biologist began July 1966 when I was hired to work with Martin Moe on taxonomy and migration of Gulf of Mexico fish. Martin and I were virtually morticians since a considerable amount of our job consisted of opening over 20,000 pickled dead fish, examining stomach contents, reproductive activity, and recording piles of data. In 1968 Martin and I got tired of the dead guys and wanted a little diversification with live guys. We initiated a small aquaculture side project to investigate the conditioning, spawning and rearing of black seabass, *Centropristes melanus.* For both of us this was our first crude attempt to culture marine fish. We were attempting to condition the fish by controlling light and temperature, factors known to trigger spawning. The conditioning experiments were conducted in a small light and temperature controlled, 300 gallon wood aquaria (Photo 1). Once conditioned, the fish were injected with HCG (Human Chorionic Gonadotrophin) in an attempt to induce spawning. Our initial experiments ended in mixed results. We achieved spawning and fertilization only by forcing the eggs and sperm from the fish and mixing them by hand. Fertilized eggs were then transferred to an elaborate, self- contained, larval rearing system (Photo 2). This recirculating larval rearing system was our first but certainly not our last over the many years that would follow. Although no seabass were raised in this unit, several species of fish were reared from fertilized eggs obtained from plankton tows in the nearby bay. The filtration, plumbing, and temperature control system of this small unit would rival many production models now being sold.

Photo 2 - Author with our first larval rearing, recirculating, system, 1969

Martin went into private industry around 1970 to spawn and rear pompano, *Trachinotus carolinus.* I continued as head biologist for the State's new aquaculture division. Our projects were diversified, including the conditioning, spawning, and rearing of pompano, black seabass, macrobrachium shrimp, and oysters. Our original conditioning tank and larval rearing unit quickly became obsolete and was replaced by seven walk-in temperature-controlled rooms, housing various aquatic recirculating systems. In addition, we had a 20,000 gallon (75.7 cu m) semi-open system which occupied a large room. While conducting this research, I hired a fellow graduate student, Chris Turk, to come to work with us. Chris had worked the previous summer with Martin Moe on his new pompano project. It was during this research that I formulated my first gelatin-based diet which is currently manufactured and distributed in the pet markets by Chris and others. The first formula in 1969 had been superseded by 75 different formulas by the time IOH closed its doors.

In 1972, Martin Moe ceased private research on pompano and concentrated his expertise on the culture of marine tropical fish. His first private marine tropical fish venture, Aqualife Research, occupied an old icehouse located directly across Bayboro Harbor from the State Marine Research Lab. The beauty of colorful clownfish compared to nondescript pompano got to Chris and he eventually went back to work with Martin.

Who And What Was Instant Ocean Hatcheries

Well, who was Instant Ocean Hatcheries, what were their roots and what did they want to accomplish? In August, 1974, I too was stricken by the beauty of clownfish and accepted a position in private research for a sister company of Aquarium Systems. I was hired to design, manage, and direct research for a private investor, Mr. Kelvin Smith, who was interested in supporting a pilot scale research facility to investigate the culture of marine ornamental tropical fish. The concept and direction came from Mr. William Kelly, who shared mutual interests and insight in realizing marine tropicals must be reared in captivity in order to save natural populations. Both investors had common early bonds with Mystic Marinelife Aquarium, Cleveland Aquarium and Aquarium Systems. I intern, hired Tom Frakes, a fellow biologist who worked with me on pompano, and my brother William Hoff.

Neptune's Nurseries (initial name for Instant Ocean Hatcheries) was established in an old ice cream factory located about 3 blocks south of Martin's facility and seven blocks south of the Marine Research Laboratory. Our plankton culture room, larval rearing room, and a portion of our juvenile rearing facilities were located in old cooler and freezer rooms. Broodstock were maintained on the second floor. A better perspective and essence of this early venture, would be to picture a new business trying to establish itself in a "bombed-out" building. We were constantly trying to accomplish our goals while the owner was slowly attempting to remove piles of trash, replace broken windows before winter, paint walls, and literally renovate and clean the entire building. It was indeed hectic, yet we managed to spawn our first clownfish by December 1974 which set the stage for the great clownfish race against Martin and his crew.

During those early years there was considerable rivalry between the staffs of Aqualife Research and Neptune's Nurseries. A close, friendly, but very competitive relationship developed between the two facilities. Virtually every Friday evening we engaged in "tongue-in-cheek" discussions at the Sunrise Bar which just happened to be conveniently located next to Neptune's Nurseries. Both parties tried to lead and bait each other into revealing just a little more information. Needless to say, excessive beer helped us both. Little did we know 20 years ago that we would become the pioneers and that these confrontations actually set the stage for a new but still poorly explored field of marine tropical fish culture.

By spring of 1975 we were spawning and rearing three species of clownfish: *Amphiprion ocellaris* (ocellaris), *A. frenatus* (tomato) and *A. clarkii* (sebae). By November 1976 we had added five additional species, including: *P. biaculeatus* (maroon), *A. ephippium* (red saddle), A. *melanopus* (cinnamon), and the first known tank-raised marine hybrids, *A. frenatus* crossed with *A. ephippium* referred to as FE, and *A. frenatus* crossed with *A. melanopus* called FM. By 1977 we added more feathers to our cap by rearing *A. akallopisos* (orange skunk), *A. polymnus* (saddleback), *A. percula* (percula), Gobiosoma oceanops (neon goby), and an additional third hybrid, *A. ephippium* crossed with *A. melanopus* referred to as EM.

We experienced the same series of trials and tribulations that all pioneers endure. Efforts to maintain microalgae and rotifer densities in each larval tank to sustain optimal feeding and water quality would be crushed by a simple pump failure or some unknown critter. Nest size would fluctuate for a given pair and eggs would hatch one time but not the next. Fish would spawn and then take three to six weeks to spawn again. Larvae were available but the rotifer cultures collapsed. Post larvae mysteriously died within a short time after being transferred to juvenile growout aquaria or went into shock while being removed from a larval rearing aquaria. Juvenile mortalities would mysteriously increase after three months in growout. Juvenile feeding initially started at eleven times per day and slowly decreased to three a day. Yet, we still obtained the same or better growth rates. Utter frustration mixed with brief periods of success is the best way to describe the early work and experimentation at Neptune's Nurseries.

In addition to solving biological problems, we concentrated on the mechanics of closed systems for holding broodstock, mass culture of larvae, mass rearing of juveniles, and continuous culture of plankton. Numerous designs of hatching tanks, larval rearing aquaria, and culture vessels were tried and rejected. Criteria such as size, shape, flow rates, flow direction, tank coloration, tanks with or without filtration, lighting, aeration, and substrate were considered tried and rejected until we finally found what worked for us. Broodstock were addressed with regards to obtaining optimal spawns and hatch by manipulation of conditioning factors including: food and feeding, tank size, habitat, water quality flow rates, moon light and size of fish. In regards to water quality, factors such as the effect of spawns, hatching, and rearing were considered. Disease problems in relation to using a recirculating system design rather than singular static aquaria were carefully weighed. All factors were considered important before deciding the next step of designing and constructing a commercial scale hatchery, Instant Ocean Hatcheries (IOH). At all times we were careful to remember that, all too often, experimental results are difficult to scale to commercial applications.

During the latter days at Neptune's Nurseries, we began working with experimental designs for greenhouses in the backyard amid our landlord's piles of trash. Our objective was to utilize Florida's abundant sunlight to help coloration, lighting, and heating during the daylight periods. After a year of trial and tribulation, it was decided that IOH would utilize a modified greenhouse type building. Site selection would be important if we wanted to utilize full benefits of the sun. After looking at 41 sites we selected land located on a southerly exposed hillside which allowed protection from northerly winds, avoided the cold low lands, and received the maximum amount of year-round light. Ironically, this site was in the "mountains" of central Florida with an extreme elevation of 210 feet and located 35 miles east of the Gulf of Mexico and 35 miles north of Tampa.

IOH was virtually designed from scratch from all that we had learned at Neptune's Nurseries and the Marine Research Laboratory. An off-center, double layered, translucent fiberglass roof maximized exposure on the south side of the building and provided a partial insulation effect. The building was split-level, being 200' by 56' (61.5 x 17.3 m) , 11,200 square feet of work space under one roof (Photos 7, 8, 9 & back cover). The upper level housed a 10,000 gallon (38 cu m) larval rearing system, consisting of 80 all-glass, 40 gallon (150 L), cube shaped aquaria (Figures 17, 30, & 31) mounted above three 25' x 4' x 2' high (7.7 x 1.2 x 0.6 m), cement block-constructed reservoirs equipped with 6" deep (15 cm) submerged gravel biological filters with eight 1" diameter (25 mm) airlifts per reservoir (Figure 17). Two additional auxiliary biological filters, 12' x 4' x 4' high

7.7 x 1.2 x 1.2 m) with 12" (30 cm) of substrate with no airlifts served as primary filters and were arranged so effluents from the larval rearing tanks flowed directly into them to remove heavy detritus. Flow from these deep anaerobic primary filters dispersed into three reservoirs located below the larval rearing aquaria. Flow to the aquaria was drawn through the submerged gravel biological filters located in the reservoirs into a diatomaceous earth filter and then into a UV filter. Flow rates depended on the larval sequence and needs and varied from 1 to 40 gallons (3.75 to 150 L) per hour The larval system received a partial water exchange of new water at a rate of 10% to 25% per month.

Also housed on the upper level of the hatchery was a 12,000 gallon (45 cu m), 180 pair capacity broodstock system. Adult pairs were housed in 40 or 29 gallon (150 or 109 L), plywood aquaria, having one window (Photo 7, and Figures 17 and 33). Each aquarium was equipped with a 2" deep (50 mm) submerged gravel filter with a single 1" diameter (25 mm) airlift and were located above six, 25' x 4' x 2' high (7.7 x 1.2 x 0.6 m) reservoirs equipped with 6" deep (15 cm) submerged gravel biological filters (Photo 8). Effluent from the tanks flowed directly into the reservoirs below. Continuous flow was drawn through the submerged gravel filters into a UV filter and then distributed to the aquaria. In addition each tank received continuous aeration.

Juvenile growout systems were also housed on the upper level and accounted for over half the floor space. Like the adult systems, rearing aquaria were mounted above twelve reservoirs which related to six closed recirculating systems consisting of two reservoirs and two rows of rearing aquaria per system. Each reservoir was equipped with 6" deep (15 cm) submerged gravel biological filters and 8 airlifts each. Growout aquaria were 40 gallon (150 L), 22" high (55 cm) cubes made of plywood with a single 12" x 12" (30 x 30 cm) glass window (Photo 8, Figures 30 & 31). Each contained a 2" deep submerged gravel biological filter with a single 1" diameter (25 mm) central airlift. In addition, each had an airstone. Each row (2 per system) contained 32 aquaria for a total capacity of 384 aquaria. Each system (6 total) was about 4000 gallons (15 cu m) with a normal stocking density of 350 to 400 fish per aquarium or approximately 24,000 clownfish per system or a total capacity of 144,000 fish based on stocking 375 per tank. Water was recirculated with a 3/4 hp pump per system. Total gallons was 24,000 (90 cu m) which relates to six fish per gallon (3.75 L). Water was drawn through the submerged filters in the reservoirs and periodically passed through diatomaceous earth and UV filters. Water utilized in these systems came from the larval area and partially from new and/or recycled water. In the event of disease, cleaning, etc. either individual adult or juvenile aquaria or an entire row of aquaria could be isolated from the system and serviced separately.

The remaining area at this level was reserved for packing and shipping, and to house a 10,000 gallon (37.5 cu m), indoor-outdoor experimental system for growing macroalgae and to house several hundred extra pairs of tank raised, subadult fish.

The lower level of the building was 20' wide by 200' long (6.2 x 61.5 m) and 5' (1.5 m) lower than the main floor. This level housed ten, 14' x 12' x 4' deep (4.3 x 3.7 x 1.2 m), 4,000 gallon (15 cu m) reservoirs. These were primarily used to increase water volume in the juvenile systems, grow marine alga, growout of larger fish, and to hold recycled water (Photo 9). In addition, there were two 3,000-gallon (11.3 cu m) reservoirs for mixing seawater which were integrated with a central manifold that serviced the entire building. Either new or recycled water could be provided to each system from this area.

Last, but perhaps the most important part of the entire operation, was the culture facility (back cover Photo M) which was located in a 20' x 25' (6.2 x 7.7 m) room on the lower southeast corner of the building adjacent to the larval rearing area on the upper level. This room was solar lighted, ambient heated and cooled by just a thermostatically controlled ceiling fan. Although this was a small area, it was sufficient to produce billions of microalgae cells and yield 20 to 50 million rotifers daily which were fed to 20 to 30,000 larva each day. This unique, simplified culture facility served as a model for many other aquaculture ventures.

Used saltwater eventually was transferred outdoors by gravity flow, to six 10,000 gallon (37.5 cu m) recycling ponds located downhill from the hatchery. Heavy aeration, additions of (specific nutrients, and inoculate of algae, triggered mass blooms of micro and macroalgae. This bloom would usually continue for three to four months thus reducing nitrates from the usual 30 to 50 ppm NO_3-N to 5 ppm or lower. At the closing of IOH approximately 70% of our water was being recycled.

Aeration to the entire building and outside facilities was supplied by a pair of lobe blowers capable of producing 160 cubic feet (4.5 cu m) of air per minute at 5 psi. For emergencies, one blower was equipped with a self-starting gasoline powered lawn mower engine triggered by a common well pressure switch. Pressure loss either by loss of electrical power, loss due to a broken pipe, or just loss of a drive belt triggered the emergency system. The battery was charged weekly by the automatic running of the main blower controlled by a timer.

In addition to the main hatchery building, there was a two-story office and water quality lab building, a separate kitchen for preparing gelatin foods, a workshop, and a home to house one employee on the premises.

Instant Ocean Hatcheries was a unique, well planned facility with a staff that accomplished a lot of work in the field of marine tropical fish farming.

Reason For This Manual

Clownfish were and continue to be the most demanded species of marine tropical fish and certainly will later be classified as the "goldfish or guppy" of saltwater tropical fish. Demand of saltwater fish and invertebrates constantly exceeds supply, and this demand will only increase in years to come. Therefore, it is prudent that constructive measures be taken now to ensure that wild fish and reefs are protected and we establish farm-raised supply of reasonably priced clownfish. This can only be accomplished by development of captive breeding programs.

The very essence of this manual is to convey to future farmers technology and data which has been well paid for and does not have to be repeated. A conservative estimate is that over the last 20 years prior to this publication in 1996, eight to ten million dollars had been spent by various investors on the controlled cultivation of clownfish. Even though data provided in this manual was recorded over 30 years ago, it still stands as the only complete commercial data available to the public. To date, only limited, ambiguous technology transfer has been provided via hobby literature. I hope this manual provides your door to the future and will eliminate the need for future investors to reinvent the wheel. Why waste a great deal of time or effort in creating something that already exists. New investments into marine tropical fish should only add to our current knowledge pool and not be wasted.

My wish is that where I and other pioneers stopped, others will soon follow and complete our endeavors in a profitable manner. I hope this manual provides the emphasis to go on. I hope it will stimulate the new field of culturing rather than capturing marine tropicals. Whether it is done in your home aquarium, den, basement, garage or in a hatchery, I sincerely feel the field of culturing marine fish, plants and invertebrates is virtually an untouched entity with extremely lucrative possibilities that are not impossible to realize.

Drawing By H. W. F. del.
Results of the Fifth Vanderbilt Expedition
By Fowler, 1944

Halichoeres stictus

TAXONOMY AND DISTRIBUTION

Taxonomy

Fishes are more numerous than all other vertebrate groups. Fossil armored plates indicate that the evolutionary process has been going on since the Ordorvician Period of the Paleozoic Era (450-500 million years). Up to the Devonian Period of the Paleozoic Era (350 million years ago) fish were evolving and the first amphibians (looked somewhat like a coelacanth) did not appear until after the Devonian Period. Fish occupy four classes, Cyclostomata (round mouth fish), Placodermi (plate skin fish), Chondrichthyes (cartilaginous fish) and Osteichthys (bony fishes). They are arranged in 50-60 orders, 500 families and 25,000 species. Florida waters have about 1,200 recorded species of which 700 are considered shore species (continental shelf region).

I have restricted this section to pertinent information which I feel might help the reader to understand more about the biology and life style of clownfish. I have summarized this section from information provided by Allen (1972). Clownfish belong to the widely diverse and well-distributed family Pomacentridae (damselfishes). There are 27 known clownfish species. They are distinguished and taxonomically separated from other damselfish by their dependence on anemones for protection. Only three other genera of fish are known to exhibit this behavior including the popular "Domino", *Dascyllus trimaculatus*. They are further distinguished from other damsels by their large (approximately 2.0 mm) capsule-shaped eggs and large larvae (approximately 4 mm) at hatch. Physically their swimming pattern consisting of exaggerated lateral flexures and alternating "paddling" of their pectoral fins, sets them apart from others in their family.

Distribution

Clownfish of the genera *Amphiprion* and *Premnas* are found throughout the vast Indo-West Pacific region. The center of abundance is around the Indo-Australian Archipelago-Philippines region (Ekman 1953). This area contains 10 known species. *Amphiprion* have been able to penetrate most of the islands of the tropical western Pacific which are numerous and generally not isolated from neighboring islands. However, where vast stretches of open water exist, such as between the Marshal Islands and Hawaii or between the Tuamotus and Easter Islands, *Amphiprion* have failed to penetrate, while species with extended pelagic larval stages, such as *Acanthurus*, have been able to make these crossing (Randall, 1955). Clownfish do not occur in the eastern fringe of the Indo-West Pacific: Hawaiian Islands; Johnston Island; Line Islands; Marquesses Islands; Pitcairn and nearby islands; Rapa; and Easter Island.

Distribution and availability of anemones are also important factors which would affect distribution of clownfish. At present, systematics and zoogeography of host anemones are poorly known (Allen 1972).

Isolation and distribution are also controlled by hydrographic factors such as deep cool upwelling waters adjacent to warm shallow lagoons. Oceanic coastal currents which are basically nutrient free and have limited live food sources. Swift currents adjacent to shorelines which sweep larva out into open deep waters. Siltation and freshwater runoff from rivers may act as coastal barriers. Many Indo-Pacific islands are nothing more than mountain peaks that emerge from the water. Adjacent to these are deep cool waters. Volcanic activity and warm underwater springs may be factors of isolation and regulators of distribution.

Amphiprion species distribution can be divided into three areas: (1) those that are widely distributed; (2) limited regional distribution and (3) those confined to specific islands or island groups. *Amphiprion clarkii* is the most widely spread species. It is found in the east to the Caroline Islands, west to the Persion Gulf, north to Southern Japan and south to the New Hebrides. Table 1 is a summation of distribution of most clownfish species.

Common and Scientific Names

Utilization of species names for common names within the pet market has created several problems that need clarification. Sebae is a common name for *A. clarkii* yet there is a true sebae species (*A. sebae*) which is torpedo shaped and has only two body bars. Ocellaris is commonly applied to either *A. ocellaris* or *A. percula* which are distinctly different in adult stages. Tomato clowns can include several species including *A. frenatus, A. rubocinctus, A. melanopus* and at times *A. ephippium*. The tomato complex is vast and confusing especially when dealing with juveniles which look very similar at this young age. Skunks can include *A. perideraion, A. akallopisos* and *A. sandaracinos*. For the sake of clarity I would like to encourage the utilization of species names and discontinue the use of common names.

More detailed taxonomy, zoogeography and colored pictures can be obtained in Allen 1972 and 1991.

Table 1 - General distribution of clownfish species based in part from Allen 1972

Species With Widely Spread Regions:

A. akallopisos .....yellow skunk......Indian Ocean
P. biaculeatus .....maroon ....Australia, East Indies, Melansia, Philippines
A. bicinctus ....?? ....Indo-Australian Archipelago; Indian Ocean
A. clarkii ....sebae ....noted in paragraph above
A. chysopterus....blue stripe..........Micronesia; central Polynesia; Fiji
A. frenatus ....tomato ....Indonesia; Philippines; east coast of Asia to So. Japan
A. melanopus......cinnamon...........Micronesia; central Polynesia; east Indo-Australian Archipelago.
A. ocellaris ....ocellaris ....Australia, East Indies, Melanesia, Philippines, Ryukyus
A. perideraion .....pink skunk ....western Pacific; Indo-Australian Archipelago
A. rubrocinctus....tomato ....central Polynesia; Fiji; north Australia
A. sebae ....sebae ....northern Indian Ocean

Species With Limited Regions:

A. percula ....percula ....Melanesia; Queensland
A. akindynos ....?? ....Queensland; New Caledonia
A. sandaracinos..orange skunk.....Philippines
A. ephippium.......red saddle ....northeastern edge of Indian Ocean
A. allardi.............?? ....East Africa; Seychelles
A. bicinctus ....?? ....Red Sea
A. calliops ....?? ....not well documented
A. latezonatus.....?? ....Lord Howe Island; No, New South Wales

Species With Highly Restricted Regions:

A. tricinctus ....?? ....Marshall Islands
A. chysogaster....?? ....Mauritius
A. chagosensis ...?? ....Chagos Archipelago
A. fuscocaudatus ?? ....Aldabra and Seychelles
A. latifasciatus ....?? ....Madagascar
A. nigripes.....................................Maldive Islands
A. mccullochi ....Lord Howe Island

GENERAL BIOLOGY & PHYSIOLOGY

Texts often lack a review of some basic aspects of fish biology and physiology. To relate problems and gain insight into what is happening, it is advantageous to consider fish as an entity that uses body language as a means of communication. Gaining basic knowledge of the function, of body parts and organs helps you to interpolate this language.

Body shapes of fish vary greatly from torpedo or bullet shaped bodies to horizontal and laterally compressed bodies. Body parts, (fins, eyes, mouths, heads, scales, skin, pigments) come in all sizes and shapes. A rule of thumb is to realize that form follows function. Therefore, adaptation over time has created these changes and you must realize that these features are important with regards to how we are going to maintain and reproduce this fish in a captive environment.

External Anatomy

General anatomical description use prefixes such as anterior for the frontal, posterior for hind end, dorsal refers to the back or upper structure, ventral refers to the under part or lower surface, and medial with regards to the central or middle body. Figure 5 and 6 provides some of the basic external and internal features of common fish. All fish have heads, bodies, gills, internal organs and skeletons and most have fins and scales. They may have varying numbers of paired and unpaired fins up to a maximum of five. Unpaired fins are the dorsal, anal and caudal or tail fin. Dorsal fins may be present or absent, united for its entire length (eels) or divided into two or three distinct parts as illustrated or further modified. Remoras (family Echeneidae) have a modified dorsal fin which has developed into a sucking disc for attachment. Other fish such as blennies (family Blennidae) and batfish (Ogcocephalidae) have modified pelvic fins that act like legs. Sea robins (family Triglidae) and flying fish (Exocoetidae) have modified pectoral fins which allow them to glide through or above the water.

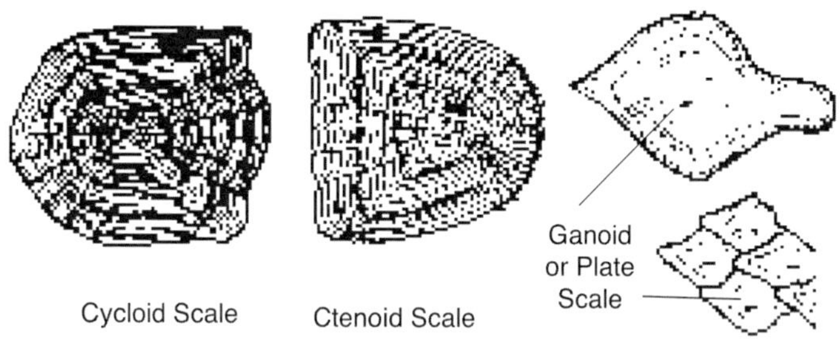

Figure 1 - Types of scales found on modern and primative fish.

Scales (Figure 1) function as a physical barrier which protect fish against injury. Fish scales basically fit into four general groups. Placoid scales are rasp-like and non-overlapping. They are typical of sharks, rays and their allies. Ganoid scales are shiny and platelike and are found on primitive species: sturgeon and gars. Most typical bony fishes have rough, overlapping, ctenoid scales which have distinct comblike teeth on their exposed posterior margins. Equally common among modern bony fishes are cycloid scales. These are thin, overlapping scales that have a smooth posterior margin. Some bony fish possess both types of scales on different parts of their bodies while others have one type while young and the other type when older.

All fish, scaled and scaless, have a protective layer of mucus or slime covering their bodies which consists of mucin, a glycoprotein which absorbs water. Mucus serves multiple purposes including; lubrication to reduce friction in water, inhibits entry of disease organisms, and chemical barrier containing enzymes (lysozymes) and antibodies (immunoglobulins) which kill invading organisms. The constant horizontal backward flow of mucus out of pores normally located around the dorsal head region provides protection against parasites, bacteria, fungi, etc. In addition, it helps to regulate osmotic balances between salts and minerals in the body versus the aquatic environment. Lungfish build cocoons out of mucus while beta or fighting fish make bubble nests.

Gills, normally four gill arches per each side, serve as the gas and mineral exchange agent.

Each has multiple blood vessels and lamellae, semipermeable membranes protected by bony gill arches. All water solutions attempt to reach equilibrium (isosmotic), waters of lower concentration move towards waters of higher concentration. The net effect is eventual equilibrium even over a selective semipermeable membrane. Most bony fish have blood and body fluid salinities of 7 to 8 ppt (parts per thousand) and a pH of 7.4 to 7.8. Freshwater fish are hyperosmotic, their body fluids contain more salt and mineral ions than the surrounding water. Saltwater fish are just the opposite, hypoosmotic. Freshwater fish fight to keep water from diluting their body fluids while saltwater fish fight to keep water from being lost. Freshwater fish drink little, but the permeable gill and skin membranes allow water to be absorbed. Copious amounts of urine are evident in freshwater fish while saltwater fish are conservative. The urine of bony fishes is either hypotonic or at least isotonic to seawater. All marine fish "drink" water and experiments on *Myxocephalus* have shown that fish swallow between 50 to 200 cc per kilo (2.2 lbs) per day (Nikolsky, 1963). Salts are removed from the blood of bony fishes in two ways: the kidneys remove mainly bivalent ions and the gills remove monovalent ions. A portion of the salts are regulated through the intestine. Stress upsets the normal electrolyte (sodium, potassium and chloride) balance which results in excessive uptake of water by freshwater fish and dehydration in salt water fish.

Taste in the majority of fishes usually consists of large number of sensing cells located on the lips surrounding the mouth. Some fish like catfish (*Siluris glanis*), cod (*Gadus morhua*) and red mullet (*Mullus barbatus*) have special outgrowths (barbels) which contain taste sensing cells. Others like gurnards actually have modified pectoral fins that have the ability to act as taste organs.

Form follows function is especially true with regards to mouth, gill rakers and teeth creations with regards to the form and consistency of prey consumed (Figures 2, 3, 4, & 4a) and where it is located. Both shape and location are indicative. Grasping mouths are typified by large, sharp, teeth on both jaw bones and often on the internal mouth vomerine and palatine bones. Fish that consume other fish and mobile invertebrates are characteristic of this group. Mouth forming a sucker, no jaws, teeth are in the form of horny tubercles, and boring teeth present on the tongue. These are typically found on semiparasitic lampreys and hagfish. Imbibing mouths are in the shape of tube of variable length, sometimes extensible, usually toothless. Typically they feed on bottom invertebrates and sometimes water column invertebrates. Species such as bream, bird wrasses and pipefish have such mouths. Crushing mouths come in the form of a beak, like parrot fish and trunk fish or platelike (skates, devil fish) or spines (catfish). This mouth type functions in crushing hard invertebrate armature found on mollusks, echinoderms and corals. Planktophagic mouths (zooplankton

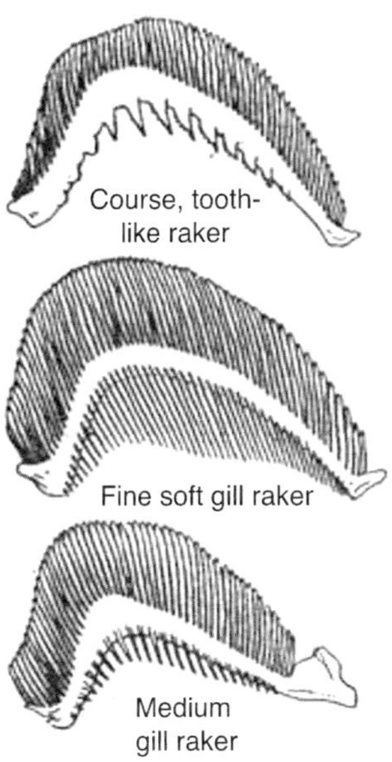

Fig 2 - Gill rakers in relation to feeding habits

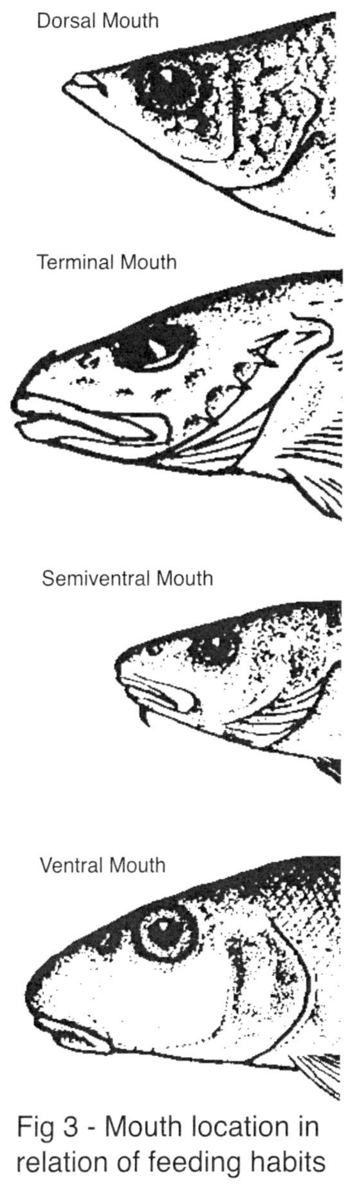

Fig 3 - Mouth location in relation of feeding habits

Figure 4 - Representative digestive tracks from fish.

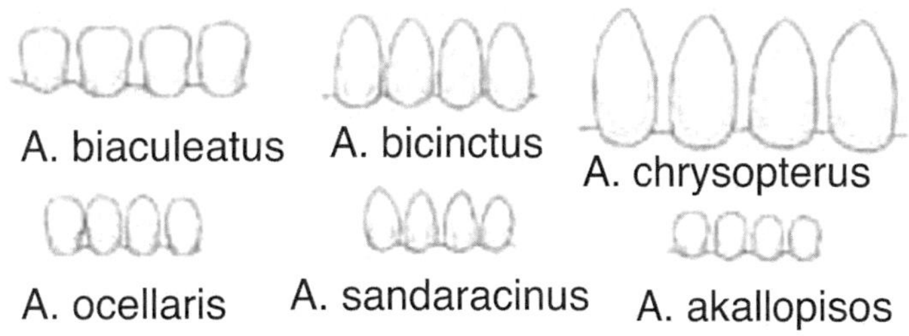

Figure 4a - Different Clownfish Teeth Scale 20X

feeders) are usually larger than other types and immobile. Teeth are small or completely lacking and gill-rakers are long. Included in this group are herring, whitefish, some cyprinids, etc. Lastly, periphyton mouths are typified on fish that feed on aquatic plants. These are ventral mouths that look like a transverse crack. Usually the lower lip has

Figure 5 - External body parts of most fish.

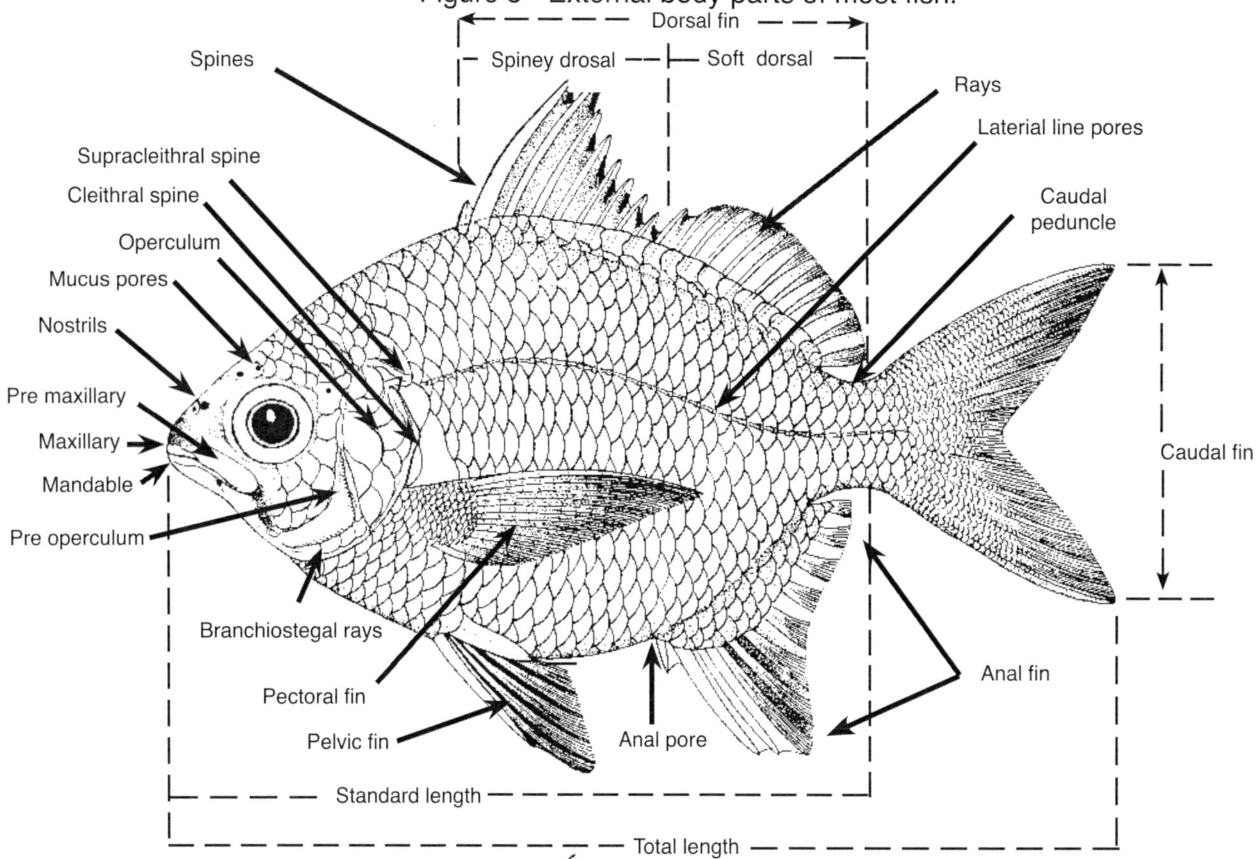

Figure 6 - Internal body parts.

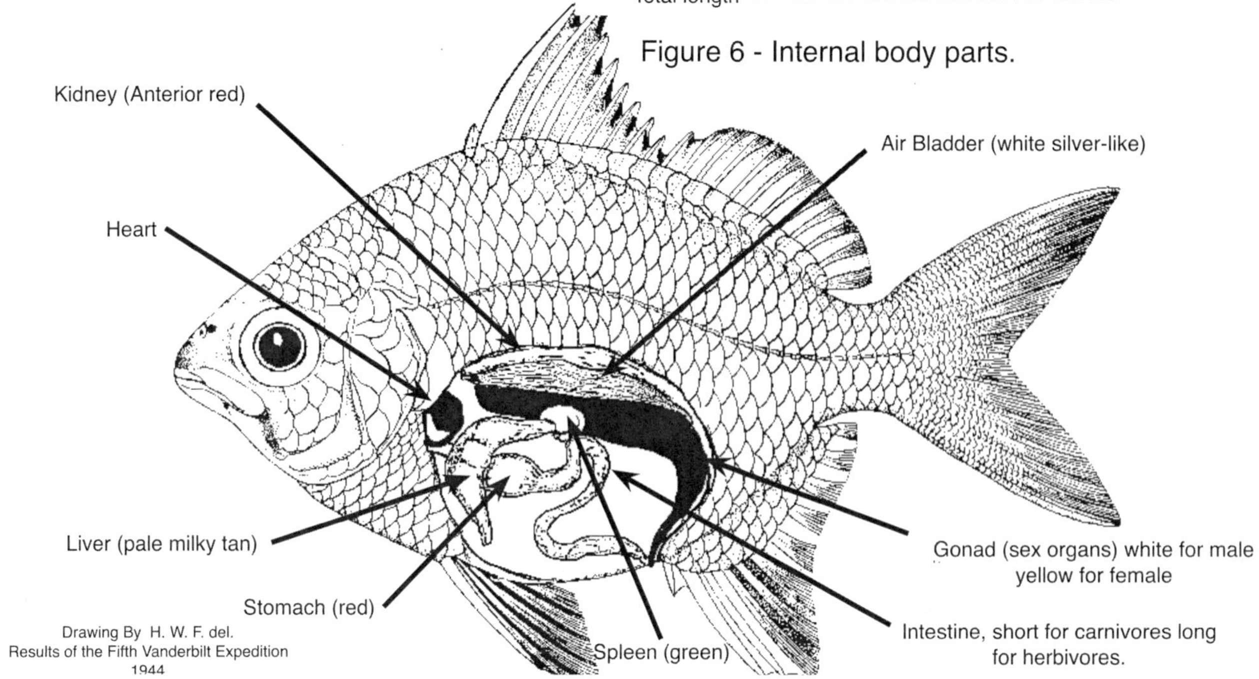

a sharp cutting edge, sometimes a horny covering and are usually lacking teeth.

Mouth dentition and form indicate food preferences but mouth location indicates where within the aquatic environment the fish feeds. Mouths may be located dorsally above the body axis; terminally, right on the body axis; semi-ventrally, lying just under the body axis and ventrally, located below the body axis. Large-mouth fish are typically predators of other fish and large invertebrates. Feeding habits are further identified by examination of the gill rakers (Figure 2). Gill rakers with extensive fine soft filaments on both the upper and lower arch are indicative of fish that feed on microalgae and small planktonic crustacea. Medium-size rakers are indicative of a plankton feeder where as those with heavy "toothlike" hard rakers are indicative of fish-eating predators.

Within the throat region or pharynx are other indicators of feeding habits. Pharyngeal teeth are often found which can be pointed, cup shaped or crushing plates. Pharyngeal teeth change during the ontogeny (development of an individual organism or anatomical or behavioral feature from the earliest stage to maturity) of the fish in relation to changes in the food of the fish (Nikolsky 1963). In larval bream and carp they are conical; in the adult bream they are hook-shaped with a masticatory platform, in the adult carp they are masticatory.

Internal Anatomy

Following the external indications of food and feeding habits, examination of the stomach and intestine are further indicators. The size of the stomach is closely related to size of prey consumed. Large stomachs are usually indicative of large prey or large portions of food during a single feeding. In predatory fishes, not only is the food bolus broken up and partially digested but the prey is sometimes killed within the stomach by the presence of hydrochloric acid (Niklosky 1963). Intestine length is closely related to normal diet. Typically, small intestines and large stomachs are indicative of carnivores and greatly elongated intestines with small stomachs are indicative of herbivores (Figure 4). Omnivore's digestive systems are found in between. Typically the intestine of herbivores comprises over 100% of the total body length. Plants are hard to digest, hence an increased digestive tract is necessary. In addition, digestive enzyme composition is quite different. In carp, amylase which digests carbohydrates, is 150 times as strong as in carnivorous pikes. Intestine tract lengths and convolutions increase as the fish grows. Many larval herbivores start off as carnivores, typified by a short straight gut and later change feeding habits. Mullet (*Mugal cephalis*) is an example.

Fish have a two chamber heart as opposed the four chamber model we have. Oxygen exchange occurs at the gills. They are cold blooded (poikilotherms) and are basically isotonic to the fluid temperature in which they live in. Heat production is low; carp weighing 105 gm produces 10.2 kcal of heat per kilogram (2.2 lbs) weight daily while starlings (birds) weighing 74 gm produce 270 kcal. Body temperature of most fish differs from their environment by only 0.5 to 1°C whereas tuna can differ as high as 10°C. Response of metabolic rates, spawning, migration etc. are closely in line with temperature. Usually, tropical and subtropical fish are stenothermal (narrow range of temperature fluctuations) than are those of boreal and higher latitudes. Tropical marine species are more stenothermal than freshwater species.

The nervous systems of fish are quite well developed and combinations of the senses of hearing, taste, smell, sight and touch are often acute.

The age of a fish can be determined by removing the otoliths (ear bones) located in the head region between and a little back of the head beside the brain. These small oval calcareous bodies in the inner ear of vertebrates are involved in sensing gravity and movement. When examined under a low power scope you see growth rings like a tree. Photo 2a on right shows about three and a half rings from a grouper fish.

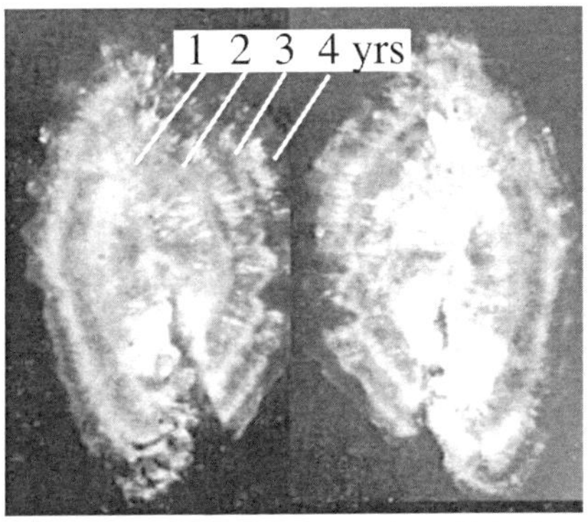

Photo 2a - Ear bones (otoliths) used for age determination)

Clownfish Biology

Anemonefish (clownfish), as their name defines, live in a mutualistic relationship with anemones. This relationship has been discussed in detail by Mariscal (1966b, 1970c) and Allen (1972). In nature, selection of preferred anemones is species specific. Specific host anemones are difficult to identify. Dunn (1981) "The Clownfish Sea Anemones" is considered a good source for identification of host anemones. Tullock (1992) is a more recent source regarding clownfish host anemones.

The thrust of this manual centers on the relationship anemones have with regards to spawning and survival of clownfish. Primary benefits to clownfish from anemone association are protection of the pair, their nests, and a portion of their progeny from predation. Reciprocal benefits to the anemone by this association are not as well defined. Graefe (1964) and Gohar (1934) indicated that the fish actually feed the anemone. Fishelson (1965) and Allen (1972) felt the fish protected the anemone from predators. Ziegler (1960), Mariscal (1970c) and Allen (1972) felt that the fish remove parasitic copepods from the anemones and vise-versa. To this array of theories, I would like to add another. Anemones are known to derive their coloration and a substantial amount of their nutrition directly from commensal microalgae called zooxanthellae which live within their surface tissues. Since nutrients are very limited in tropical reef situations, it is possible that the urea and fecal matter excreted by the pair of clownfish serves as a basic nutrient source which inturn enhances growth and survival of the zooxanthellae which in turn provide nutrients to the anemone.

How clownfish can survive the stinging cells within the tentacles of anemones is covered by Lubbock (1980). He found that *A. clarkii* achieves protection from stinging by means of the development of a special external mucus layer. This layer appears to be three to four times thicker than that of related fish species that do not utilize anemones. It consists of glycoprotein containing neutral polysaccharides. The fish become covered with a small amount of anemone mucus as a result of contact with its host which is perhaps beneficial in that it increases the thickness of the inert mucus layer covering the fish. Yet, there is no evidence that *A. clarkii* is primarily dependent upon anemone mucus for protection or survival.

In captivity, necessity and dependence on anemones diminishes since the possibility of predation is greatly reduced. Environmental substitution of alternative "security blankets" for captive pairs consists of proper housing, confinement, and isolation as non-biological substitutes for anemones. This relationship will be discussed in more detail under the conditioning section. Although anemone fish seem to prefer specific types of Pacific anemones, we have successfully introduced the Atlantic anemone, *Condalactus,* to the tomato species (*A. melanopus, A. frenatus, A. ephippium*), sebae (*A. clarkii*) and maroon clownfish (*P. biaculeatus*). We have had limited success introducing Atlantic short tentacle anemones to percula (*A. ocellaris, A. percula*) or saddleback (*A. polymnus*). Compatibility with Atlantic anemones further demonstrates that they are utilized more as "security blankets" or for nest site protection than for some physiological / biological need. Perhaps some clownfish species can adapt to a wide range of nematocysts while others cannot. It would appear, that use of nonspecific anemones does not alter a pair physiologically, but is more appealing for other reasons. In the wild, selection of specific anemone species may simply be based on location of specific anemones in relationship to currents, light, reef habitat, and specific food sources. It is sort of asking what came first, the chicken or the egg.

It is evident physiological alteration occurs when tank-raised clownfish are introduced to an anemone for the first time. Although they show almost immediate curiosity, they will not swim into or adjust as quickly to anemones as wild caught fish. Initially, they nip the tips of anemone tentacles and immediately display shock, stress, heavy breathing, locked jaw and often turn a pale whitish color. This conditioning continues for several days and eventually they casually immerse themselves in the anemone. However, our data has substantiated that these obvious physiological changes are apparently not necessary for anemone fish to sexually mature and spawn. We have successfully spawned and reared clownfish to three generations without using any anemones. Yet, I still feel there might be some relationship between acquired immunity to diseases and perhaps some nutritional disorders by not being able to undergo these physiological changes during their life. This is only theory but, has some validity based on observations made over the nine years of raising clownfish. This physiological change clownfish undergo in order to adapt to anemones might be the main reason why so many larger adult fish break down and die quickly after capture and transport. Removal from the anemone and transport is perhaps too radical and quick for them to adapt back to a healthy, non-anemone, physiological condition.

Pairing

Clownfish appear to be monogamous, pairing for life, which usually ends with the demise of the oldest individual, the female. There is also a possibility that some species may be polygamous. Moyer and Bell (1976) noted that populations of *A. clarkii* off southern Japan may at times have a male and two females. They also observed two nests laid within a three day period on a single site. Whether polygamy is constant or intermittent is not clear. They further stated that a nest site is used by the same pair for several years. Observations by Allen demonstrated that wild pairs of *Amphiprion* mate for life. Of 14 pairs he observed over a period of 2.5 to 8 years, in all cases but one monogamy occurred.

Spawning and Hatching

Spawning within the central regions of clownfish concentrations occurs about once a month. Both Allen (1972) and Randall (1961) indicate a higher percentage of spawns occurs within six days prior to and after the full moon. Within captive broodstock maintained for over nine years we could not substantiate this observation. In the wild, there is no incidental bright light source at night except from the moon which reaches a high output on full moon of about 0.03 lumens. Yet, within a hatchery, even the street light located outside a window could create an artificial "moon light" effect as high as 2 to 25 lumens.

Of the spawns that Allen observed in the wild he indicated the female took the lead role and they normally spawned in the early afternoon. He further speculated as to why spawning and hatching occurs during the period of the full moon Lunar periodicity may facilitate nocturnal egg care. Lunar light might also benefit newly hatched larvae since they are photopositive. Low light may stimulate larva to swim up into the water column immediately after hatch and allow greater dispersal of hatchings since surface tidal currents are greater during the period of the full moon. Korringa (1947) noted that many marine invertebrates are known to spawn during the period of the full moon thus newly hatched larva would benefit from the available food sources. The latter point is significant in a tropical reef situation since phyto- and zooplankton are normally limited and the timing of hatch would be certainly advantageous.

Ross (1977) observed nineteen pairs of cinnamon clowns, *Amphiprion melanopus,* for a year. He stated that spawning occurs 2-3 hours after sunrise and lasts 1.5 hours. Hatching occurred 1.5 hours after sunset on the seventh or eighth day of incubation. Spawning activity peaks near the first and third quarters of the lunar cycle. Consequently, hatching activity peaks near full moon and the new moon at the time of the day when high spring tides also occur. Nocturnal hatching on high spring tides is interpreted as a mechanism for reducing predation on hatchlings due to the inactivity of some planktivores at night and the more effective dispersal of hatchings away from the nest during high tides. He found there was no seasonal periodicity in spawning activity. Average was 2 spawns per lunar month per pair with an estimated annual fecundity of 7,200 eggs per year (300 eggs/nest).

Sex Changes

An interesting relationship occurs in nature which has some important applications to a breeder of clownfish. Pairs accept and allow a small percentage of juveniles to live in conjunction with them. At first this relationship seemed strange but later we realized that clownfish had the ability to change sex at a later age. To our knowledge a large percentage of most clownfish species initially start off as males and can later change to females if necessary. Sex changes from male to females are not reversible and are referred to as protandrous hermaphroditism. Within fish that pair for life, this ability to change sex is important to continued survival of a specific gene pool. If one were to die, let's say the female, the surviving male of the pair retains the ability to change to a female and one of the associated juveniles assumes the new male position. Whether these associated juveniles are the parent's progeny or a mixture from other pairs is not known.

During early development, a fish larva is, phenotypically, neither male or female in that it does not possess ovaries, testes, or other characteristics associated with the reproductive systems. Instead, an embryo possesses embryological precursors of ovaries and testes (primordial germ cells), and at this stage an embryo is "totipotent" because it could develop into either a male or female. At a specific time during embryological development (it is different for each species), a chemical signal originates from a gene or set of genes, and this signal "informs" the totipotent tissue which way to develop (Tave, 1993). It is during this stage that breeders often attempt to hormonally alter populations into mostly male or mostly female fish depending on what is desired. With regards to clownfish, these procedures could be advantageous to produce mostly female populations since they grow larger and faster than males. Sex reversal with regard to pairing of clownfish is discussed in more detail in the broodstock pairing section.

Cross Breeding

Within tropical oceans, shallow reefs and atolls are usually tops of isolated underwater mountains or volcanos. Waters adjacent to these shallow reef areas are extremely deep, colder and have strong directional currents. The combination of these factors tied to the fact that clownfish are basically sedentary, tends to isolate specific populations over long periods of time. These environmental isolating factors allow specific subspecies to slowly develop by inbreeding and to eventually diverge from the original parent coloration and, in some cases, body configuration, feeding, and spawning habits.

Another interesting point to breeders is that isolation of specific reef species does not imply that they cannot cross breed. We have successfully documented cross breeding within captive stocks of IOH. Although this is not clearly documented in nature, I have occasionally seen photos in popular magazines and texts in which I strongly question the species described. After examining several and comparing them to the crosses which we made within the tomato complex, I am sure cross breeding of subspecies of clownfish occurs often in the wild. For example, *Amphiprion ephippium*, (red saddle clown) which is found in a highly restricted range encompassing the Andaman Sea and coasts, primarily the Indian Ocean around Sumatra and Java, will easily cross with various tomato species. We have successfully crossed this species with *A. frenatus* (tomato clown) and *A. melanopus* (cinnamon clown). Yet, *A. frenatus,* mainly inhabits the reefs of the South China Sea, southern Japan and Korea. *Amphiprion melanopus* is mainly confined to the eastern portion of the Australian Archipelago and islands of the western Pacific. The ease with which these fish cross bred, without artificial hormonal inducement, strongly indicates that the "isolation factor of species" in reef fish is probably the main reason why more color variants or strains are not found. Over time this isolation may indeed develop different color patterns and body shapes but essentially they are just geographical variants. Genetic chromosomal comparisons of these three species of fish showed no significant difference (unpublished data) but exterior coloration and bar patterns are completely different.

Although we never completed the work, there is a strong indication that the sebae complex could be crossed within their specific group variants or against species from the tomato complex. At the closing of Instant Ocean Hatcheries we had several sebae/tomato possible pairs set up which appeared to be very capable. These would indeed make a very interesting looking fish.

Feeding Habits

Anemone fish are classified as omnivorous, opportunistic feeders. This means they consume a wide variety of food items consisting of plant and animal matter and do not travel long distances to pursue their food. Feeding is generally confined to the surface area and water column immediate to their host anemone. Tidal flow and wave action supply bits of floating debris, mostly algae, into their water column. Largest feeding territory recorded by Allen (1972) covered 20 square meters in a relatively predator-free environment. Most feeding ventures are along bottom substrates and do not expand beyond three to five meters from the host anemone.

Allen (1972) examined the stomach contents of 74 specimens from Eniwetok. Species examined included *A. chrysopterus* (bluestripe), *A. melanopus* (cinnamon), *A. tricinctus* and *A. perideraion* (pink skunk). By volume, 91.8% of their food consisted of an average of 34.8% copepods, 33.0% algae, 9.7% worms, 6.4% crustacean fragments, 5.2% tunicates, and 2.7% amphipods. The remaining 8.2% consisted a wide range of food items including; isopods, barnacle nauplii, *Amphiprion* eggs, shrimps, crustacean larvae, barnacle appendages, unidentified eggs, crabs, nematodes, ostracods and polychaetes. It is important to note that copepods and macroalgae constituted 67.8% of their diet.

Copepods consumed were not specific but widely varied including: calanoids, *Centropages sp.*, cyclopoids, harpactacoid, *Oithona sp.*, *Paracaudacia truncata,* and *Tisbe furcata.* The most common copepods consumed were the calanoid *Paracaudacia truncata* and the harpactacoid *Tisbe furcata.*

Algae was also widely diversified including: diatoms, red algae *Ceramium sp.*, *Ceramiales*, *Hypnea sp.*, *Polysiphonia sp.*, blue greens, *Schizothrix mexicana*, brown algae, *Dictyota friobilis*, *Debesia sp.*, and green algae, *Caulerpa peltata.* The most common algae were the red algae, *Hypnea,* and the blue green algae, *Schizothrix mexicana.*

Of the species examined, only *A. perideraion* (pink skunk) consumed more algae than copepods (43.5% to 25%). This may indicate a preference of algae over meat or plankton. Dentition also differs from other species in that they have more incisiform teeth which appear to be partly adapted for algal feeding. Whereas, conical-type dentition is more common in most clownfish. Divergence of *A. perideraion* dentition may have evolved due to habitat preference or niche feeding habits.

It is important to remember that feeding preferences in clownfish are probably forced upon them by their tendency to remain isolated and somewhat sedentary. No doubt many other foods can be substituted in lieu of items normally found in the stomachs of wild individuals. Yet, I feel it is very important to point out that most frozen diets on the market consist mostly of carnivore (meat type) foods and really do not adequately address the general specific needs of clownfish. Certainly few frozen foods contain 33% algae. (see Food and Feeding)

GONADAL CYCLING & EGG DEVELOPMENT

A mature haploid male or female germ cell (gamete) that is able to unite with another of the opposite sex in sexual reproduction to form a zygote (oocyte/egg). What occurs within the sexual organs of maturing male and female fish? Gonadal cycling is a term used by aquaculturists referring to the formation and regression of maturing eggs (oocytes) within the ovary (egg producing organ). Egg (oocyte) formation is cyclic and is dependent on maturity, external environmental and internal physical factors which influence development or regression. Once the fish is sexually mature and the pathway is properly stimulated a sequence of events take place. Figure 7 (Roberts 1987) outlines what is going on inside the various organs of the body. In addition, natural processes can be stimulated or overridden by the injection of hormones at the points indicated by hexagons. During oocyte (egg) development, several distinct conditions or states can be recognized by taking in-vivo (live) samples.

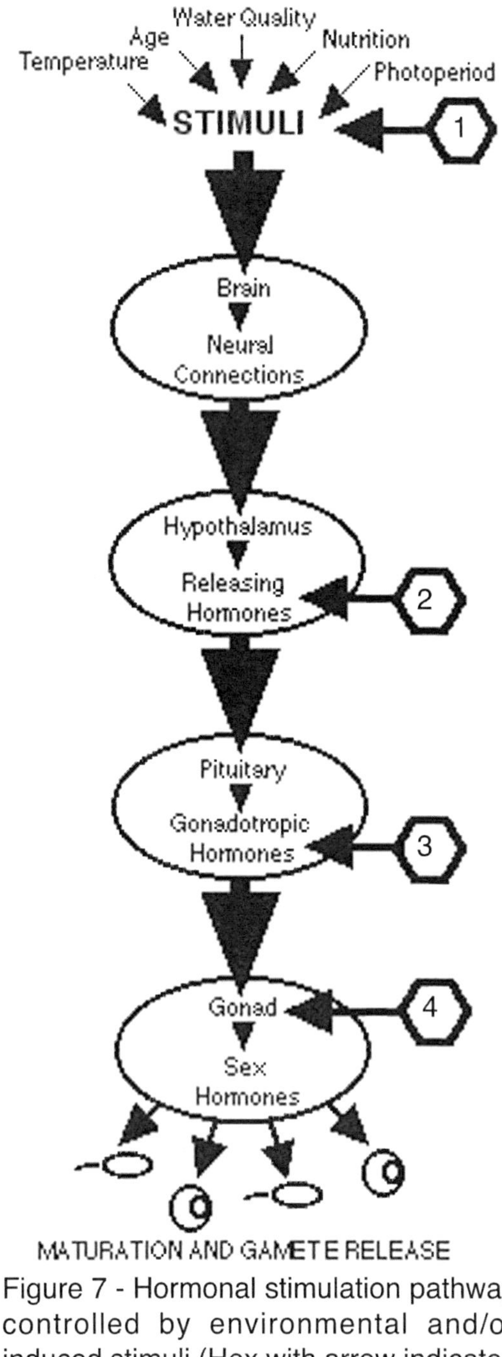

Figure 7 - Hormonal stimulation pathway controlled by environmental and/or induced stimuli (Hex with arrow indicates where artificial stimulation can be used with injected hormones). Modified from Roberts 1987.

Oocyte Sampling

Removal of live samples can be injurious and difficult to utilize on small fish like clownfish. Yet, on larger tropical fish such as freshwater sharks (*Labeo sp*), marine pompano, seabass, grouper, and angelfish, this method has been routinely employed without permanent injury. The following technique was employed on pompano, and seabass in 1970 (Hoff 1973) and is still widely practiced. Live oocytes were removed using a simple suction device made of a 5 or 10 cc plastic syringe, equipped with a stiff rigid flat tip needle. A 10 to 12 cm x 1 mm ID polyethylene tube is mounted on the blunt end of a stainless steel needle (Figure 8). A small hole is drilled at the base of the syringe chamber and used to release the vacuum, thus preventing the sample from being sucked into the syringe chamber. If properly managed, the egg sample will remain in the sampling tube and can be forced onto a slide by plugging the hole again and pushing the plunger down. Following each sample, the syringe sampler is filled with sterile saline solution to flush out remaining eggs

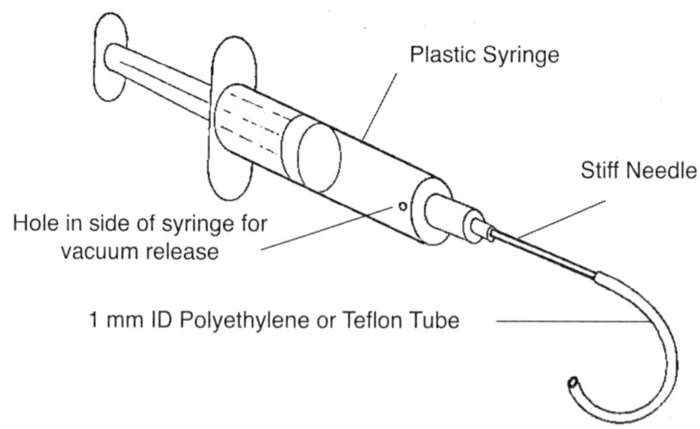

Figure 8 - Oocyte (egg) sampling device, Hoff 1972

and the sampling end (tubing) is rinsed in ethyl alcohol to maintain asepsis.

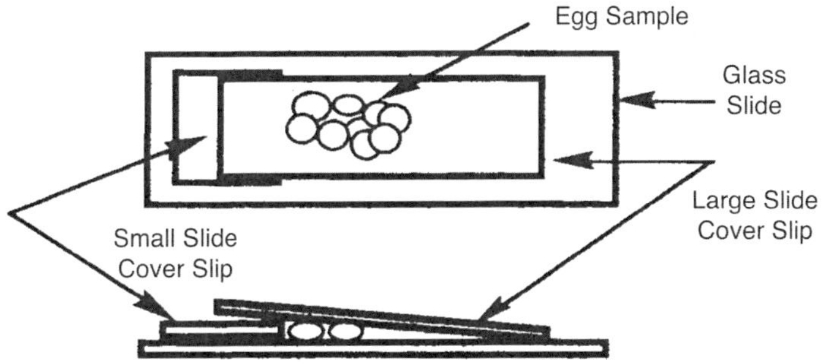

Figure 9 - Egg sample preparation for microscope viewing.

Insertion of the sampler must be done carefully. First it is important to recognize the ovaduct (egg duct or gental opening) and not tear any external tissue (Figure 11). Fish are usually anesthetized with MS-222 or quinaldine prior to insertion to relax them. They are then placed on a wet pad or work table. The sampling tube is inserted gently until a resistance is felt. Then it is slightly forced into the ovary. Once in the ovary a suction is created by placing your thumb over the hole and pulling the plunger. While removing the tube, release the suction by uncovering the hole.

Egg samples are then squirted onto a standard microscope slide. Adjacent to the sample, a small glass cover slip is glued to the slide (Figure 9). Both eggs and mounted glass cover are then covered with a larger cover slip. A small amount of sterile seawater is injected under the slide cover to keep the sample moist and improve resolution. The object of the lifted cover is to prevent smashing and adverse distortion of the eggs yet flatten them enough so that you can easily see through them. Ripe or ripening eggs have considerably larger diameters and often need two small slide covers glued onto the slide to elevate the final larger cover slip (Figure 9).

If necessary, live samples can be stored overnight. Prepared slides are covered with sterile saltwater-dampened paper towels, placed in a plastic bag, sealed, and refrigerated. Numerous comparative examinations of pompano and seabass eggs showed that this storage method did not radically alter the size or stage of egg samples up to 24 hours.

Anesthetizing Fish

When sampling larger fish or even during treatment or handling smaller fish, it usually is necessary to anesthetize them. Two drugs are commonly used: Quinaldine (5 ml per 38 liters) or MS-222 (5 gm per 38 liters). Of the two, I prefer Quinaldine since MS-222 often created an adverse nervous reaction and high agitation during recovery. Let the fish remain in the anesthetizing solution only long enough to become completely limp and easy to handle. Fish that are about ready will roll on their sides, lay on the bottom, and respire very slowly. They can be revived more quickly by manipulation of the mouth

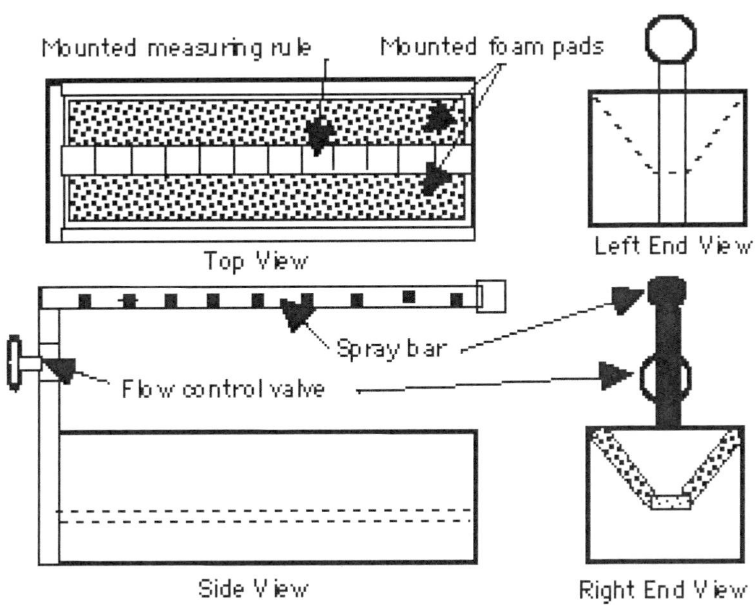

Figure 10 - Wet table for handling, measuring, in-vivo sampling and injection of large fish

To aid in handling anesthetized fish and help prevent external eye and scale/skin injury during sampling, a "wet-table" was designed (Figure 10). The "wet-table" consists of a slightly "V" shaped trough made of epoxy coated boards and covered with 1/2" polyfoam. A measuring rule is mounted in the bottom of the trough. At the head board of the trough is a spray bar connected to a water supply from the brood tanks. Water is continuously sprayed on the pads to keep them very moist. A strong word of caution is necessary. If a wet table is used, make sure the foam pads are always well soaked before placing the fish on them and do not press the fish tightly against the pads nor release it quickly. Pressing the fish displaces water from the pads and a quick release can actually suck mucus and scales off many smooth skinned fish. Another word of caution; always handle fish with wet hands.

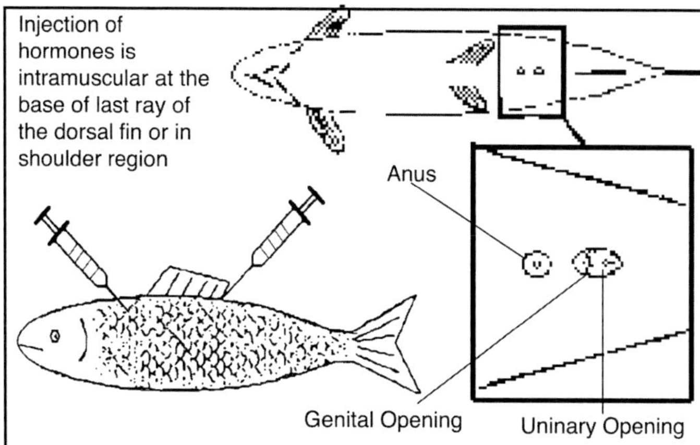

Figure 11 - Egg samples are obtained by insertion of sampling tube into the genital opening. Modified from Rottmann et al 1990.

Gonadal Development

Since the bulk of this manual is concentrated on smaller marine fish such as clownfish, the aquaculturist must rely on external appearances and actions to determine the reproductive status of the female. Several months prior to spawning, females often display gonadal abdominal swelling (Figure 40). Swelling is created by development of maturing oocytes (eggs) within the gonad. Initially, swelling followed by a decrease and no resulting spawns may be evident. Gonadal cycling without spawning may go on for several weeks or months and is more evident in younger, sexually maturing individuals. What actually happens is that a percentage of the eggs within the ovary are developing to almost a spawning state but do not completely mature, break down and are reabsorbed. Non-spawning gonadal cycling is basically due to immaturity and/or poor environmental conditioning.

Initial spawns are often hard to detect. Usually they are sporadically spread upon the substrate, not well fertilized, consist of few eggs of poor quality and are often consumed by the parents within two days. Attempts to hatch and rear initial spawns is usually a useless endeavor. If these eggs survive incubation, they often hatch late, resulting in a few weak, inferior larvae. It usually takes several months of consistent spawning of two nests per month to reach full potential and quality. Gonadal cycling which results in no spawn is common and nothing to be alarmed about. It can be compared to puberty in humans. It is a growing up process, a training period, a new change within and without the body. Yet, it is important to realize that if this condition continues longer then two months other factors may be inadequate.

Gonadal cycling is a natural occurrence of maturing and mature fish. However, this natural cycling is strongly influenced by improper environmental conditioning procedures. If conditions are not adequate, cycling can continue for extensive periods of time, resulting in no spawns. Factors such as diet, feeding, lighting, temperature, habitat, and general water quality, are all involved and have direct influence on gonadal cycling. Diet and feeding habits are the most important factors to obtain quality spawns including the number of eggs, percent hatch and larval response. Whereas, temperature and light are the most important factors in triggering spawn time, hatch and maintaining sexual organs in a ready spawning state. Water quality is important with regards to health, egg quality, egg hatching, and sperm activity. Often just one of these factors can trigger poor quality spawns. More detail is provided in the conditioning section.

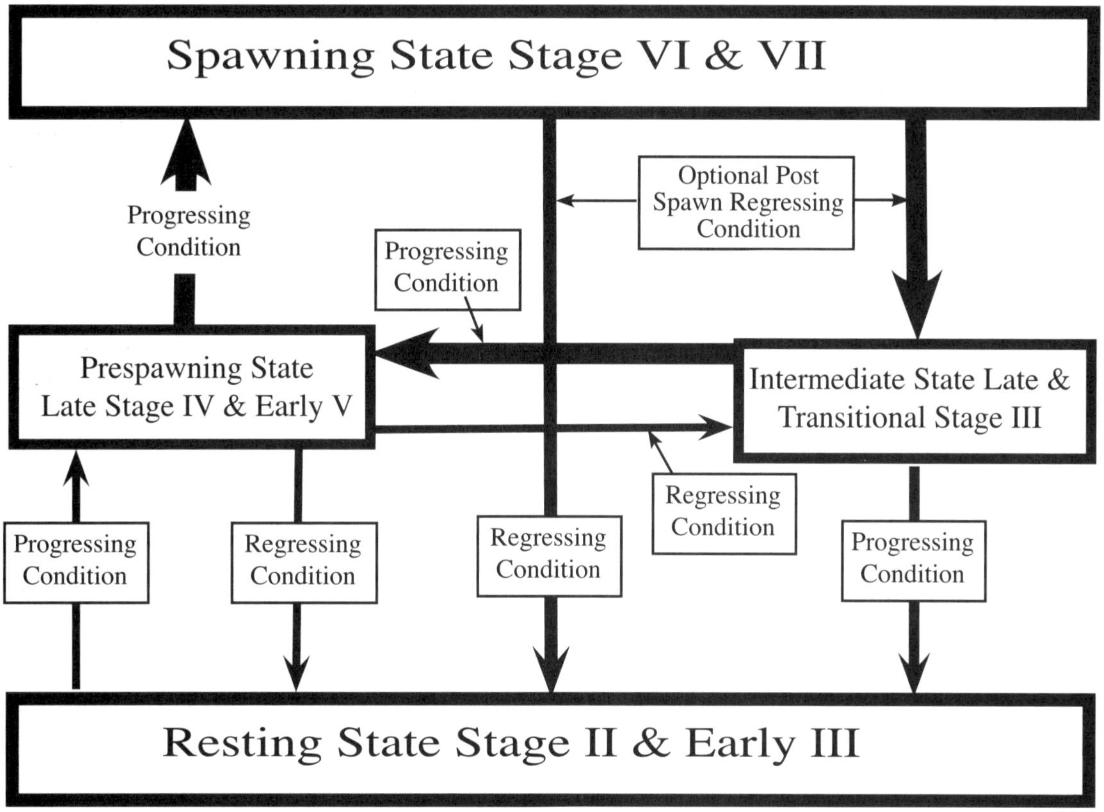

Figure 12 - Gonadal conditions and states associated with gonadal cycles which are controlled by environmental conditions and physical health. Heavy arrows indicate the normal pathways for a protracted spawner like clownfish.

Egg Development

Well, what is going on within the gonad? What do the eggs look like, what is causing the swelling, and what do the eggs consist of? A typical pattern for pelagic spawning fish (those that spawn free floating eggs) is shown in Figure 12. Although this chart is patterned after a pelagic spawning fish, the same cycle occurs in benthic spawning fish (those that spawn on bottom substrates) as well. As mentioned, these states of development are highly dependent and regulated by internal and external environmental conditions.

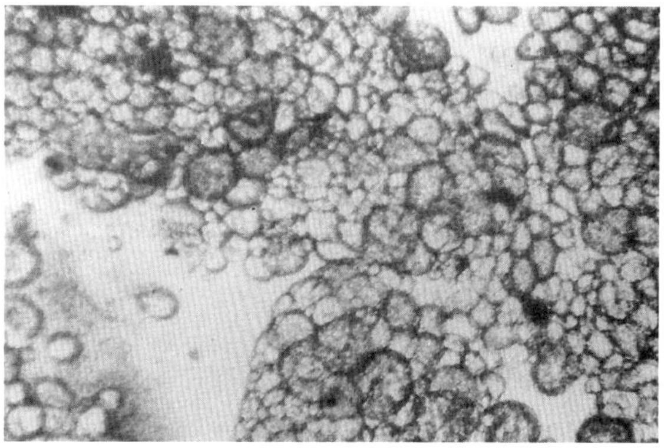

Photo 3 - Resting State, top
Photo 4 - Pre-spawning State, below

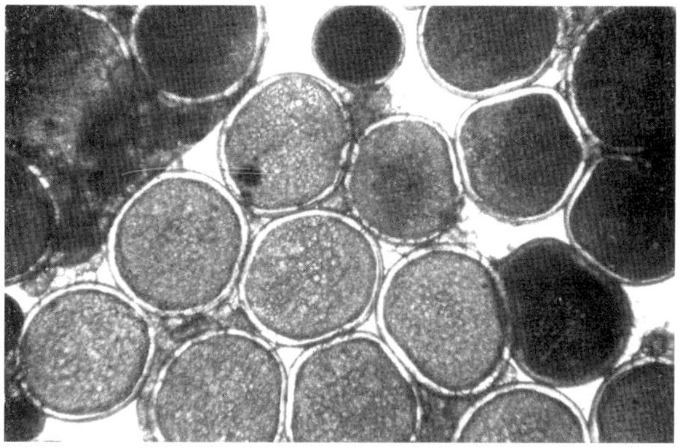

Immature fish or adult fish not in their normal spawning season contain mainly resting eggs (Photo 3). Resting eggs are characterized as being small and very uniform in shape. As environmental parameters, mainly light and temperature, change toward optimal conditions, a specific group of eggs begins to mature and the gonad reaches a prespawning state (Photo 4). At this point, a sudden change in the physical environment, habitat, or other forms of induced stress can severely alter the pending spawn. For example, if you decide to move fish to another tank or perhaps change habitat in the tank, you can actually cause enough stress to force developing eggs to regress back to a resting or intermediate state (Photo 5). I have often seen fish farmers remove fish in a pre-spawning state from their ponds and place them in a completely different tank, water supply and light condition and then expect them to spawn several days later. Usually attempts to spawn these fish, even in conjunction with subsequent hormone injections, is futile. Although the eggs obtained might look normal, they usually result in inferior larvae. Basically, what has happened is that the stress they induced starts to effect the maturing eggs as soon as the fish is captured. If the fish is not spawned within 24 hours or less, the eggs will regress to an intermediate state. If the stress is minimal and not prolonged, the eggs will revert back to an intermediate state (Photo 5, Figure 12). If stress continues, the entire gonad will regress to a resting state (Photo 3, Figure 12). Intermediate state eggs are mid-stage in development and can remain in this holding sequence for extended periods of time. If all stress is removed within a specific time frame and all environmental triggers are optimal, eggs will progress to a pre-spawning state. If forced to regress to a resting state, development back to a pre-spawning state will take considerably longer regardless of environmental conditions.

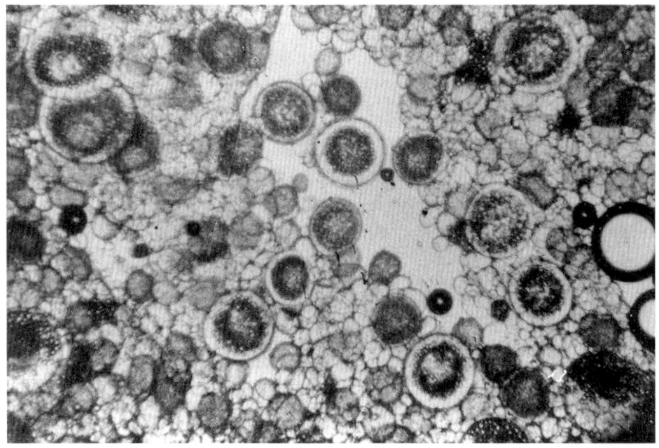

Photo 5 - Intermediate State

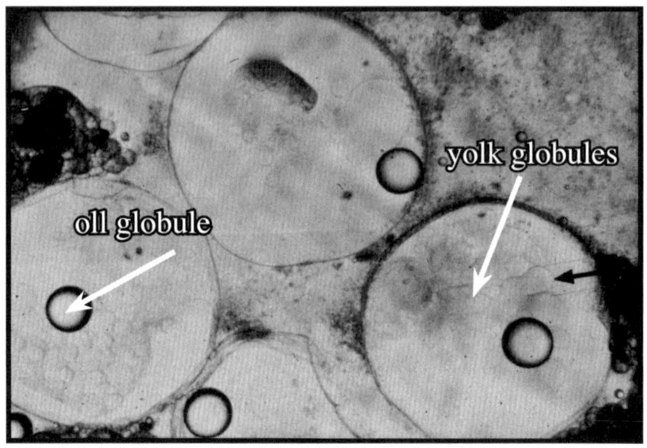

Photo 6 - Spawning State

Once oocytes reach a pre-spawning state (also referred to as vitellogenesis), hormones are often used on larger fish. Currently many commercial farmers are injecting a combination of hormones including HCG with carp pituitary, synthetic LH-RH analogue (LRH-A) used with reserpine. Generally they are injected into the fleshy upper shoulder region behind the head immediately after capture. These are designed to block the release of enzymes which in turn destroy the developing eggs. A further injection is often used to induce a spawning state. The hormone of choice over many years has been human chorionic gonadotrophin (HCG). Dosages vary but usually one and sometimes two are administered at a rate of 50 to 500 IU/Kg.

If stress remains minimal, eggs enter a pre-spawning state and will quickly develop to a spawning state (Photo 6, Figure 12). Once the eggs develop to a pre-spawning state, a spawn is inevitable but the resulting egg quality can be good or poor. On larger commercial aquaculture species, injection of hormones is commonly administered usually to only females in a pre-spawning state.

Once the gonad is triggered, either by induced hormones or natural environmental conditions, there is no turning back regardless of any induced stress. If stress does occur during a spawning state or if the eggs were not completely ready to be force-induced by injections, spawning will occur but egg quality will be affected. Depending on environmental conditions after spawning, ripe or ripening eggs remaining in the gonad will be reabsorbed and the gonad will regress to a intermediate state or a resting state. Which state the gonad regresses to depends on the species

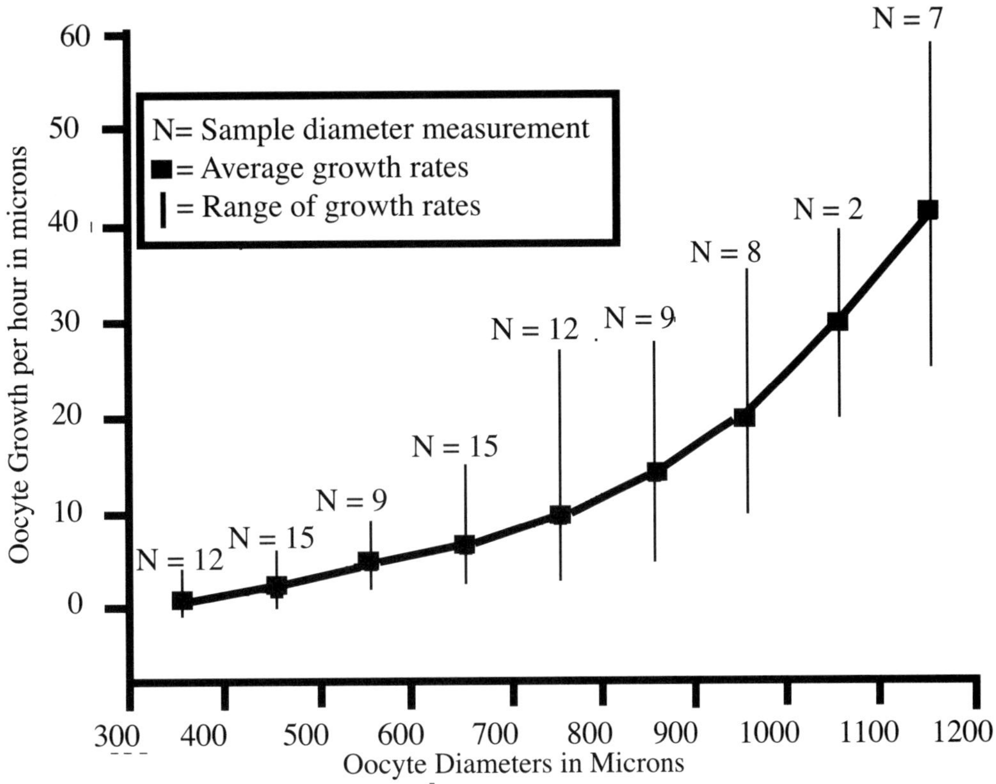

Figure 13 - Oocyte (egg) growth rate per hour from an intermediate state to a spawning state after hormonal injections (Hoff 1976).

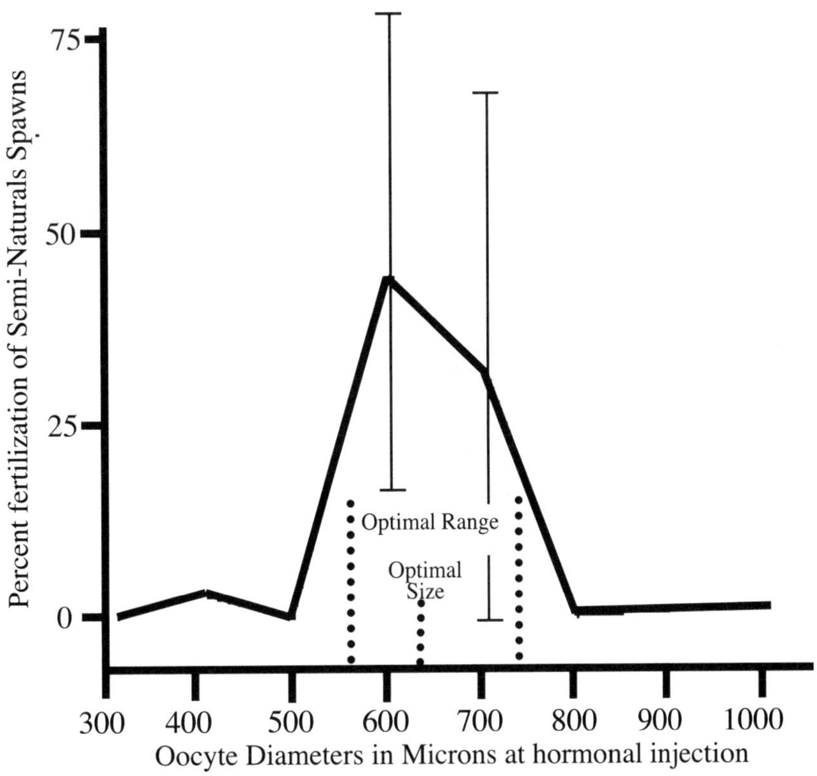

Figure 14 - Influence of oocyte diameter at initial injection on percent of fertilized eggs. Note optimal oocyte range for injection.

fish after spawning and/or the environmental conditions. For a complete oocyte (egg) development photographic series including prepared and live oocytes see the egg sampling kit and chart from Florida Aqua Farms Inc. (1996).

Growth of eggs from a pre-spawning state to a spawning state, referred to as hydration, is very quick (Hoff 1978). Growth rates (Figure 13) were determined from live egg samples removed from pompano (*Tracinotus carolinus*). From a pre-spawning state to a spawning state, eggs can develop at a rate of 25 to 30 microns per hour (1000 microns = 1 mm). Once hydration is triggered, spawning to some degree is inevitable.

The optimal size of oocytes prior to injection directly influences the percent of fertilization. Immature or overripe oocytes yielded lower fertilization rates (Figure 14). Oocytes within the pre-spawning range of 550 to 750 microns (Photo 4) are ideal.

Mature males undergo developmental gonadal changes as well but usually only from a pre-spawning state directly into a spawning state. If you examine a section of male gonadal tissue, there are pockets tightly packed with numerous tailed sperm. Sperm are actually held in a mature, ready state, but lack hydration. When you press the side of a non-hydrated male to test for sperm expression, it is a white, thick, pasty consistency. Hydrated males express a free-flowing thin milky fluid that actually squirts out. Hydration in male fish is triggered by pheromones released by a ripening female fish. It can be triggered by utilizing hormones but this is usually not necessary. Simply by allowing males to remain in the same tank with ripening hydrating female fish is usually all that is necessary.

Conditions regarding water quality, temperature and light play a important role in the quality and fertilization of spawned eggs. This is further discussed under conditioning.

Centroprites melanus
Black Sea Bass

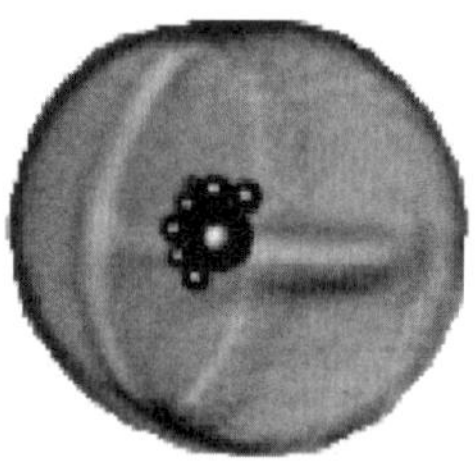

Early Neurula

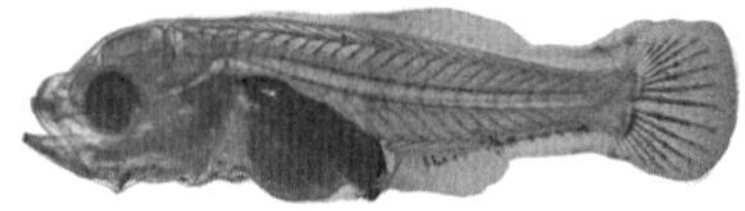

Sea Bass early true juvenile 11 days, TL 6.95 (see pg 74)

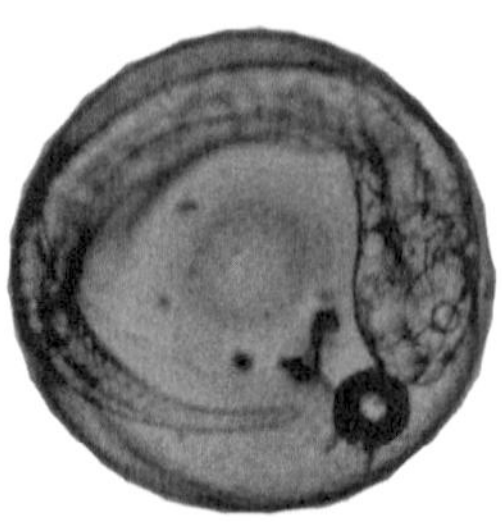

Late Neurula

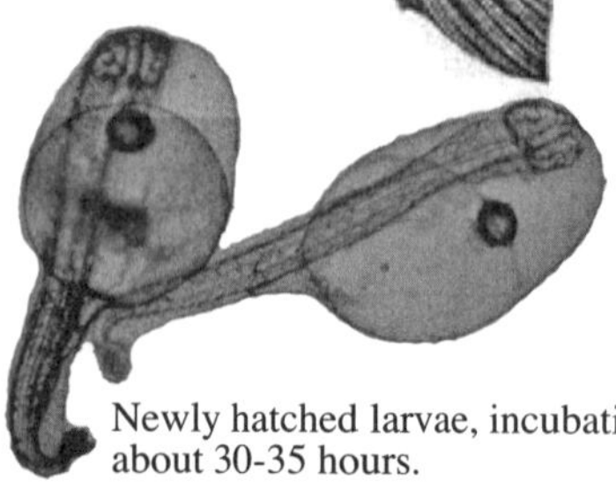

Newly hatched larvae, incubation about 30-35 hours.

Juvenile 17 days, TL 11.7mm, SL 9.7mm

Juvenile 45 days old, 5.4cm TL, SL 4.7 cm Wt. 2.54 gm.

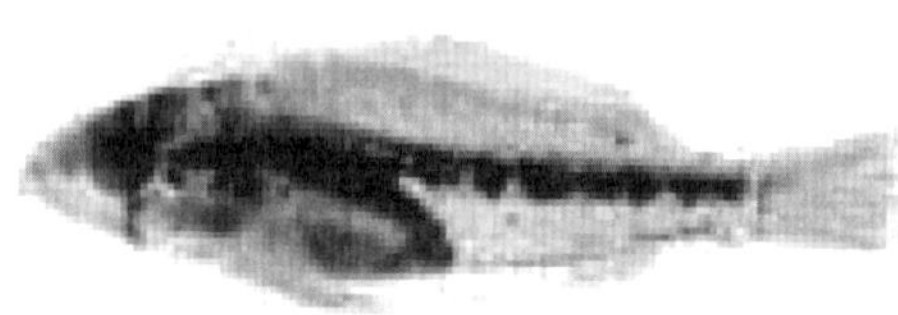

uvenile 22 days, TL 16 mm, SL 13 mm

SPECIES SELECTION

Although this manual is basically directed toward the culture of clownfish, it is important to understand why this group was selected or why any species is selected. Correct selection of the species you wish to culture is important toward success in a reasonable amount of time. Even consideration of feeding habits and swimming habits can be important factors. For example; constantly swimming pelagic species usually live in large open waters and are not the easiest to spawn in closed confined aquariums or tanks. Normally this type of fish is a pelagic spawner which release eggs freely into the water column. Few seem to spawn in average, well maintained home aquaria or even under the best laboratory conditions without the aid of hormone injections. Those that are exceptionally large at adulthood may be difficult to handle and spawn especially if hormone injections are utilized. Table 2 lists the criteria for species selection with regard to biological aspects and economics.

There is a fine line between species selection and the amount of biological data and your expertise. If you like challenges, have time, reasonable expertise, and can tolerate long term failures, then select species that have never been reared before. If you are the type of person who needs regular incentives and reasonable ease of success, then select species with considerable background data like clownfish.

Based on data and biological knowledge, I have listed several species that I feel have exceptional potential for the marine tropical fish market (Table 3). I have also rated these as to what I feel will be their degree of difficulty. Although this rating is only a guess, it is based on years of experience in marine fish breeding. The degree of difficulty is rated from 1 to 10 with the higher number being the easiest. I have also indicated that published literature exists regarding the specific species or closely related species. This table is further expanded by noting that unpublished work exists on this species or closely related species. This table is only a guideline to future endeavors with regard to species. Further information on commercial feasibility is available in the last chapter.

Table 2 - Criteria for specie selection for commercial culture.

1 Demand - is there a current strong demand and short supply situation in the pet market?
2. Value - is there sufficient value in selecting a particular species to sustain your input?
3. Space requirements - can you obtain sufficient space to meet the requirements of the particular species you select?
4. Biological Literature - is there sufficient literature and/or biological data regarding the selected fish or a related species?
5. Biological quirks - is there something unusual with regards to conditioning, spawning or rearing the selected species which would hamper success in a reasonable amount of time?
6. Sustaining time - do you have sufficient time and money to invest to reach your goal?
7. Marketing - can you or someone else market your production on a constant basis?
8. Many of these factors would be the same for culturing food fish.
9. Major differences occur in the scale of a facility designed for raising food fish, including tank sizes, space, water demands, growth time required to obtain a marketable fish and farm gate prices compared to small topical fish.
10. However, there is a limit on just how many fish species you can produce and marketing efficiently.

Table 3 - Criteria for species selection of marine tropical fish based on previous published and unpublished data. Lower numbers indicate the'fish that are currently more easily raised and usually more articles and unpublished articles exist.

Family/Species	Common Name	Difficulty	Articles	Unpub
Aluteridae				
Eubalichthys mosaicus	mosaic leatherjacket	1	--	--
Antennariidae				
Histro histro	sargassum fish	1	--	--
Aulostomidae				
Aulostomus chinensis	trumpetfish	1	--	--
Callionymidae				
Synchiropus splendidus	mandarin fish	5	X	X
Carrangidae				
Alectis indica	threadfin	3	--	X
Selene vomer	lookdown	3	--	X
Chaetodontidae				
Chelmon marginalis	margined coral fish	1	--	--
Chelmon rostratus	copper-banded	1	--	--
Chaetodon lunula	racoon butterfly fish	1	--	--
Forcipiger flavissimus	long-nosed	1	--	--
Damselfishes				
Pomacentrus coelestis	blue devil	4	--	--
Glyphidodontops talboti	??	1	--	--
Glyphidodontops hemicyaneus	??	1	--	--
Glyphidodontops cyaneus	??	1	--	--
Eupomacentrus aureus	??	1	--	--
Ambiyglyphidodon aureus	??	1	--	--
Abudefduf saxatilis	sargent major	6	X	X
Dascyllus trimaculatus	domino	3	X	X
Dascyllus melanurus	??	1	--	--
Chomis analis	??	1	--	--
Premnas biaculeatus	maroon clownfish	8	X	X
Amphiprion sandracinos	orange skunk	7	X	X
Amphiprion perideraion	pink skunk	5	X	X
Amphiprion tricinctus	??	5	X	X
Amphiprion polymnus	saddle back	6	X	X
Amphiprion frenatus	tomato	10	X	X
Amphiprion melanopus	cinnamon	8	X	X
Amphiprion percula	percula	8	X	X
Amphiprion ocellaris	ocellaris	10	X	X
Amphiprion latezonatus	??	3	X	--
Amphiprion clarkii	sebae	8	X	X
Amphiprion chrysopterus	??	3	X	X
Amphiprion tricinctus	??	3	X	X
Diodontidae				
Atopomycterus nicthemerus	porcupine fish	1	--	--
Grammidae				
Gramma loreto	fairy basslet	3	--	X
Gramma melacara	blackcap basslet	3	--	X
Lutjanidae				
Lutjanus sebae	red emperor	4	--	X
Monocentridae				
Cleidopus gloriamaris	knight fish	1	--	--

Family/Species	Common Name	Difficulty	Articles	Unpub
Ostraciontidae				
Anoplocapus lenticularis	high-backed boxfish	2	X	X
Canthigaster cinctus	black-saddled puffer	2	X	X
Ostracion lentiginosus	blue boxfish	2	X	X
Lactoria fornasina	blue-spotted cowfish	2	X	X
Opistognathidae				
Opistognathus aurifrons	yellowhead jawfish	6	X	X
Plectorhynchidae				
Plectorhynchus polytaenia	ribbon sweetlip	2	--	X
Pomacanthidae				
Centropyge hispinosus	two-spined angelfish	4	X	X
Centropyge flavissimus	lemonpeel angelfish	4	X	X
Centropyge loriculus	flame angelfish	5	X	X
Euxiphipops navarchus	blue-gridled angelfish	1	--	--
Pomacanthus semicirculatus	blue angelfish	2	X	
Pomacanthus annularis	blue-ringed angelfish	2	X	--
Pomacanthus imperator	emperor angelfish	2	X	--
Pomacanthus semicirculatus	semicircle angelfish	2	X	--
Pomacanthus paru	french angelfish	7	X	--
Pomacanthus arcuatus	gray angelfish	7	X	--
Holacanthus tricolor	rock beauty	2	X	--
Holacanthus ciliaris	queen angelfish	3	X	--
Plotosidae				
Plotosus anguillaris	striped catfish	3	--	X
Sciaenidae				
Equetus lanceolatus	highhat	6	X	X
Scoprpaenidae				
Brachirus zebra	butterfly lionfish	3	X	--
Pterois volitans	fire lionfish	3	X	--
Serranidae				
Cephalopholis miniatus	coral cod	4	--	--
Hypoplectrodes nigrorubrum	black-banded perch	3	--	X
Hypoplectrus clorurus	yellowtail hamlet	4	--	X
Hypoplectrus gemma	blue hamlet	4	--	X
Hypoplectrus gummigatta	golden hamlet	4	--	X
Hypoplectrus indigo	indigo hamlet	4	--	X
Serranus annularis	orangeback bass	3	--	X
Serranus tigrinus	harlequin bass	3	--	X
Serranus totugarum	chalk bass	3	--	X
Syngnathidae				
Phyllopteryx taeniolatus	weedy sea-dragon	5	X	X
Hippocampus erectus	common sea-horses	7	X	X
Triglidae				
Paratrigla vanessa	butterfly gurnard	1	--	--
Family?				
Calloplesiops altivelis	comet	6	X	X

Sourcing Literature And Unpublished Data

Sourcing literature is very important when starting out in anything you want to do. There is absolutely no need to re-invent the wheel. When speaking about freshwater fish culture there are extensive amounts of published data in popular and scientific literature, yet marine literature is scarce. The techniques and experiences found in freshwater literature should not be disregarded since considerable data is applicable to marine culture. A fish is basically the same with regard to needs and social interaction. Species quirks exist, which are the problems you want to solve.

Popular literature is probably the most abundant and easiest to obtain with regard to culturing tropical aquarium fish. Yet, care should be taken when reading or utilizing data from this literature since hobbyist authors often publish based on only one or two events and have very little back-up data to support their theories and speculations. Often data on temperatures, water quality, lighting, feeding, tank size, etc. are written in such a manner that the data should be considered as gospel. Often they lack supportive bibliographies. When reading this literature, be flexible and glean what is useful to your needs but keep in mind these experiences are based on little data. Yet, often this literature is all that is available and

On the other hand, scientific literature often becomes narrow with regard to reporting all the parameters. Often it dwells on nitty-gritty details which do not have apparent, immediate, practical applications. Yet, what is obtained should be based on multiple data, past literature, and conducted by rigid scientific protocol. Additionally, scientific papers require the author to conduct a literature search prior to publishing, which helps you quickly locate other related sources.

Popular literature sources include local aquarium clubs, index printouts available from popular aquarium magazine publishers, and direct contact with the authors of these articles. Publishers can provide you with addresses and/or phone numbers to contact the authors. Book listings are available either from the publishing company or through local book vendors. Internet computer sources have current data which may help. Even though your pet store may lack a comprehensive book selection, wholesale suppliers of pet supplies have a wide selection. Ask your pet shop dealer to take you with them to the wholesaler or request a specific book from Amazon. Some local public libraries may have a fair selection of aquarium books. There are also several book suppliers listed in tropical fish magazines which have a wide variety of books from various sources and can help you obtain what you are looking for. I have found many good books in used book stores.

Scientific literature is available through a multitude of sources. University libraries have access to virtually any published scientific literature via their inter-library exchange system. Copies of literature not within their facility can be obtained for a nominal fee. Many of these libraries are computerized, which eliminates a lot of work. Federal and state aquatic extension agents are available to the public within your county government agencies. Consult your phone book or the department of agriculture agencies within your state. Marine and freshwater fish hatcheries and research labs often have extensive libraries which you are allowed to use. Try the Internet, it is filled with information from hobbyists who have had limited success.

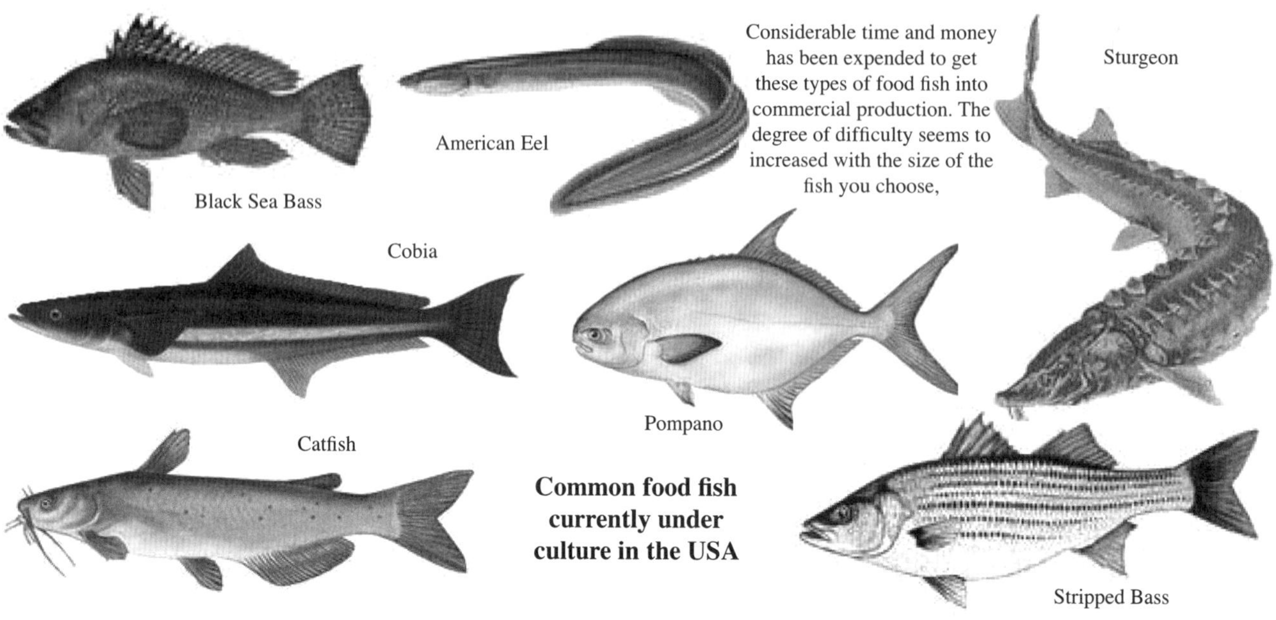

Considerable time and money has been expended to get these types of food fish into commercial production. The degree of difficulty seems to increased with the size of the fish you choose,

Common food fish currently under culture in the USA

SYSTEM DESIGN AND FILTRATION

System Versus Singular Tanks

I have always preferred centralized, closed, recirculating systems over single tank systems or open-water systems. "Diseases will spread in a central system from one fish to another" is absolutely not true if the system is properly designed and maintained and the fish are not stressed. From years of experience, I have found that diseases within adult broodstock systems, occurred on an individual (tank to tank) basis and normally did not spread to every tank via the common water source (see Disease section). Recirculating systems are particularly prone to disease problems when too high a density and eutrophic conditions exist. Yet, distinct positive advantages exist when dealing with daily environmental parameters that must be controlled and maintained equally from tank to tank.

Placing broodstock pairs in individual aquaria is like placing each pair in a different aquatic environment. Only light and sometimes temperature are equal from tank to tank. Dissolved organics, phosphates, nitrates, ammonia, trace nutrients, vitamins etc. are all different on a per tank basis. No doubt these differences have a direct bearing on the performance and nest quality of each spawning pair. Trying to maintain each tank even close to equal is virtually impossible. In commercial applications, utilization of individual static tanks is burdensome to maintain, clean, check for individual water quality, and make water exchanges. Extreme maintenance of individual aquaria becomes costly and time consuming.

For the average home aquarist, single tanks equipped with subgravel or sponge filters are certainly suitable. Even small systems consisting of a remote wet/dry filter, fluidized bed biofilter, foam skimmer and UV and several small holding aquaria can be used.

In designing filtration systems, initial costs and space requirements should not become limiting factors. Filtration is a main key to success and will "make or break" a commercial venture regardless of whether it is an open, semi-open, or closed system. Flexibility and simplicity in design is required so maintenance, repairs, and improvements can be made at minimal costs and ease of operation.

Types of Systems

There are three types of systems used in aquaculture, "closed, semi-open, and open". Open systems utilize either surface waters (salt or freshwater) from adjacent rivers, streams, lakes, bays and oceans, or from deep or shallow wells. Within these facilities water is constantly flowing in and out of the tanks. Prefiltration is normally only mechanical, utilizing sand filters and at times followed by UV or ozone sterilization. Some use a series of holding reservoirs where enough water necessary for one day's operation is prefiltered, retained, sterilized by UV, chlorine and/or ozone, then neutralized and utilized the following day (Figure 15). Open intakes are usually retained in a large box equipped with removable screens for filtering out large particles. Water is then passed through a series of sand filters then directly into culture tanks or reservoirs. Other intakes consist of a manifold equipped with a series of shallow well points buried close to the shore lines (Figure 16). Sandy shoreline conditions are best but porous plastic containers filled with sand can be buried into rocky shorelines. This arrangement allows the surrounding sand to act as a sand filter thus eliminating normal particle contaminants. Of the two arrangements, I prefer the latter. Others utilize deep salt or freshwater wells. Keep in mind freshwater floats on top of saltwater so in many places deep saltwater aquifers are available. Although this is a tempting situation, water obtained from deep wells is virtually sterile, is often not well balanced with regards to dissolved minerals, metals, contains hydrogen sulfide and lacks oxygen. Elaborate preconditioning of deep well water is often necessary.

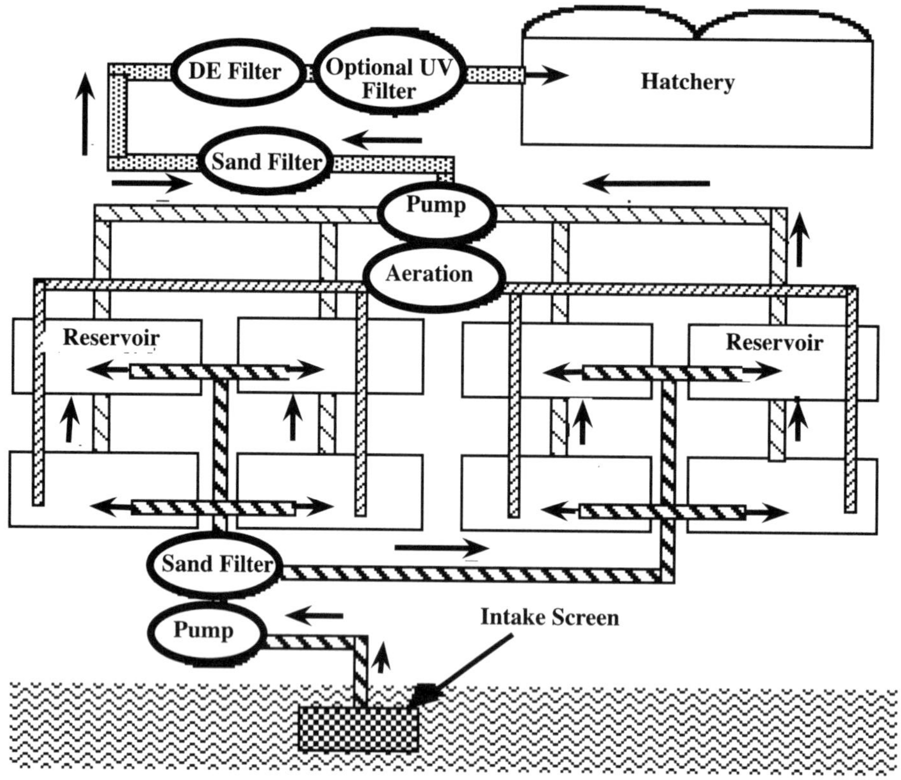

Figure 15 - Alternating reservoir open-water exchange system. One or more reservoirs are used for daily water exchanges. Newly refilled reservoirs are chlorinated over a 24 hour period and neutralized.

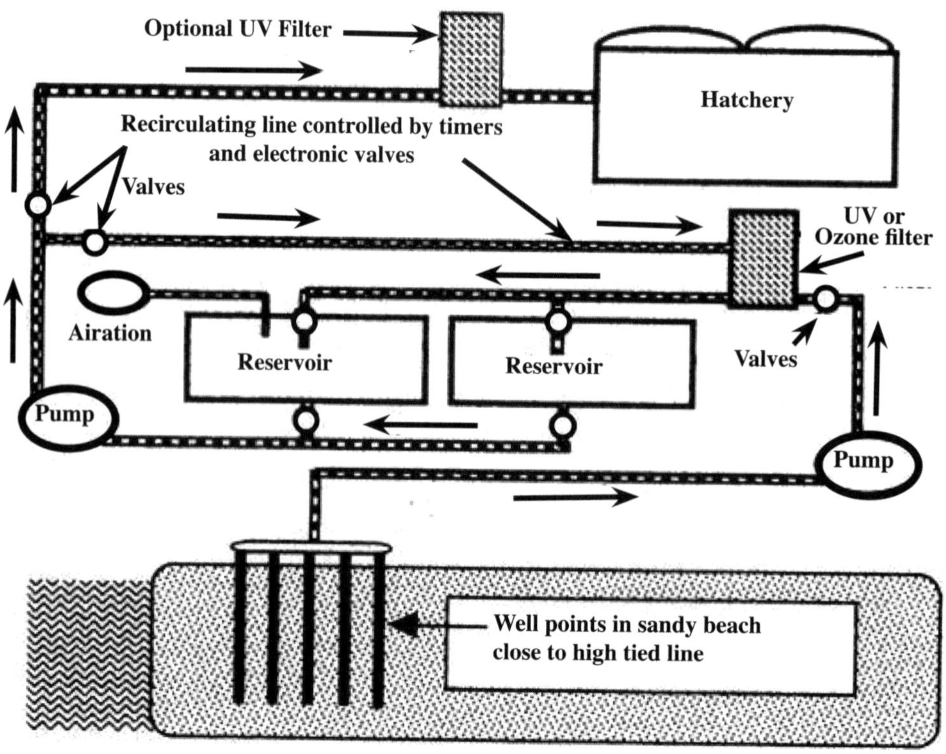

Figure 16 - Open water system utilizing well points and conditioning reservoirs.

In open systems a specific amount of water is constantly being exchanged in the culture tanks. Usually one to four complete exchanges are made per tank in a 24-hour period. A distinct disadvantage of utilizing water from either adjacent surface waters or using shallow well points is chemical and temperature changes which occur seasonally. Consistency of surface waters changes seasonally due to rain, terrestrial runoff, dissolved nutrients, inorganics and organics, light intensity, temperature, and current changes. Physical and chemical changes can have a pronounced effect on microalgae, rotifers, and larval culture. These changes are usually less pronounced with regard to broodstock and juvenile culture. Temperature can be easily rectified at some cost but changes in dissolved nutrients, metals, organics etc. is harder to detect and rectify.

Open-systems utilizing deep-well water, either fresh or saltwater, present problems if water exchanges per day are excessive. Exactly where that point occurs depends on individual differences in water sources, how the water is pretreated, how it is introduced into the tank and the particular species being cultured. Deep-well water is basically "raw water", not much different from newly mixed artificial seawater. It is basically stagnant, low oxygen, sterile water. Once water mixes with surface oxygen and other gasses, chemical changes and exchanges occur. Often dissolved metals and minerals, such as excessive iron, phosphate, and silicate precipitate out when contacting oxygen. In addition, hydrogen sulfide, ammonia and at times nitrate are present in deep well water. As dissolved organics are added, well water becomes balanced and more suitable for culture work. Subjecting aquatic animals to excessive amounts of "raw water" is not natural and can cause constant stress which in turn leads to disease, etc.

Semi-open systems are becoming more popular in older open-flow hatchery facilities. This consists of exchanging a set volume of water each day, usually 10% is maximum. Between exchanges, water is either maintained within the tank or recirculated through a central system consisting of various filters. Exchanges are made by either draining a set amount of water and then refilling the tank or system or by flowing a small set amount of water into the tank or system on a continuous basis. Of the two, draining and filling is best. Simply flowing water into a filled tank or system only partially dilutes nutrient levels since a percentage of exchange water is also lost directly to the drain. Excessive exchanges, higher then 10% per day, may be detrimental to sensitive aquatic animals.

Closed systems utilize water exchange but it is usually 20% per month. A distinct advantage of semi-open or closed systems is water stability. Water stability and balance is necessary within all aquatic bodies to keep animals growing, sexually mature, and healthy. Recirculating systems offer this security and balance.

System Arrangement At IOH

Instant Ocean Hatcheries (IOH) entire broodstock section was interconnected into three central closed systems. Each system (Figure 17) was designed so that individual tanks and entire rows of tanks could be isolated from the central system. This allowed us to treat, repairs and/or clean a tank or an entire row of tanks without influencing or interrupting the entire system. Each tank was equipped with a submerged gravel biological filter with a single centrally located air lift pump. There were 6 rows of adult broodstock fish (204 tanks) which were all connected together. Tank walls were opaque on all sides except the front. The stand-pipe, for maintaining water depth, was located on the outside of the tank (Photo 7, Figures 17 and 31) to eliminate excess internal plumbing and aid in routine maintenance and capture. Each was provided a plastic spigot drain on the bottom for washing and draining. The juvenile growout systems were basically the same consisting of 12 rows of tanks with 2 rows comprising a single system. The larval rearing area consisted of one large 10,000-gallon system but had the same basic flow design as the broodstock and juveniles.

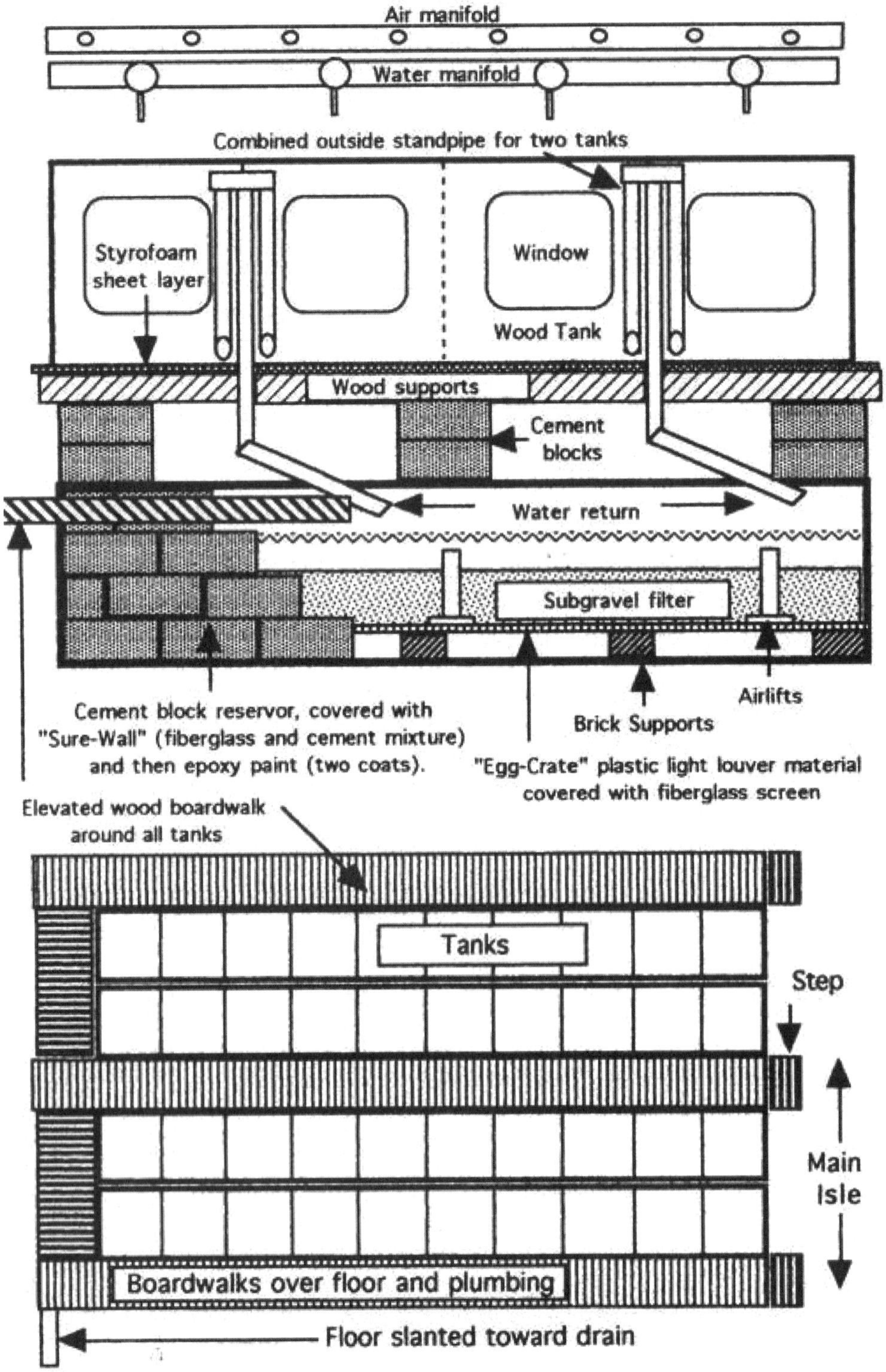

Figure 17 - Arrangement of adult and juvenile tanks of IOH. Elevated boardwalks helped keep workers dry , covered floor and allowed for easy tank servicing.

Photo 7 - Exterior view of tanks located above reservoirs and showing exterior stand pipe. Elevated boardwalk is between the reservoirs for ease of draining and cleaning.

Photo 8 - Interior view upper level showing main access isle (left) and rows of tanks mounted above reservoirs.

Interior Views of Instant Ocean Hatcheries

Photo 9 - Interior view showing lower level with pumps on boardwalk, DE filter on far right and UV filters mounted above pumps on wall. DE filters are mounted along side of large auxillary reservoirs.

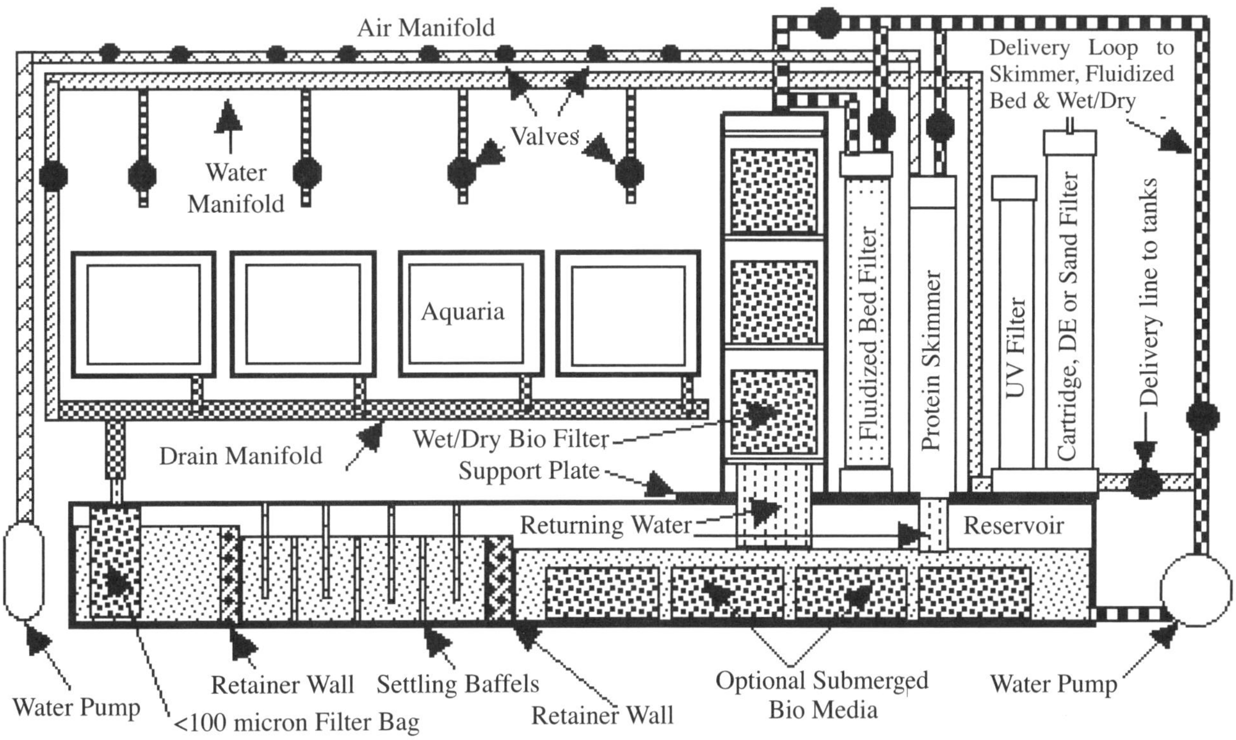

Figure 18 - Recirculating system suitable for marine or freshwater tropical fish culture.

Instant Ocean Hatcheries' entire operation was housed in a 200' x 56' (61.5 x 17.3 m) insulated cement block building with a double layered whitish-clear fiberglass roof. Lighting was ambient with low-level, indirect, timed, lighting located between the two layers of roof panels. Temperature was controlled by a combination of heat sources including solar daylight heat, forced hot air, and the heat sink created by the large volume of water (110,000 gal. = 413 cubic meters) retained within the hatchery (Photos 7, 8, 9, and back cover).

Key Points For Hatchery And System Designs

Simplicity, efficiency, and ease of maintenance are key issues with regard to hatchery and system designs. If asked to design another commercial hatchery for marine tropical fish it would not be exactly like Instant Ocean Hatcheries (IOH) but similar to the ones I helped design for C-Quest in Puerto Rico (Figures 63, 64, 65 & 66) and Hidden Lake Hatchery in Lakeland, Florida. Although the overall design, tank layout, location, size, water and air circulation, etc. were quite adequate at IOH, the filtration systems were not. Central systems for either broodstock or juvenile growout should contain the following;

1. A large capacity reservoir.
2. Particulate filter systems capable of removing particles to 5 to 10 microns in size.
3. Remote fluidized bed biofilter.
4. Remote, large, wet-dry biological filtration system.
5. Aeration and water systems which deliver to each tank.
6. Maximum UV filtration.
7. Remote protein skimmer.
8. Remote carbon filters for occasional use.
9. Simple, removable, biological sponge filter for each tank.

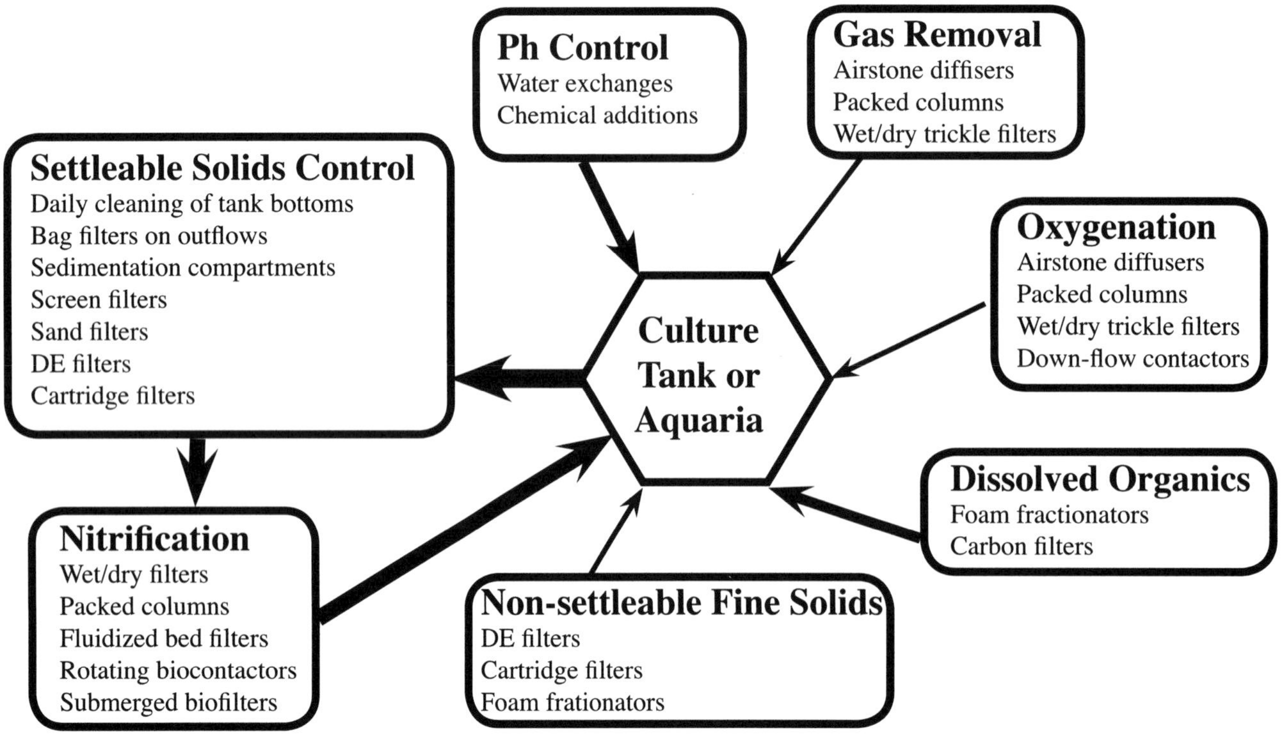

Figure 19 - Typical components used in recirculating production systems (Modified after Losordo et al 1992). Single directional arrows indicate units which are usually plumbed with a separate pressurized water source and not part of the main flow system.

10. Sufficient, easy to use, system water exchange.
11. Ability to provide variable volumes of water and air to each tank.

Remote indicates that these filters are attached to a seperate delivery loop and not directly to the line feeding the aquaria. Figure 19 further outlines the types of filters and their function and use within closed recirculating systems.

Estimating Rate of Water Exchange

Maintenance of satisfactory water quality in a recirculating system is affected not only by management but by system design. In an open-system, water quality is directly proportional to the rate of exchange in each tank. In a closed or semi-closed recirculating system other factors drastically affect water quality. The ratio of tanks to filter volume (R_a), time (T) it takes for all the water to pass through the filter, exchange rate (E) or percent of new water added each day, retention time of water in culture tanks (R_t) and retention time in the filters (R_f) are all related to water quality. Basically the larger the filter, or the lower the tank/filter ratio, the better water quality should be. Faster turnover of recycled water and shorter retention time in the tanks results in improved water quality. As retention time in the filter increases, and as the percent of new water (E) added per day increases, the water quality (Q) increases. These relationships are expressed in the equation 1 below

$$Q = \frac{R_f \times E}{R_t \times T \times R_a} \qquad \text{(Equation 1)}$$

Water quality (Q) increases proportionally with new water exchange (E) and flow rate, but varies inversely with retention time per tank and tank to filter ratio. Using this equation we can analyze any given recirculating system and determine if it would be more practical to increase new water exchange, water flow into the tank, or the size of the filter to improve water quality. Equation 2 predicts the carrying capacity in open-flow raceways for fish of any given length. W = total weight in the system of the carrying capacity, F = is the loading factor for the system, L = average length of fish in inches and I = is the gpm (gallons/minute) new water exchanged.

$$W = F \times L \times I \qquad \text{(Equation 2)}$$

Since carrying capacity cannot be directly related only to gpm in a recirculating system, this equation must be modified. Substituting Q (water quality) for I (new water flow).

$$W = F \times L \times Q \qquad \text{(Equation 3)}$$

Using Equation one we develop

$$W = \frac{F \times L \times R_f \times E}{R_t \times T \times R_a} \qquad \text{(Equation 4)}$$

These equations only indirectly evaluate the filter efficiency of a recirculating system, but they do so on empirical terms. When any factor (ammonia, DO, pH, BOD, CO_2, etc.) becomes limiting, the carrying capacity will have been reached under those conditions and the loading factor can be determined. To evaluate the effectiveness of different methods of water purification, types of filter media, or perhaps the effect of foam strippers in any one system, respective carrying capacities can be established and the resulting loading factors compared. This equation has its limits, as any system has a finite carrying capacity; however, this equation can be used to evaluate the design of recirculating systems.

For example, to predict water quality in a recirculating system consisting of a 3,000 gallon tank, a 1,000-gallon filter, a pump rated at 60 gpm, and a 10% exchange rate, equation one is used.

$$Q = \frac{R_f \times E}{R_t \times T \times R_a}$$

$$Q = \frac{1000/120 \times 0.10}{3000/60 \times 4000/60 \times 3}$$

$Q = 1.67 \times 10^{-4}$

If you double the pumping rate to 120 gpm

$$Q = \frac{1000/120 \times 0.10}{3000/120 \times 4000/120 \times 3}$$

$Q = 3.33 \times 10^{-4}$

If you double the new water exchange

$$Q = \frac{1000/60 \times 0.20}{3000/60 \times 4000/60 \times 3}$$

$Q = 3.33 \times 10^{-4}$

If you double the filter size

$$Q = \frac{2000/60 \times 0.10}{3000/60 \times 5000/60 \times 1.5}$$

$Q = 5.35 \times 10^{-4}$

By reversing the fish culture tank from 3,000 to 1,000 gallons and using 3,000 rather then a 1,000 gallon filter.

$$Q = \frac{3000/60 \times 0.10}{1000/60 \times 4000/60 \times 0.33}$$

$Q = 1.34 \times 10^{-2}$

Based on experimental data, a theoretical system with a Q of 1.34×10^{-2} (1:3 ratio) might produce 2,400 lbs of fish per 4,000-gallon (476 ft^3) system or 2.4 lbs per tank gallon. Whereas, a tank to filter ratio of 3:1 with a Q of 1.67×10^{-4} might produce only 644 lbs per system or 1.35 lbs/ft^3 or 0.22 lbs per tank gallon. The carrying capacity and water quality of recirculating systems increase as the filter size, flow rate, or exchange rate increase. The most valid method for comparing loads in a recirculating system is by expressing them by total pounds per cubic foot of the system.

Another example of determining open water exchange is proposed by Losordo et al 1992. Assume that a 5,000-gallon culture tank is gong to be maintained at a density of 0.5 pounds of fish per gallon of culture tank volume. If the 2,500 pounds of fish are fed a 32% protein feed at a rate of 1.5% of their body weight per day, then 37.5 pounds of feed would produce approximately 1.1 pounds of ammonia-nitrogen per day (approximately 3% of the feed becomes ammonia-nitrogen). Additionally, if the ammonia-nitrogen concentration in the tank is to be maintained at 1.0 mg/L, then a mass balance calculation on ammonia-nitrogen indicates that the required flow rate of new water through the tank would be approximately 5,600 gallons per hour (93 gpm) just to maintain the desired ammonia-nitrogen concentration.

When dealing with small tropical fish, it is hard to relate to pounds of fish since it may take 100 to 400 fish to make a pound. But basically, the bigger the filter size, exchange rate and overall volume the less chance of failure due to water quality problems.

Particle Accumulation Versus Filtration

The absolute key and very heart of any closed or semi-open system are the mechanical macro and micro particulate filters, followed by biological filtration. Biological filters are not and should not be utilized to remove particulate matter. The main objective of biological filters is to support aerobic bacteria colonies which digest and convert dissolved organics and nutrients. Sludge (total suspended solids = TSS) discharged within recirculating systems is composed of partially stabilized excreta, uneaten food particles and biofloc (particles derived from bacterial biofiltration) and is controlled by mechanical filtration. Particulate matter is further defined by Losordo et al 1992 as: (1) settleable solids, those that settle out of the water within one hour under calm conditions; (2) suspended solids, those that will not easily settle out and, (3) fine and/or dissolved solids, particles $< 30\mu$ (microns). The first two categories are usually controlled by settling and sand filters (Figures 18 & 20). Often the last category is ignored yet can contribute up to 50% of the total suspended solids in a recirculating system.

Virtually all sludge wastes (solid and dissolved) generated in a recirculating system stem from the feed. Using a feeding ratio of 1 to 2 lbs of feed/100 lbs of fish, and disregarding uneaten food, results in 80% of the feed volume (on a dry basis) to be excreted as solid or dissolved fish wastes (Hopkins and Manci 1989). Sludge volume is derived from uneaten feed, excreted waste, chemically created precipitate, incoming suspended particulate matter and/or biologically created biofloc and is the major factor to consider when designing a system.

TSS excretion rates for trout range from 0.40 to 0.53 kg/kg of feed (Chen *et al* 1991). Catfish range from 0.18 to 0.69 kg/kg of feed. Calculated rates of 0.3 to 0.5 appear to be typical (Chen *et al* 1991). Murphy and Lipper (1970) reported soluble BOD (biological oxygen demand) at 58% and 42% in particulate matter for channel catfish. Whereas Wimberly (1990) reported 23% of the BOD excreted as dissolved and 77% as particulate. Variance in feed conversion is highly dependent on temperature, species, feed consistency, and feeding rate.

Suspended solids derived from bacterial biofiltration (biofloc) depend on the growth of bacterial growth during BOD removal and nitrification (Chen *et al* 1991). Considering ammonia nitrogen excretion rates of 1.8% to 4.6% of the feeding rate, biofloc production due to nitrification is estimated at 0.03% to 0.09% of the feeding rate (Chen *et al* 1991). Biofloc production due to dissolved BOD consumption is more significant. For example, if 23% of the BOD (0.23 kg BOD/kg of feed) is produced in soluble forms as reported by Wimberly (1990), the total soluble BOD production will result in a biofloc production which averages 5% of the feeding rate. This portion of BOD is absorbed during biofiltration, producing TSS levels equivalent to 9% of the feeding rate. Assuming the dissolved BOD has a composition similar to typical municipal sludge, values for biofloc production of 8% to 12% can be considered (Chen *et al* 1991).

Most fish feeds contains 7.2% to 7.7% nitrogen by weight. Of the nitrogen in feed, 67% to 75% will be lost to the environment as nitrogen gas or dissolved nitrogen (Iwama 1991). The total excreted nitrogen rate of fish averages 5% of the feeding rate. Of the total nitrogen excreted, approximately 20% to 50% is particulate.

Particulate matter becomes trapped and clogs biological filters making them inefficient and contribute to the BOD (biological oxygen demand). Therefore, standard subgravel filter plates commonly used in aquaria are not as suitable for exactly this reason. Basically this type of filter provides both particle filtration and biological digestion. For the average home aquarium, where all we are trying to accomplish is to maintain low densities of fish and invertebrates with minimal feeding, a subgravel biological/mechanical filter is fine. In these situations the bioload (amount of food and living material) remains minimal. In commercial operations, where high levels of food and animal density are required, more particle waste is produced. Under these conditions, subgravel filters have a tendency to become overloaded, collapse, and in turn toxic. This is particularly true in juvenile growout tanks where the main objectives are to grow fish to marketable size in the shortest time possible, within the smallest space and at highest possible density. Martin Moe (1982) has referred to this collapse as the "toxic tank syndrome."

Mechanical Filters

Photo 10 - Absolute best, cheapest initial primary particle filter, made from reusable polyester felt

Sufficient mechanical (particle) filtration is the most important part of a system with regard to long term continuous operation and optimal water quality. Approximately 25% of the BOD (biological oxygen demand) in a system is attributed to sludge decomposition. Therefore, continuous removal of all particulate material is essential. Particle filters are referred to as clarifiers or mechanical filters and can consist of simple settling tanks, flow-through screens, bags, pressurized sand filters, DE filters or cartridge filters. The size of a clarifier can be determined by two methods: (1) hydraulic loading and (2) solids loading (Chen *et al* 1991). Assuming the design hydraulic loading for the clarifier is 16.3 $m^3/m^2/day$ (EPA 1975), the clarifier surface area needed per 1,000 kg of catfish is calculated to 0.004 to 0.04 m^2 Assuming a solids loading of 72 $kg/m^2/day$ a clarifier of 0.054 to 0.088 $m^2/1{,}000$ kg of catfish is needed (Metcalf and Eddy 1979). Primary clarifiers (settling tanks, gravity fed filter screen, and filter bags) should always be based on the highest calculated value or larger. After primary clarification, a total suspended solid level of 3 to 7% and a BOD/TSS ratio of less than 0.2 is acceptable. Effluent from secondary clarifiers (sand, DE, or cartridge filters) should be 10 to 30 mg/L (EPA 1975). Even though as demonstrated, a small area will filter a large amount of water you should design oversized filters (primary and secondary) to reduce maintenance, provide for possible future growth, or unexpected problems.

Particle filtration should be utilized on all delivery and return lines. On return flows, primary particle separation down to one micron can be cheaply achieved by using simple removable, polyester filter bags on return outlets (Photo 10 & Figure 18). If arranged properly, filter bags can be quickly removed and replaced with clean ones. In addition, they can be soaked, chlorinated and cleaned repeatedly. A second means of removing particles is to install baffles in the reservoir (Figure 18). In addition, upwelling, sand, and/or a DE (diatomaceous earth) filter, and/or a cartridge filters can be installed on the pressure end of the system after the pump.

DE (diatomaceous earth) filters are common in the swimming pool industry but not widely utilized in aquaculture (Figure 21). If properly coated, a DE filter can remove particles down to 5μ (microns). They have been effectively used to control ectoparasites and protozoan infestations. Like sand filters, DE filters can be back-flushed to remove trapped particles. DE filters should not be considered a primary particle filter but utilized only as a polishing filter, in lieu of, or included with cartridge filters and installed before a UV sterilizer and/or water delivery to culture tanks. (Figure 18)

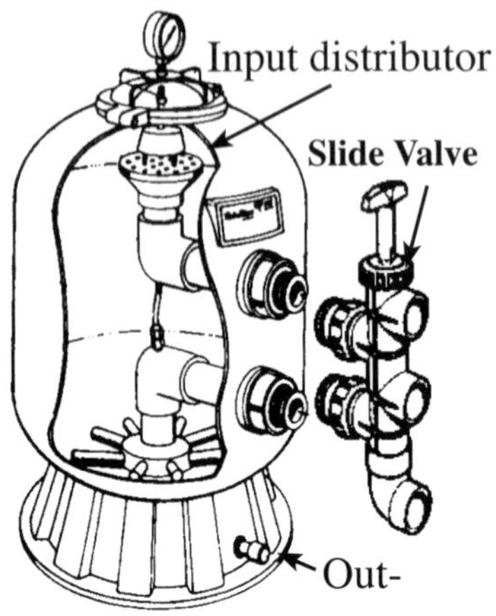

Figure 20 - Typical sand filter assembly.

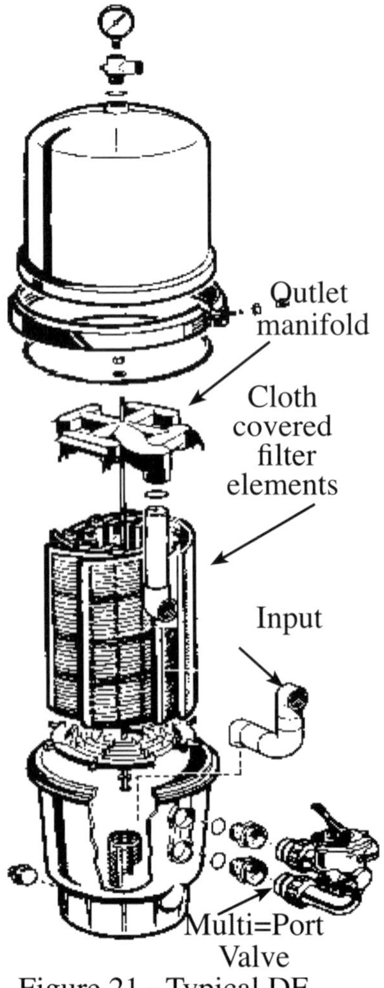

Figure 21 - Typical DE (diatomaceous earth) filter.

Sand filters (Figure 20) commonly strain out particles as small as 30 to 35μ if properly maintained. Many users prefer split-designed sand filters that can be unbolted for easy access and media changes. Often sand filters are filled with activated carbon, zeolite etc. for removing specific dissolved organics, metals etc. Sand filters are also biological filters but unlike fluidized bed filters the capacity is usually restricted to the top surface area of the sand bed rather than the entire sand column.

Standard cartridge filters that utilize 2.5" diameter cartridges are capable of removing particles less than 1μ depending on the cartridges used. Cartridge filters are capable of virtually removing bacteria. In addition, ultrafiltration cartridges are available that can be used to remove pyrogens and macromolecular compounds. Cartridge filters offer the highest degree of versatility by offering optional alternatives including special chemical retaining cartridges for carbon, zeolite, etc. Under high detrital conditions, sand or DE filters should be used prior to cartridge filters. Basically, the lower the particle size you want to remove, the slower the flow rate per unit area. To maintain optimal flow rates and remove small particles, the filter must be oversized. Keep in mind, these are mechanical filters and provide little biological filtration. Calculated pressure drop per filter utililized is estimated at 1 to 3 psi.

Foam Skimmers

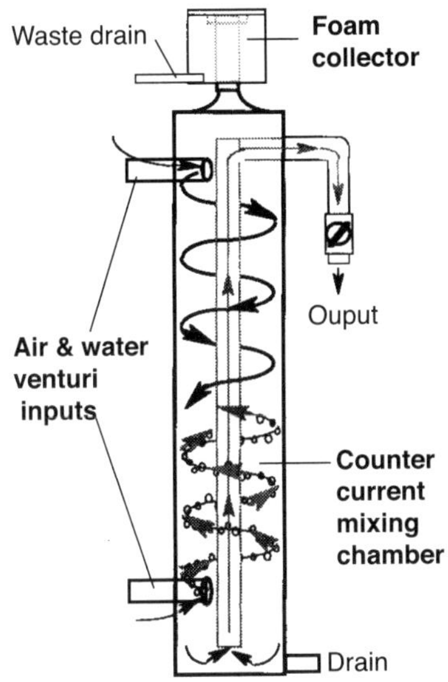

Figure 22 - Counterflow foam skimmer using venturies (drawing by Top Fathom)

Foam skimmers (protein skimmers) are evident in nature when viewing turbulent water and the resulting foam that washes onto the shore. Foam skimmers are actually a form of chemical and fine particle filtration. These filters create bubbles which in turn coalesces into a thick foam. The electrostatic attraction to the bubble surface causes dissolved substances to become entrapped. Efficiency is based on the size of the bubbles, preferably between 0.5 and 0.8 mm diameter. Small bubbles are formed by using fine airstones made of wood or fine silica or by using a venturi which consists of a stream of water passing through a constricted port causing a pressure drop and suction into which air is injected and mixed with the water. Many variables affect the efficiency including air flow, water flow, contact time, turbulence, temperature, pH, surface tension, specific gravity, column height and width, and mode of operation (Delbeck & Sprung 1994). Usually foam skimmers work best in saltwater or other water with heavily dissolved organics or inorganics, at a high pH and temperature range of 21-27°C (70-80°F). Production is usually higher during the late evening and early morning, tapering off during the day.

Delbeek & Sprung (1994) have provided an extensive review on the utilization of only foam skimmers which are popular in aquaria with live hard coral. They recommend as a guideline that the entire volume of the aquarium pass through the foam skimmer at least once an hour. When used in conjunction with other filters I feel that a only 20% of the total volume needs to pass through per hour.

Basically there are four types of skimmers. Smaller models use fine airstones located at the bottom of a column of flowing water. Simple models are open-bottom and placed in a body of water which is drawn up the column in a common airlift fashion. Others fit inside or outside tanks and utilize water flow provided from the pump. More advanced popular models use a separate water pump that passes pressurized water through a venturi, creating a suction into which air is injected into the flowing water before it enters the mixing chamber (Figure 22). These are usually located outside the tank and do not utilize airstones. Skimmers work best under constant flow and pressure since efficiency adjustments demand constant conditions. Venturi models require a separate water pump that can deliver 3 to 12 PSI (7' to 27' head pressure) for 120 to 2,500 GPH models. Connecting foam skimmers on lines from main pumps is difficult unless some sort of pressure regulator is installed.

Principles of Biological Filtration

Nitrification (biological filtration) rate depends on total ammonia, nitrogen, dissolved oxygen concentrations, and diffusion rates (Parker et al, 1990). Concentration ranges at which the two parameters become rate limiting depend on biofilm (bacterial slime, micro algae, diatoms, particulate matter, etc.) properties and diffusion rates (Okey & Albertson 1989). Normally, nitrification rates will increase as the concentration of dissolved oxygen increases and as penetration of oxygen into bacterial film is improved (Harremoes 1982). It has been suggested that nitrification rates are limited by diffusion rates into the biofilm rather than by low ammonia concentrations normally found in aquaculture effluents as opposed to sewage effluents (Brune & Gunther, 1981. Brune & Piedrahita, 1983). Work by Bovendeur et al (1990) showed a substantial reduction in nitrification rates at relatively low dissolved organic loadings.

Types of Biological Filters

Considerations for selecting and sizing a biofilter should include the following: operating temperature, oxygen requirements, water source, feed fed per day, type of feed, type of fish being reared, cost, and maintenance. Types of biological filters include: rotating biological contacters (RBC), wet/dry trickle filters, packed columns, upwelling fluidized beds, and submerged beds (Photo 11, Figures 23, 24, 25, & 26).

Photo 11 - Rotating biological contacter RBC (Marineland)

All are designed for gas exchange and biological digestion. Yet, trickle filters, packed columns, and RBCs are the most efficient for both activities. Dissolved oxygen consumption, total ammonia, nitrogen, and carbon dioxide production rates may be expressed as functions of feed input as 0.2, 0.03, and 0.28 kg/kg of feed respectively (Colt 1986). The main excretory products of fish are carbon dioxide and ammonia. Dissolved gas in water is a balanced situation. If oxygen is lowered, carbon dioxide and ammonia increase. Excess carbon dioxide depresses pH at different rates depending on alkalinity. High levels of alkalinity suppress carbon dioxide absorption but, at low alkalinity, pH values drop rapidly. Safe level of carbon dioxide is 30 mg/L. Nitrifying bacteria are sensitive to pH. Optimal activity is 7.0 to 8.0 for *Nitrosomonas* and 7.5 to 8.5 for *Nitrobacter*. Outside optimal ranges, decreases as high as 50% efficiency occur over less than 0.5 pH units (Grady & Lim 1980).

Water normally contains 3 to 8 ppm oxygen depending on biological oxygen demand (BOD), dissolved organic and inorganic elements, temperature, and altitude. Cold, clear, freshwater retains the highest oxygen levels whereas waters with high salinity, organics, dissolved gases and/or high temperature hold less. Atmospheric air at sea level contains 20 ppm oxygen. The principle of these filters is to flow thin sheets of water over irregular shaped media maximizing water/air contact which in turn accelerates gas exchange and biological digestion and conversion by bacteria. Aerobic bacteria, as the name implies, thrive on and utilize oxygen to digest dissolved nutrients and gasses into less harmful gasses and elements. By maximizing surface substrate, flow patterns and flow rates, oxygen levels are elevated and bacterial efficiency is increased per unit area.

Flow rates (hydraulic loading) are dependent on the type of media, filter design, oxygen levels, nutrient particulate loading and dissolved nutrient loading. Lu and Piedrahita (1993) tested 5.5" (14 cm) ID diameter by 117" (3 m) tall packed columns filled with 3/4" (1.6 cm) diameter Pall™ rings (surface area 341 m^2/m^3) at flow rates of 0.53, 5.3, and 10.7 gal/min (2, 20, and 40 L/min) for efficiency of nitrification. Compressed atmospheric air (10 psi) was injected from the bottom of the column equal to the water flow rates thus creating a counter-flow situation. Their results showed that nitrifying trickle filters can be operated at very high flow rates (40 L/min) corresponding to a hydraulic loading rate of 3,765 m^3/m^2/day or a specific hydraulic loading of 3.66 m^3/m^2/day. Hydraulic loading rate is defined as the flow rate divided by the cross-sectional area of the filter column. Specific hydraulic loading rate is defined as the flow rate divided by the surface area of the media in the filter.

Artificial Biological Media

Determining the right surface area and media for efficient biological filtration is very confusing especially if you listen to all the claims by various manufacturers. In reality all media work. What is required is to understand what these media are designed to do. Ideally you want to provide the most surface area possible yet not restrict the water flow, bacteria flushing, or gas exchange. Media cost is fundamentally based on surface area but not necessarily on efficiency. Cost should not even become a limiting factor when designing a filter since this is the "heart" of your grow-out system. If you short change the filter, you jeopardize your chance at consistent success.

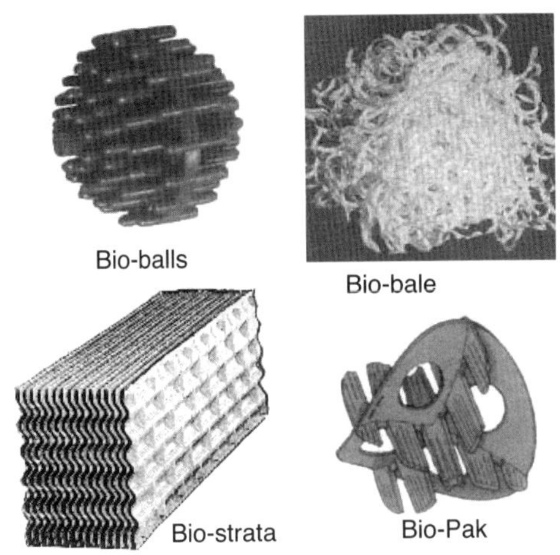

Photo 12 - Various biological substrates used as contact surfaces for bacteria in wet/dry and trickle filters.

Media selected for a wet/dry filter are usually based on space requirements, configuration of your filter, desired flow rates, efficiency, availability, bioload, and lastly price. More surface area per unit volume does not directly reflect the efficiency of the media and usually does not make it better. When selecting a media consider the "actual active area". This means that a poorly designed filter or selection of the wrong media may actually create dead areas where little degassing or biological activity occur and biological bacterial sludge accumulates. In general, the higher the surface area the less void space and lower ability for degassing. Dense media restrict air flow, air-water mixing, do not disperse water, are prone to canalization, and clog faster with biofloc. DLS (double layered spirals) consist of a fine dense polyester mat layered with a semirigid, three dimensional, fused mesh material. This popular type of media is initially efficient in biological filtration but less suitable for degassing and tends to clog quickly. DLS may be suitable for small scale systems, but for high intensity, high flow situations it is less desirable. Another popular media, sintered glass, has a high initial porosity but quickly clogs with biofloc thus drastically reducing usable surface area.

Cheap media such as rocks, crushed coral, "plastic packing peanuts" etc. are readily available but the surface area is low, therefore more space is required to achieve the desired filtration. In addition, this media is heavy, harder to clean, and has a tendency to nest tightly thus restricting flow and efficiency. Photo 12 shows four distinctly different plastic media used in wet/dry or submerged filtration. Bio-Bale™ is a convoluted strain of PVC material that is "spongy" in consistency. This media claims to have 250 square feet (23.23 M^2) of surface area per cubic foot (0.03 M^3) and is more suitable for irregular shaped, shallow depth filters with medium flow rates.

Bio-Pak™ is a rigid, more complex, media originally designed by Dow Chemical 30 years ago for degassing columns. This unique design allows maximum flow rates and "nests" together creating interlocking networks that increases "active surface area". It comes in four sizes (0.75", 1.00". 2.00" and 3.75") with a surface area from 150 to 5.25 square feet per cubic foot. Size selection is strictly based on the flow rates and / or the amount of solid particulate material. Bio-Balls are the most popular media in the home aquaria market. This media does not nest as well as Bio-Pak™ but does provide good random water flow patterns. The last media, Bio-Strata™, is a laminated material that looks like corrugated plastic roofing material glued together at alternating angles. Surface area ranges from 110 to 42 square feet (10.22 to 3.9 M^2) per cubic foot (0.03 M^3). This is a rigid media more suitable for very high flows and for use in tall towers.

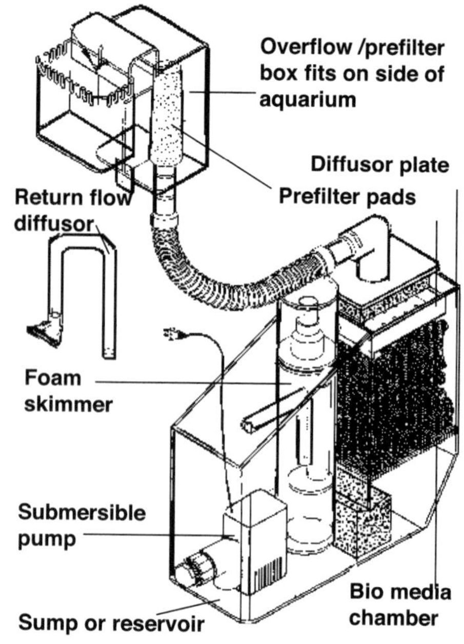

Figure 23 - Common acrylic, wet/dry filter frequently used in the home aquarium industry.

All media are prone to biofilm buildup consisting of particle waste, diatoms, bacteria, micro and macro algae which are deposited or grow on the structures. Biofilm is what you want since this is the active part of nitrification. As this film becomes thicker it becomes less efficient and if not routinely removed it tends to clog the filter. Media like Bio-Pak™ and Bio-Strata are designed for sluffing-off a percentage of this buildup as water flows through the media. Whereas, fine pore media trap biofloc particles and quickly clog, thus quickly reducing the effective biological surface area. Differences in media performance are attributed primarily to differences in biofilm-substrate contact time, which is related to liquid retention time. Contact time is affected by liquid redistribution characteristics of the media and the tendency of media to plug (Richards 1986). Void space and redistribution design actually are more important then surface area with regard to efficiency of biofilm development, water flow, and biofloc removal. In addition, efficiency usually increases as depth increases.

Wet/Dry Filters

Figure 23 shows a typical wet / dry filter normally utilized on home aquaria. Often these units are large enough to service several aquaria. Figure 18 shows a simple central system suitable for broodstock, larvae, and juvenile growout which utilizes a simple wet / dry filter. All filtration media should be calculated to a higher capacity than needed. Wet / dry biological filters should be considerably larger than required. Commercially, a base figure of 300 square surface feet (28 M^2) per 100 lbs. (46 kg) of biomass (fish plus food etc.) is often recommended. In a closed system used for growing tropical reef fish I feel this is minimal and should be at least doubled to 600+ ft^2/100 lbs (56 M^2/46 kg). Granted, 200 pairs of broodstock clownfish weigh less than 50 pounds. But, you must take into consideration the excess food you are going to feed, biosludge buildup, and the long term existence of a system. Commercial food fish calculations are usually based on a single growout period with routine sizable water exchanges (semi-open). These systems are used seasonally, then restarted, whereas our objective is to keep a system continuously going for several years with minimal water exchange and breakdown cleaning. The most common mistake I have seen in system designs has been inadequate filtration. Space limitations allocated for filtration in lieu of a few more tanks of fish should never be considered.

Trickle (wet / dry) and packed tower biological filters are simple and cheap to construct. Common aquarium units are made of clear acrylic which are often "more show than go" Ironically, these filters should be enclosed in an opaque, not clear, housings which allow excessive light.

These filters should not be exposed to light since this only enhances excess macro and microalgae growth which expedites clogging of filter pores. Although algae growth is beneficial with regard to nitrate removal, it is must be routinely harvested to be effective. Trying to remove algae from a wet/dry filter is labor intensive. Opaque containers help prevent excessive overgrowth. Aerobic bacteria function in either light or dark. Filter designs can consist of space-saving towers located at an end of a row of tanks or shallow trays placed in the reservoir below a row of tanks. We have found simple stacking styrofoam shipping boxes to be excellent, cheap, filter retainers (Figure 24). The bottom is cut out leaving a one inch support rim around the bottom. Plastic "egg crate" (light louver material) is cut so that it fits into the bottom. Bio-media is then placed in the box and others are stacked on top creating a cheap, waterproof, opaque enclosure that can be stacked to any height.

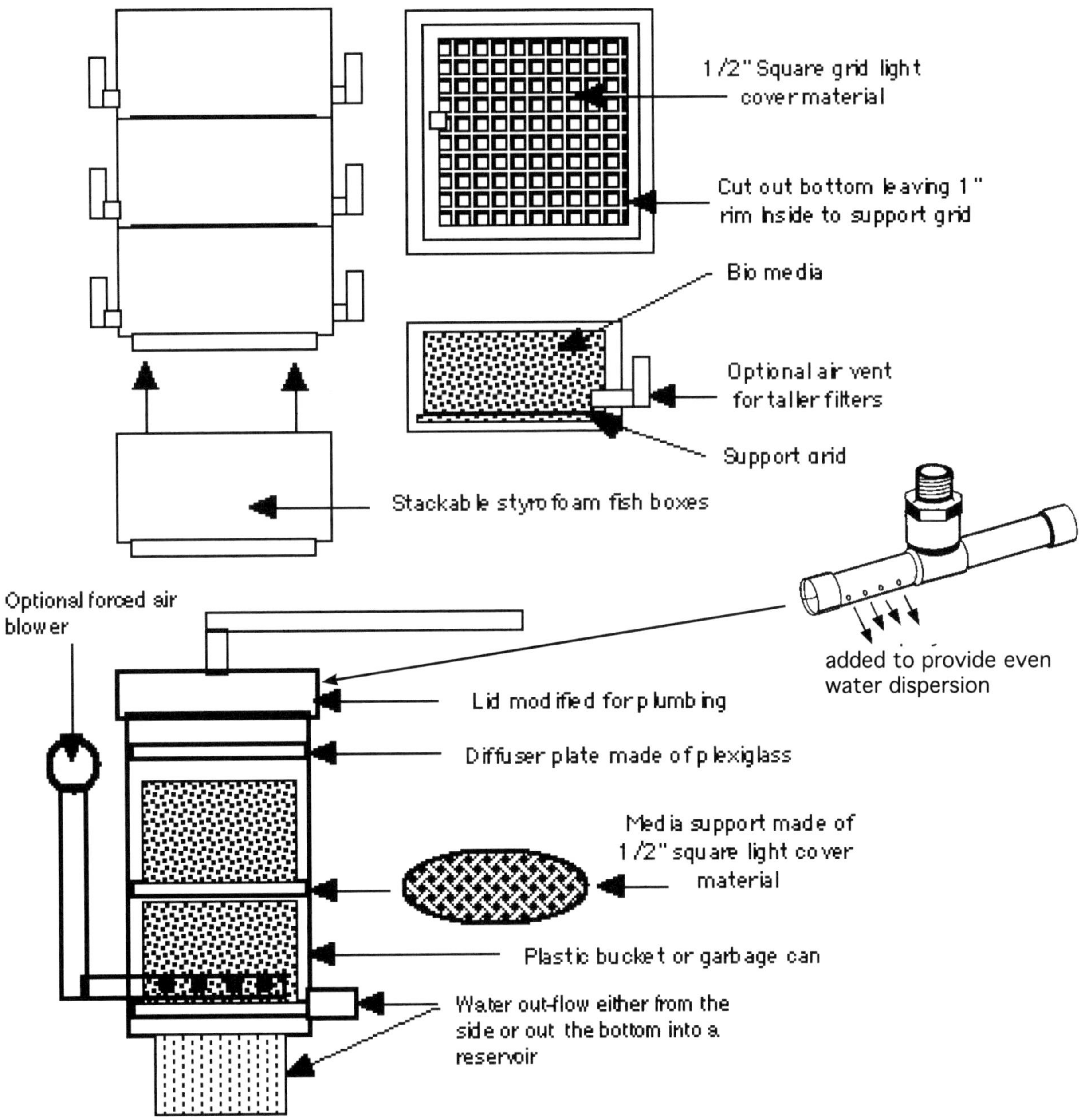

Figure 24 - Wet/dry degasser biofilter made of stackable styrofoam shipping boxes, plastic buckets, or garbage cans

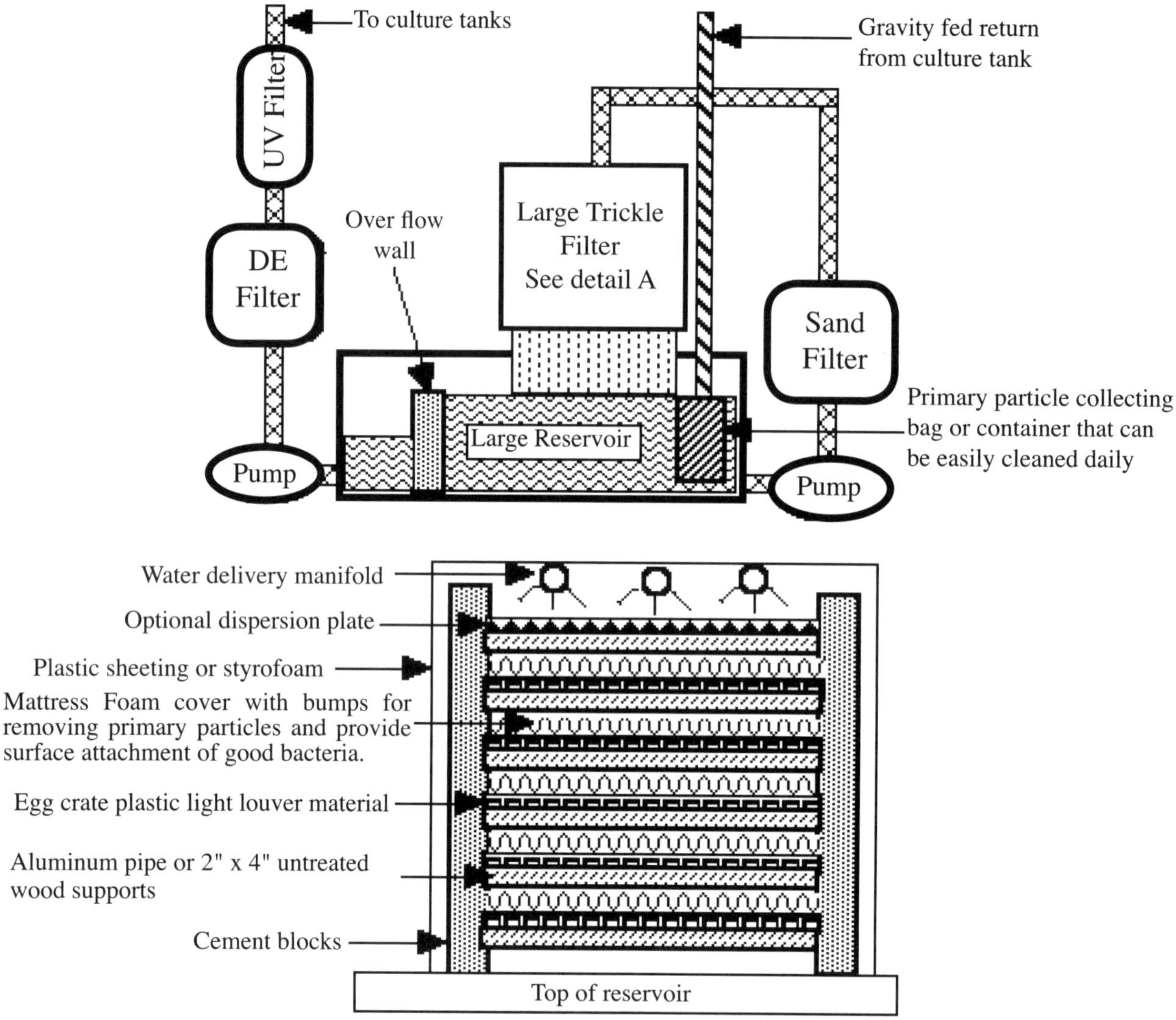

Figure 25 - Large recirculating system utilizing two pumps and a large trickle filter as the main biological filter. Sand filter installed for removing primary particles before water is sprayed on biomedia. Used on several freshwater facilities for years.

A key factor to efficiency is constant water delivery to all the media. All media should be constantly bathed with a thin film of flowing water. Commonly, water is delivered using a rotating spray bar, diffuser manifold or diffuser plate located at the top of the column. In tall columns, air or pure oxygen can be injected along the column or at the bottom to improve efficiency. Even larger units can be constructed of layers of foam placed on racks (Figure 25)

Keep in mind that a wet/dry filter is not a mechanical filter for removing detritus but only to biologically convert dissolved nutrients into gases and nitrate. As a precaution a polyester pad can be placed on the top to trap any particles that were missed.

Flow rates are important with regard to efficiency and biofloc removal. The level and consistency of the biofilm that develops on the filter media is an important factor with regard to the efficiency of these filters. Low hydraulic loading results in relatively thick biofilm which results in low diffusion rates of ammonia and oxygen into the biofilms, limiting nitrification rates (Okey & Albertson, 1989 ab; Siegrist & Gujer, 1987; Miller & Libey, 1984; Kaiser & Wheaton, 1983;

Harremoes, 1982). In the above experiment, Lu and Piedrahita (1993) found that at 2 L/min, biofilm development was scarce whereas at 20 L and 40 L/min a heavy growth of biofilm developed. In their experiments, higher flow rates were the most efficient.

Upwellers Or Fluidized Beds

Recently, upwelling sand filters and fluidized bed filters have been introduced as high surface (6,200 sq ft/cu. ft surface area) biological filters. In wet/dry biological filters, the media for bacterial colonization is sedentary or rotating and not totally submerged whereas, in fluidized beds, the media is in constant motion and submerged. In both cases the media is being bathed in constantly flowing oxygenated water except one is submerged and other is not. Even though fluidized beds are subject to lowering oxygen conditions, the constant replenishment of up flowing new water through a relatively small amount of media keeps it well oxygenated within a very restricted tube. Apposed to old submerged subgravel biological filters, waste conversion rates and bacterial growth are very rapid, yielding a highly more efficient filter with no channelization. The dynamic rolling sand assures full utilization of the entire bed. In this mildly abrasive environment the bacterial slime layer around the sand grains is constantly eroded, thereby eliminating clogging problems or stagnation, common problems of undergravel filters. Like all biological filters, fluidized beds should never be considered particle filters. In fact, if used as a particle filter, they actually take larger particles and break them down into even smaller particles which only adds to the BOD demand. Under normal operation they create more particles than wet/dry filters since the accumulated biofloc material is constantly being removed as the media rubs against itself but this is not overly significant or a big problem. If used in an auxiliary loop line (Figure 18) which allows the effluent to return to the reservoir, a filter bag can be used if desired.

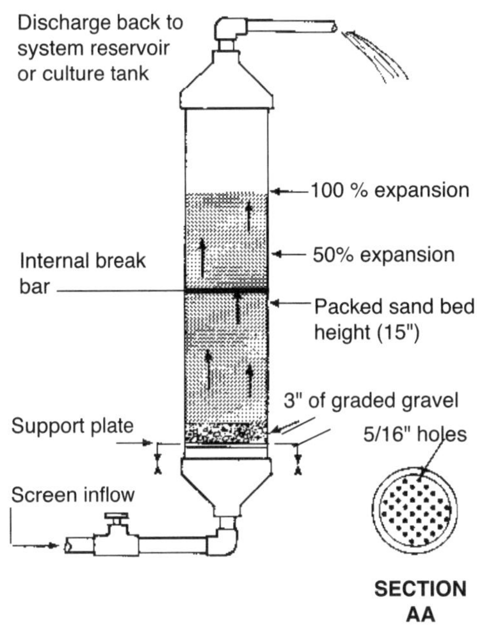

Figure 26 - Typical design of a fluidized bed filter

Two types of fluidized bed filters are manufactured: non-pressurized and pressurized (Figure 26). Non-pressurized units are more economical but require higher flow rates to completely fluidize the entire bed at 100% capacity. Pressurized units are smaller, compact, and require less than 30 gpm. There is a trade off with regard to oxygen need and utilization. Pressurized units are not open to ambient air and therefore depend on dissolved oxygen content in the water flowing through the unit. Some units allow you to inject oxygen into the chamber to further increase efficiency. An ideal installation of a pressurized fluidized bed biofilter is to pass the effluents into a wet/dry/degases tower which flows back into the reservoir (Figure 18). Since water coming out of a fluidized bed is lower in oxygen, passing it back through a trickle filter will reoxgenate the water.

Malone and Burden (1988) discuss the use, design, and carrying capacity of submerged rock filters, upflow sand filters, and fluidized columns in a blue crab shedding system. They found upflow sand filters robust, rapidly adjusting to changes in loading, while water quality remained consistently superb, providing at least 50% of the sand was allocated to upflow. Nitrification power of the upflow sand filter reflects two major factors. First, the intermittent removal of bacteria and trapped solids from the filter bed significantly reduces the amount of decaying material in the recirculating system. Second, the oxygen demand in the upflow sand filter is only one-third that of the submerged rock filter. This means that two-thirds of the wastes, either excreted by the crabs as solids or created by bacterial growth, are physically removed from the system. Flushed wastes can be collected and physically removed from the system, reducing the BOD and aeration demand on the system. Solid removal effectively reduced the total safe system volume to one gallon/crab compared to a normal rate of two gallon/crab.

The carrying capacity for an upflow sand filter exceeds that of the submerged rock filter by over 20 times (660 crabs versus 33 crabs/cubic foot of filter media) The carrying capacity for the illustrated fluidized column bed filter is 30 times more than that of the submerged rock filter (1,000 crabs versus 33 crabs/ft^3 filter media). Fluidized beds use only the expanded mode of operation where 25 to 100 percent of the bed is always expanded (fluidized). They have a much lower pressure loss, thus require less energy to operate efficiently than sand filters. Table 4 can serve as a guideline for determining carrying capacity and necessary flow rates (Malone and Burden 1988).

Table 4 - Approximate carrying capacity utilizing the fluidized column design shown in Figure 23

Filter Diameter	Sand Volume	Carrying Capacity	Expansion Flow
4"	0.11ft^3	110 crabs	6 gpm
6"	0.25 ft^3	250 crabs	13 gpm
8"	0.44 ft^3	440 crabs	23 gpm
10"	0.68 ft^3	680 crabs	36 gpm
12"	1.00 ft^3	1000 crabs	51 gpm

Like the wet/dry biofilters, non-pressurized fluidized beds should be run on a auxiliary loop line coming out of the pump with outflow going back into the reservoir (Figures 18 & 19). Basically, pressurized water are introduced through the bottom through a dispersion device into a cylinder containing sand. The velocity and volume of water is adjusted so the entire bed swells or expands and moves like boiling water. Effluent comes out of the top and back into the tank or reservoir. Care must be taken not to overexpand the bed so that it starts to flow out of the filter. Like protein skimmers, they are subject to flow fluctuations and should be connected to their own pump or pressure regulated valve for optimal efficiency. Like wet/dry filters and protein skimmers, it may not be necessary for 100% of the water to pass through the filter each time to reach efficiency based on bioload and volume.

Reservoirs

Although reservoirs are not actually considered filters, they are important contributors to maintaining water quality and eliminating quick fluctuation in water conditions. I prefer reservoirs with a capacity to hold water from all tanks when drained. A large reservoir provides several advantages, including: a large heat sink which helps modify hatchery temperatures from abrupt changes, keeps water quality from changing abruptly, and allows recirculated water to flow slow enough to settle out most of the particles. Reservoirs should not be deep since gas exchange becomes limited in deep water. I prefer 12" to 18" deep units.

Using a series of baffles inside the reservoir (Figure 18) slows and mixes return flows and allows more particulate matter to settle out. Baffles should be arranged so they are easily accessible and positioned so that debris efficiently settles out in distinct areas. Pockets of deposited matter must be routinely siphoned out.

If this is a semi-open system, the reservoir is the best place to automatically drop exchange water and refill with new water. This can be accomplished using a split stand-pipe from which a section is removed allowing water to draw down to a specified level (Figure 45). Another option is an external stand-pipe with is bent down to a measured level. A third option is use of timed solenoid valves. Exchanges should be moderated from 10 to 15% to reduce system shock. Multiple exchanges are possible during the day but should be spaced from 6 to 12 hours apart.

Ultraviolet Sterilization

This is an important part of a system with regards to parasitic disease and bacteria control. In closed systems, UV (ultraviolet sterilizers) with flow ratings at 10 GPM per 1000 gallons (37.5 L/266 L) of water are considered optimal for depuration (bacterial cleaning) of oysters and clams within 24 to 48 hours. Over sizing is recommended when choosing a UV. Prices of a UV can range from outrageous to extremely reasonable. Bottom line is how effective they are in killing and/or reproductive disruption of bacteria, yeast, protozoa etc (Table 5). There is absolutely no reason to spend thousands of dollars on a fancy stainless steel unit with buttons, alarms and flashing lights, when a PVC or polyethylene unit 99% effective and costs 1/4 to 1/3 the price. Keep in mind, the essence of a UV sterilizer is the power of the bulb; the distance from the bulb and flow rate that determine the effective qualities of a particular UV unit. Those that utilize 3" (75 mm) diameter cylinders are rated the most efficient based on wave lengths emitted by normally available UV bulbs.

The quality of the UV bulb is the most important component. Germicidal UV bulbs are measured in micro watt seconds/cm^3 and refer to both intensity and exposure time. The most efficient bulbs are made of quartz, mounted inside quartz sleeves and emit a wave length of 254-265 nanometers (2537-2647 angstroms) of radiation. Recommended flow rates for units are

Table 5 - KILLING DOSE of UV IN MICROWATT/ SEC/CM^2

BACTERIA	
Aeromonas salmonicida	20,000
Bacillus anthracis	8,700
Bacillus megatherium sp (veg)	2,500
Bacillus megatherium (spores)	5,200
Bacillus paratyphosus	6,100
Bacillus subtilis (mixed)	11,000
Bacillus subtilis (spores)	22,000
Clostridium tetami	22,000
Corynebacterium, Dephtheriae	6,500
Dysentery bacilli	4,200
Eberthella typhosa	4,100
Esherichia coli	6,600
Eberthella typhosa	4,100
Escherichia coli	6,600
Micrococcus candidus	12,300
Micrococcus piltonensis	15,000
Micrococcus sphaeroides	15,400
Mycobacterium tuberculosis	10,000
Neisseria catarrhalis	8,500
Phytomonas tumefaciens	8,500
Proteus vulgaris	6,600
Pseudomonas aerugenosa	10,500
Pseudomonas fluorescens	6,600
Salmonella sp.	10,000
Salmonella enteritidis	7,600
Salmonella typhimurium (ave)	15,200
Sarcina lutea	26,400
Serratia marcescens	6,160
Shigilla paradysenteriae	3,400
Spirillum rubsum	6,160
Staphylococcus albus	5,700
Staphylococcus aureus	6,600
Staphylococcus hemolyticus	5,500
Streptococcus lactis	8,800
Streptococcus viridans	3,800
YEAST	
Saccharomyces ellipsoideus	13,200
Saccharomyces sp.	17,600
Saccharomyces cerevisiae	13,200
Brewer's Yeast	6,600
Baker's Yeast	8,800
Common Yeast Cake	13,200
MOLD SPORES	
Penicillium roqueforti	26,400
Penicillium expansum	22,000
Penicillium digitatum	88,000
Aspergillus glaucus	88,000
Aspergillus flavus	99,000
Aspergillus niger	330,000
Rhisopus nigricans	220,000
Mucor racemosus A	35,200
Mucor racemosus B	35,200
Oospora lactis	11,000
VIRUS	
Bacteriophage (E. coli)	6,600
Tobacco Mosaic	440,000
Influenza	6,800
PROTOZOANS	
Costia necatrix	318,000P
Paramecium	200,000
Ichthyophthirius tomites	100.000
Trichodina sp.	35,000
Trichodina nigra	159,000
Myxosoma cerebralis	35,000
Ceratomyxa shasta	30,000
WORMS	
Nematode Eggs	92,000
ALGAE	
Chlorella vulgaris	22,000
FUNGI	45,000
Saprolegnia hyphae	10,000
Saprolegnia zoospores	39,600

normally based on an intensity level of 15,000 microwatt seconds/cm^2. Retention or exposure time is very important to the effectiveness of UV sterilizers. By oversizing the unit and/or slowing the flow, exposure time is increased, making it more effective. For example, simply reducing the recommended flow rate 50% doubles the exposure time to 30,000 micro watt seconds/cm^3 thus increasing the "kill power." Colored water, particles, high dissolved salts, or slime coated bulb sleeves reduce the effectiveness of the unit. Pre-filtering the water for particles, carbon and protein skimmers for dissolved organics, routine cleaning of the sleeve, slowing the flow and over sizing the unit all increase effectiveness. Within a closed system, the effectiveness is not 100% since there is continuous reproductive replacement going on within each tank which is continually being fed back into the system. It has been calculated that usually about 88% is a more realistic measurement in closed systems which is well within the accepted levels. Bulb potential decreases with time and the bulb should be changed yearly. Table 5 provides a list of aquatic organisms and the "kill power" needed to effectively destroy them.

Ozone

Ozone is another sterilizing option instead of or in addition to UV. It has been successfully used in the sewage treatment industry in Europe since the early 1900's. It is more effective than UV when sterilizing colored, cloudy, or turbid water. Ozone is a special form of oxygen (allotropic form) which has three oxygen atoms instead of two for oxygen. It is produced by generating a very high electrical discharge called a corona and passing air through this energy field or by passing air by a high energy UV radiation source. Ozone output is expressed in ppm (parts per million), gms/hr (grams per hour), percent and still others in grams per meter cubed. Usually percent or milligrams per hour (mg/hr) is used. Ozone concentration in water is usually measured and reported in mg/l (ppm). When air is passed through the charge only a percent of the air is oxidized. Corona discharge units have a higher output than UV generators, about 10 to 300 times more (Mouw 1990). Ozone production is controlled by air flow rate, water vapor, and heat. Generally as air flow is increased the relative rate of ozone production increases until it reaches optimum, flattens out and starts to drop off. Water vapor in the air fed into a corona unit is more important than in an ozone UV bulb unit. Decreases of 30% are typical as the dew point drops from -40°C to -20°C (Mouw 1990). Ozone has a very high power expressed as millivolt and redox potential. It amounts to 2076 millivolts or 2.076 volts which is stronger than chlorine (1360 millivolts). It is very unstable and breaks down rapidly, releasing a single atomic oxygen. This single oxygen is very reactive and can effectively break down complex organic wastes that are not commonly removed by mechanical filtration. Ozone can oxidize ammonia into less harmful products and is effective against ammonia surges that can occur in high density closed systems.

Extreme care should be used, not only for personal safety but, if not properly installed ozone can kill fish, invertebrates, and biofilters. Residual levels of ozone 0.02 mg/l or greater are toxic to aquatic animals. However, Wedemeyer et al (1979) advised that a very low residual permissible level of 0.002 mg/l is not harmful to fish and may prevent disease from reoccurring. Toxic levels to fish vary within the literature (Lawson & Merry, 1991). Table 6 summarizes several articles that report on toxic levels to fish.

Table 6 - Reported toxic ozone levels on fish and humans.

Source	Organism	Level	Effect
Arthur & Mount (1975)	flathead minnows	0.2-0.3 mg/l	toxic
Roselund (1974)	rainbow trout	0.01-0.06 mg/l	gill damage
Wedemeyer et al (1978)	salmon	0.05 mg/l	gill damage
Sutterlin et al (1984)	salmon smolts	1.62 g/Kg/hr	no damage at 8 hour
Sutterlin et al (1984)	salmon smolts	1.62 g/Kg/hr	toxic at 15 hours
Honn & Chavin (1976)	nurse sharks	0.132 mg/l	no toxic effect noted
Owsley (1988)	humans	0.1 mg/l	max conc./8 hour day

With regard to humans, caution is recommended. Keep in mind that if you can smell ozone, the input is dangerously high or the mixing chamber is not adequate. The odor is usually described as "sharp" and "fresh" and often noted after a lighting storm. Humans can smell ozone in the range of 0.05 mg/l. Maximum safe concentration for an eight-hour work day is 0.1 mg/l (Owsley 1988).

Ozone must be carefully introduced to a saltwater system since it can produce toxic brominated and chlorinated compounds. In seawater, the half life of some of these toxic compounds is 24 hours, therefore dwell time is very important. In addition, certain trace elements such as iron, manganese (Wheaton 1977), and calcium (Moe 1989) are precipitated out and may have to be routinely replaced. Decomposition of ozone is rapidly accelerated in the presence of high pH of 8.0 or above and seems to be more effective at 7.0 or lower. However, Colberg and Lingg (1978) reported that the oxidative capacity was greater in high pH. Therefore, when used in saltwater, higher input levels may have to be used to achieve maximum effect. In freshwater, ozone is very reactive, less stable, and decays with few residual effects.

In a closed system I feel the input of ozone should be directly introduced into the primary effluent line just after solids removal since it is not effected by turbidity, color or small particles. Others (Lawson & Merry 1991) show introduction of ozone prior to solids removal in a semi-closed systems. This does not seem to be logical since ozone breaks down solids into shorter chain molecules which are more easily degraded by bacteria. Breaking down primary solid effluents into more soluble nutrients is not what we want to achieve in a closed recirculating system. Our initial goal is to remove any particles that will over time break down into dissolved nutrients and smaller particles. Ozone can also be installed in a closed auxiliary loop to the system, as UV units are often placed (Figure 18). In this loop, a portion of the water is passed into the ozone reactor unit, through carbon, then back to the reservoir. The main concern is removal of residual ozone to less toxic levels prior to introduction to culture tanks. Ozonated water should never be added directly to the housing aquaria.

On open systems or semi-open systems, ozone should be installed on the incoming exchange water especially when using natural surface water. The water is pushed through the reactor into a separate heavily aerated reservoir and allowed to release the toxic compounds. After 24 to 48 hours it is then brought into the hatchery for water exchanges.

Ozone is often injected into the venturi on protein skimmers which allows for maximum mixing. Kynar plastic venturi are usually necessary since raw ozone gas can destroy PVC or poly plastics.

When comparing UV and ozone I feel that for small recirculating systems less than 10,000 gallons, an oversized ultraviolet sterilization unit is the safest and easiest to control. However, Rodriguez (1993) feels that UV and ozone are not in competition with each other with regard to choice. He states that even when ozone is selected as the primary disinfection mechanism, ultraviolet has a potential role to play as an ozone destruct device for residual ozone levels in water. Ultraviolet ozone destruction units produce complete removal of ozone by catalyzing the decomposition of ozone to oxygen. Therefore, both should be used on any size system especially when fear of residual ozone is possible. Activated carbon and degassing columns (trickle towers) are also used to remove residual ozone.

Aeration

Oxygen content is the key to the efficiency of biological filtration. The overall rate of oxygen consumption for a system is the sum of the respiration rate of the fish, oxygen demand of the bacteria breaking down organic wastes and uneaten food (also referred to as the Biochemical Oxygen Demand or BOD), and the oxygen demand of nitrifying bacteria in the filter. A "rule-of-thumb" is that for every pound of feed added to the culture tank, approximately 0.25 pounds of oxygen is consumed (Losordo et. al. 1992a). To maintain adequate DO (dissolved oxygen) levels in the culture system, the input of oxygen must be at a rate equal or better than the consumption rate of the fish and bacteria. In warmwater applications, saturation of atmospheric oxygen rarely exceeds 8.75 mg/L at >20°C (68°F).

Carbon dioxide (CO_2) is a by-product of respiration of fish and bacteria, and can accumulate within a system. Elevated CO_2 concentrations are not usually toxic to fish when sufficient oxygen is present. However, CO_2 levels should be maintained at less than 20 mg/L.

Aeration is a word normally used to denote the dissolution of oxygen from the atmosphere into water as opposed to adding pure oxygen which is referred to as oxygenation. Many methods are used restore oxygen levels in recirculating systems.

Airstones or air diffusers utilize low pressure atmospheric air or pure oxygen to disperse coarse (6 mm diameter), medium (3 mm diameter) or fine (1 mm diameter) bubbles into a water column. Coarse bubble stones require the lowest air pressure and are more resistant to clogging, but are about a third as efficient as medium bubble stones in transferring oxygen into water (Aquatic Eco Systems, 1996). Fine bubble stones are obviously superior but do not justify their use since they are more prone to clogging and require higher pressure to operate properly. These are typically used with pure oxygen or for administering ozone. Clogging is normally due to fine particles in the air or calcium carbonate deposits in hard or salt water. In high intensive fish culture systems, airstones are limited. Based on calculations by Losordo *et al* (1992b) at 28° C (82° F) a tank that receives 29.75 pounds (13.5 kg) of pelleted feed per day would have an oxygen demand of at least 0.31 pounds O_2/hr (0.14 kg O_2/hr). This tank would require a 0.33 hp (0.25 kilowatt) air blower supplying 21 cubic feet (0.60 M^3) of air per minute, equivalent to 17.5' (532 cm) of airstone diffusers. Ironically, if you were forced to use this much diffuser material, the turbulence may be more than the fish could tolerate. Obviously this is a high amount of food to be added to a typical tropical fish culture system on a daily basis. Typically you would not expect more than 1 to 4 pounds of feed for 15,000 to 22,000 fish. Therefore, air diffusers would be suitable. In calculating how and what size to use, I offer Table 7 which is modified information from Aquaculture Supply USA and Aquatic Eco Systems catalogues. Note the increase in oxygen transfer between the two nine inch stones with regard to depth. As the depth increases, contact time and pressure increases, forcing more oxygen into the water.

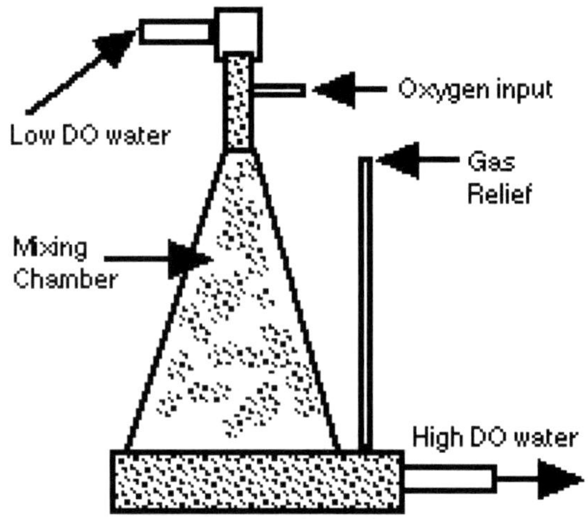

Figure 27 - Downflow bubble contactor after Colt & Watten, 1988.

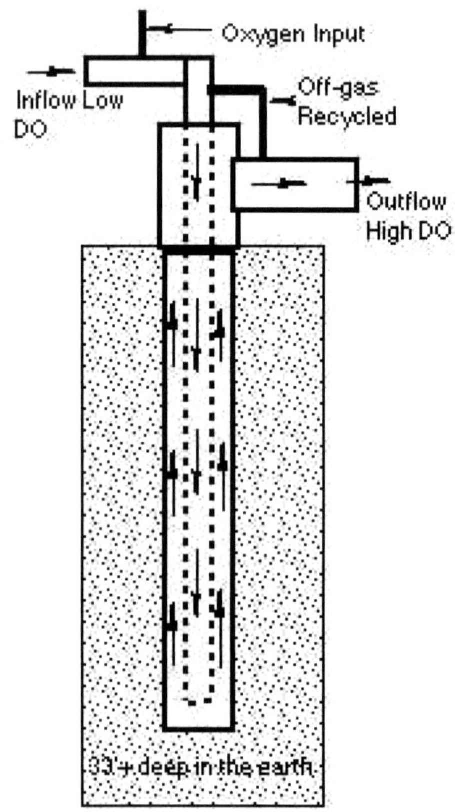

Figure 28 - U-tube oxygen concentrator.

Table 7 - Oxygen transfer using atmospheric air, stones with medium air bubbles. Items with (*) are actual data, remaining data is extrapolated and intended to be only used as a guideline.

Length	Width	CFM	Gms O_2/Hr	Lbs O_2/Hr	Depth
1.50"	0.50"	0.05	0.5	0.001	2'
1.50"	0.75"	0.10	1	0.002	2'
2.00"	1.00"	0.20	2	0.005	2'
*1.50"	1.50"	0.25	2	0.005	2'
3.00"	1.00"	0.30	3	0.015	2'
3.00"	1.50"	0.35	6	0.013	5'
6.00"	1.50"	0.50	11	0.025	5'
*9.00"	1.50"	0.75	10	0.022	3'
*9.00"	1.50"	0.75	17	0.037	5'
*12.00"	1.50"	1.00	49	0.108	

Mechanical aeration consists of splashing water into the air to increase contact. The larger the contact area and period the higher oxygen exchange. Paddlewheels and submerged pumps with a propeller are often used in large ponds or extensive systems. In comparison with airstone diffusers, vertical pump aerators, on the average, are 50% more efficient at transferring oxygen.

Packed column and wet/dry trickle filters (Figures 24 & 25) are efficient to use in most tropical fish culture applications. In systems using submerged or fluidized bed filtration, the concentration of dissolved oxygen will most likely be lowest and carbon dioxide highest at the outflow of these filters. Effluent from these filters can be passed through these columns to restore and elevate oxygen levels and remove carbon dioxide as a gas. Losordo *et al* (1992b) provide an example. With an input of DO of 4 mg/ O_2/L an effective oxygen transfer rate of 1.25 pounds O_2/hp hr (0.75 kg O_2/kw hr) can be attained. While not as high as other methods, the energy costs are minimal. If the column is to be used for carbon dioxide stripping, a low pressure air blower can be used to force at least five times as much air (by volume) as water that passes through the column.

Pure oxygen is used in high intensity recirculating systems when the rate of oxygen consumption within the system exceeds the maximum feasible rate of oxygen transfer through normal aeration methods. Sources of oxygen include compressed gas, liquid oxygen, and on-site oxygen generators. Like the use of atmospheric oxygen, the use of pure oxygen through diffusers is not the most efficient but is usually is the cheapest method. At best, the efficiency of such systems is less than 40% (Losordo *et al* 1992b). A number of specialized methods have been developed.

A downflow bubble contactor is capable of transferring in excess of 90% of the oxygen mixed. Figure 27 is a downflow bubble contactor oxygenator called a DFBC, bicone, or Speece cone. It consists of an inverted cone-shaped reactor with a water and oxygen input port at the top. As water and oxygen bubbles move down the cone, the flow velocity decreases until it equals the upward velocity of the bubbles. This type of configuration and mode of operation provide a long contact time between the water and bubbles and nearly 100% absorption. Water outflow can be as high as 25 mg/L given a system pressure of approximately 10 psi.

A counter flow diffusion column is a simple device which utilizes low pressure water. A tall vertical reactor column, 10' high, allows low pressure water to slowly flow downward from the top while oxygen is injected from the bottom through a diffuser. The rate of rising oxygen bubbles is slowed by adjusting the downward water flow velocity, attempting to allow a minimum of bubbles to reach the top. This device has the advantage of being simple but not as efficient as the Speece contactor above.

U-Tube diffusers are advantageous when high operating pressures are available. It consists of a contact loop, usually a pipe within a pipe (Figure 28), that is buried in the ground to at least 33 feet depth (the height of water required to add one atmosphere of pressure = 14.7 psi). Oxygen is mixed with the water at the entrance to the U-tube and travels with the current to the bottom of the water column. Additional pressure from the water column accelerates the rate of oxygen adsorption. Advantages of this system are the low energy requirements for oxygenating large flows and the resistance of clogging with particulate solids. The disadvantage is the initial well drilling costs. Oxygen transfer is generally below 70% with resulting concentrations up to 250% higher than atmospheric saturation of 15 to 20 mg/L.

Other methods, including pressurized spray towers and fluid filled pressurized packed columns are also used, but I suggest reading Boyd and Watten (1989) for more information.

Carbon Filtration

Carbon filtration is utilized to remove dissolved organics which appear as a yellow coloration in the water. Once this is accomplished I feel further use is not necessary for several weeks. Although I like to use carbon, I am not an advocate of constant use. There is charcoal and then there is activated carbon. The former is simply obtained by burning wood, bone or coal at normal atmospheric conditions. The latter is obtained by burning the media under high pressure and temperatures thus driving out impurities and forming a more porous media. Carbon works by entrapment (absorption) and polar chemical bonds (adsorption).

Carbon can be very selective for certain essential trace metals and nutrients. Removal of iodine and iron are examples and may have to be supplemented in closed recirculating systems. Other minor trace items once considered nonessential may indeed prove to be very important, but are being removed by constant use of activated carbon. In recent years the recognition of the minor trace metal, selium, is an example.

Many aquarists utilize carbon as a mechanical, biological, and chemical substrate rather than just a "chemical treatment" for removal of dissolved organics. Overall efficiency of carbon is quickly depleted when used this way, but it does provide a suitable bacterial substrate for biological activity and if tightly packed can easily serve as a mechanical filter. Carbon block cartridges of 1 micron or less and impregnated spun wound polyester cartridges are optional methods over typical granular carbon.

Particle size is important, the smaller the particle the more active it is. The more active it is the less you need. I prefer brewery grade material (12 x 40 mesh) 0.8 to 1.5mm. Recently carbon pellets (made of powdered carbon) have been introduced and are good for fast flow applications. Dosage will vary but we found 7 liters for 4,000 gallons was sufficient for at least 6 months of intermittent use. Periodic use to remove dissolved organics is recommended for one week on followed by two to three weeks off-line. Figure 29 is a unique 5 gallon carbon filter utilized at Instant Ocean Hatcheries.

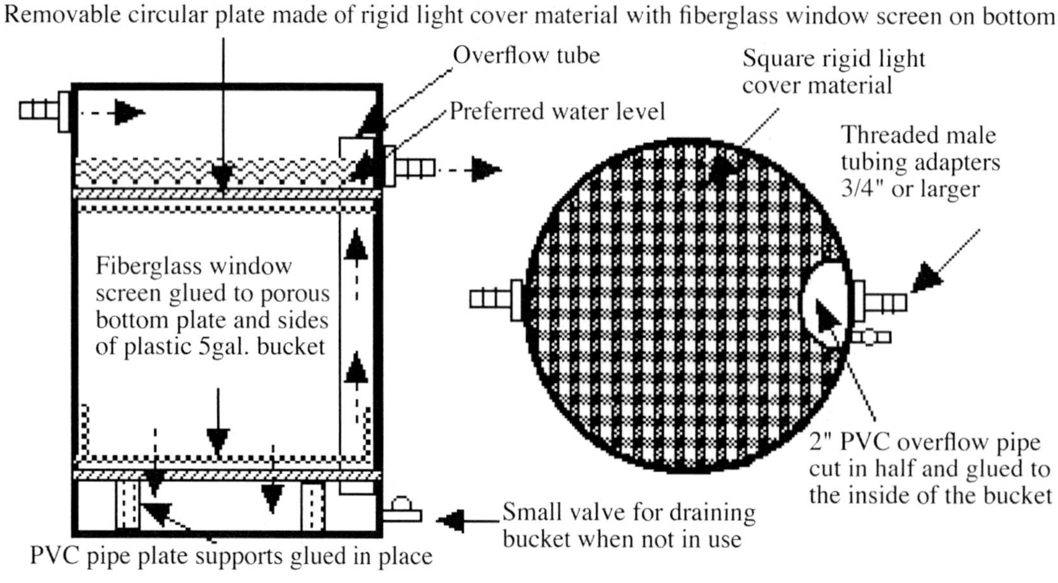

Figure 29 - Easy to construct carbon filter made from a common 5 gallon plastic bucket.

BROODSTOCK HOUSING

Tank Selection

Selection of the proper aquarium is important, but not as significant as people lead you to believe. Over and over we read that you need a certain size tank for a certain size species of fish. This is simply not true and is often more wrong than right. I do not understand who decided what size tank was necessary since so few marine fish have actually spawned in captivity. With regard to tank size, it is far less important than being able to meet the activity and ecological needs of the fish. Fish that swim constantly, will obviously need larger, taller aquarium. Large, normally active, swimming fish usually do better in a circular tank rather than a square or rectangular one. Large sedentary fish can be housed in smaller, shallower, aquaria. In general, constantly swimming fish need more vertical area than sedentary fish, such as clownfish which need more bottom area.

With regard to clownfish, *P. biaculeatus,* (maroon), *A. clarkii,* (sebae), and *A. polymnus* (saddle-back) are more comfortable in a standard 29 gallon (109 L) tank or larger. Smaller tanks such as 20 gallons (75 L) can be used for *A. frenatus,* (tomato), *A. melanopus,* (cinnamon), and *A. ephippium* (red saddle). Smaller species such as *A. ocellaris,* (ocellaris), *A. percula,* (percula), *A. akallopisos* (orange skunk), and *A. perideraion* (pink skunk) can be housed in 15 gallon (56 L) tanks. The sizes suggested are not set since we have successfully spawned many fish species in tanks as small as 5 gallons (19 L).

For commercial applications, custom, equal-size, tanks are recommended. Ideally the tank should be a cube which is not a common design in standard manufactured aquaria. At Instant Ocean Hatcheries, we used 3/4" (1.9 cm) thick "plyform" plywood tanks with 12" (30 cm) square windows (Photo 7 - Figure 17). Marine plywood, although expensive, is also suitable. Currently pressure treated plywood is available and is recommended over standard untreated wood. Two tank sizes were used. Larger species were housed in 20"L x 24"W x 20"H (50 cm x 60 cm x 50 cm). Inside volume was 42 gallons (158 L). Smaller species were housed in 20"L x 16"W x 20"H (50 cm x 40 cm x 50 cm) with an inside volume of 28 gallons (107 L). If properly designed, four to six of these tanks can be constructed in one unit cut out of 2 sheets of plywood (Figure 30). These tanks were joined together with stainless steel staples, painted with FDA approved aqua green epoxy paint (2 coats) and sealed with a bead of silicone in the seams. Holes for windows and plumbing were made before the tanks were painted and all edges were painted. Only the top three inches of the outside of the

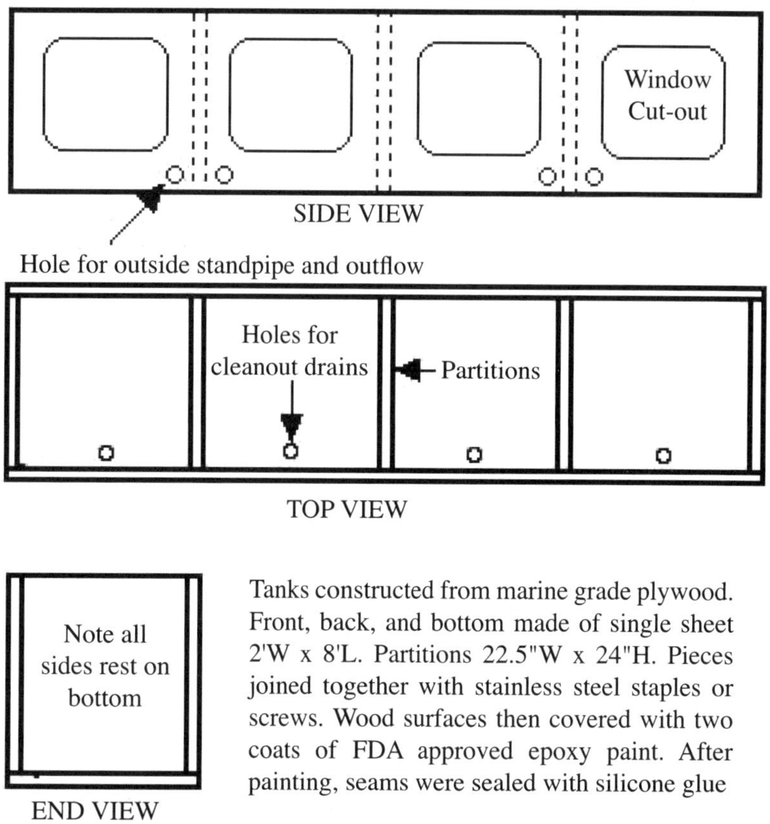

Figure 30 - Construction of four tanks from two sheets of marine grade 3/4" thick plywood

tank was painted in order to allow the wood to breathe. Windows consisted of 1/4" plate glass glued on with silicone after the tank was painted. Make sure silicone used is a non-toxic variety recommended for aquaria such as GE-35. Mounted plumbing included a mounted external standpipe and drain (Figure 31).

Currently, heavy molded plastic laundry sinks can be easily modified into durable long-lasting tanks. They are equipped with slanted bottoms and threaded drains which help eliminate detritus buildup. Just cut out an area for a window, epoxy paint to desired color and glue 1/4" (6 cm) plate glass in place with aquarium grade silicone or "goop" glue.

Special glass tank initially designed for C-Quest, can be constructed (Figure 32). These are unique in that two to four tanks could be assembled using the same base and sides. All inner tanks overflow into a box equipped with a common 1.5" (3.75 cm) drain at the back of the tank. This box can be equipped with a removable sponge or polyester pre-filter. Drain holes are equipped with bulkhead fittings. Glass is custom cut from 1/4" (6 cm) tempered plate stock and drilled prior to assembly. All sides must rest on the 48" x 20" (120 cm x 50 cm) base plate and be sealed with the appropriate silicone glue. The end and inner plates are 19.5"W x 20"H (14.75 cm x 50 cm) with one 1.5" x 6" (3.75 cm x 15 cm) overflow notch in the corner. In option three, two additional plates are needed, a 14"W x 20"H (35 cm x 50 cm) divider and a 28"W x 20"H (70 cm x 50 cm) divider with overflow cut-outs. As with the plywood tanks, the sides and bottom should be opaque except for the front.

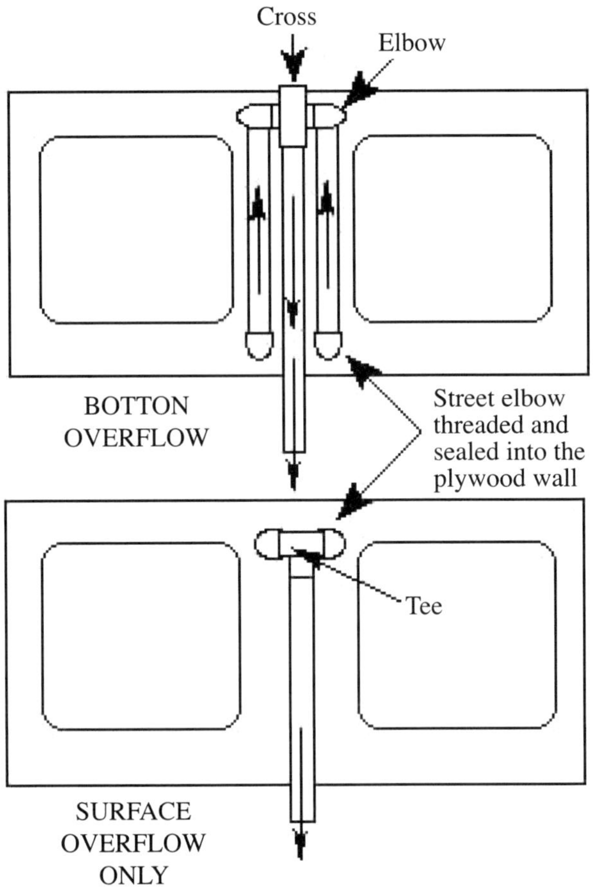

Figure 31 - External overflow arrangements which allow less inside obstructions when cleaning or capturing fish.

If space is not a limitation, I feel the larger custom tanks similar in size to those used at Instant Ocean Hatcheries and C-Quest are easier to maintain, clean, and routinely service. Standard glass aquarium tanks are usually too long and not easy to access. Tanks should be located so that they can be easily serviced. We prefer to have the top of the tank come to just below the breast of a 5.6"(165 cm) tall person. Care should be taken to not have excess plumbing on the floor or on the outside of the tank where a person will stand. Stacking tanks above the reservoir and filtration system conserves space and utilizes the entire vertical height of the building. Providing a slotted pressure treated wood boardwalk along the tanks (Photos 7, 8, 9, Figure 17) allows the technician easy access, dry feet, and provides easier plumbing arrangements. If a hatchery is designed from scratch, a tilting cement floor, 2" drop /35' (5 cm drop/10.5 m), with a common open floor drain located along the wall will provide easy drainage, as well as cleaning of tanks and entire systems (Figure 63). See last section of this book for more discussion of hatchery design. Water and detritus simply drop through the boardwalk onto the tilted floor and drain out of the building. Occasional yearly cleaning of the floor might be necessary due to buildup of siphoned waste detritus that becomes entrapped on the floor. Instant Ocean Hatcheries was designed so that a technician could walk on the boardwalks from one row to another without stepping back onto the cement floor. An added feature is the ability to locate pumps on the boardwalks to help eliminate electrical shock. If starting from scratch, make sure the cement floor is "rough brushed" prior to setting in the direction of the flow.

Tank Filtration

At IOH, we utilized submerged, subgravel bottom filters in each tank. As mentioned, this type of filter is susceptible to detrital build-up and if not properly maintained can break down and become toxic. For home use these are fine, but for commercial applications I feel a simple airlift biological sponge filter is best. These can be suspended above the bottom to facilitate ease of bottom cleaning. They can be easily removed and routinely cleaned without greatly disturbing the pair. I am against not providing individual tank filtration even when tanks are on a central recirculating system. I feel that each tank should be capable of standing alone in case of water recirculation failure. Aeration supply is paramount and the system should be equipped with a automatic backup. A 1/2" (12 mm) layer of coarse calcium carbonate substrate can be added which also serves as a biological filter and helps calm the pairs. Substrate should be coarse, 3/8" to 3/4" (9 to 19 cm) in size, to allow for siphoning of loose detritus without sucking up gravel.

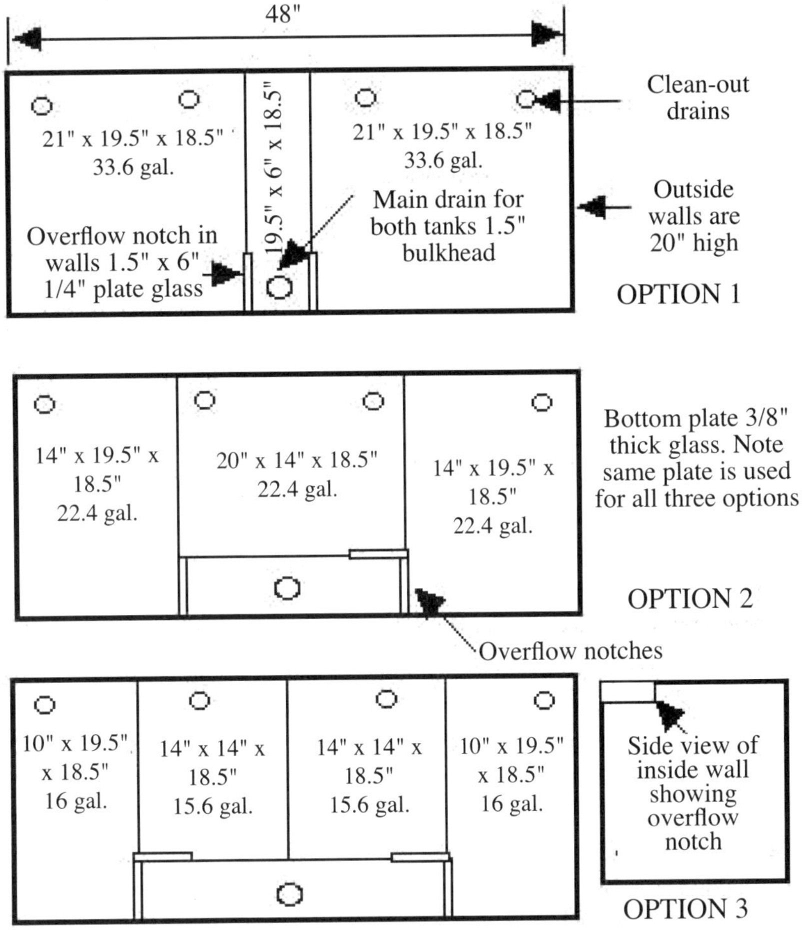

Figure 32 - Unique, versatile design of all glass tanks originally designed for C-Quest. Note all three designs use a standard, 3/8" thick, 48.5" x 20.5", base plate. All outside and interior walls are glued upon this base plate.

Spawning Habitat

More important than aquarium size is the amount and type of habitat. Absolutely no other competitive fish or invertebrates should be kept with spawning pairs. Habitat in the form of rocks, plants, coral, and excessive filter substrate should be avoided or kept to a minimum. Spawning surfaces, for benthic spawners such as clownfish, should be fairly large and consist of a hard opaque surface such as red clay flower pots, glazed flower pots, large diameter PVC pipe, natural slate or PVC rectangular plates placed vertically in the tank. Ideally, habitat should serve several purposes including: a suitable spawning substrate, artificial protection of the nest and a place where the adults can reside. Spawning substrates should be aged in saltwater prior to use to leach out any detrimental chemicals. In addition, they should be removed and cleaned monthly and replaced immediately with a clean, dry substrate. This should be done regardless of whether the fish are spawning or not. I prefer two 4" to 6" diameter (10 cm to 15 cm) new red clay flower pots per tank.

Since nests are often removed for hatching, spawning substrates should be placed in the center of the tank (Figure 33). This helps discourage the fish from spawning on the walls of the aquaria. Excess habitat, over what is specified, allows them to hide nests which becomes a nuisance when trying to locate or remove the nest for hatching or routine servicing. Occasionally a pair insist on spawning on the side of the tank regardless how you arrange the spawning surfaces. Placing clear acrylic plastic or glass strips, 4" x 9" (10 cm x 22 cm), against the side where they normally spawn permits easy removal of the nest. It has been suggested that keeping the walls of the tanks clean helps discourage spawning on the walls (Lichtenbert, 1992, personal communication).

If you are working with pelagic spawning fish this same tank design can be used. In this case, the eggs would flow over the overflow wall and be collected in a device that allows water to pass into the sump but retain eggs in a moist or wet state (Figure 34). It is important that the collected eggs are not subject to constant extreme water flow which might damage them. A large submerged collection screen or mesh bag, is suggested. To aid collection water input into the tank may have to be directed toward the overflow wall to create a positive current.

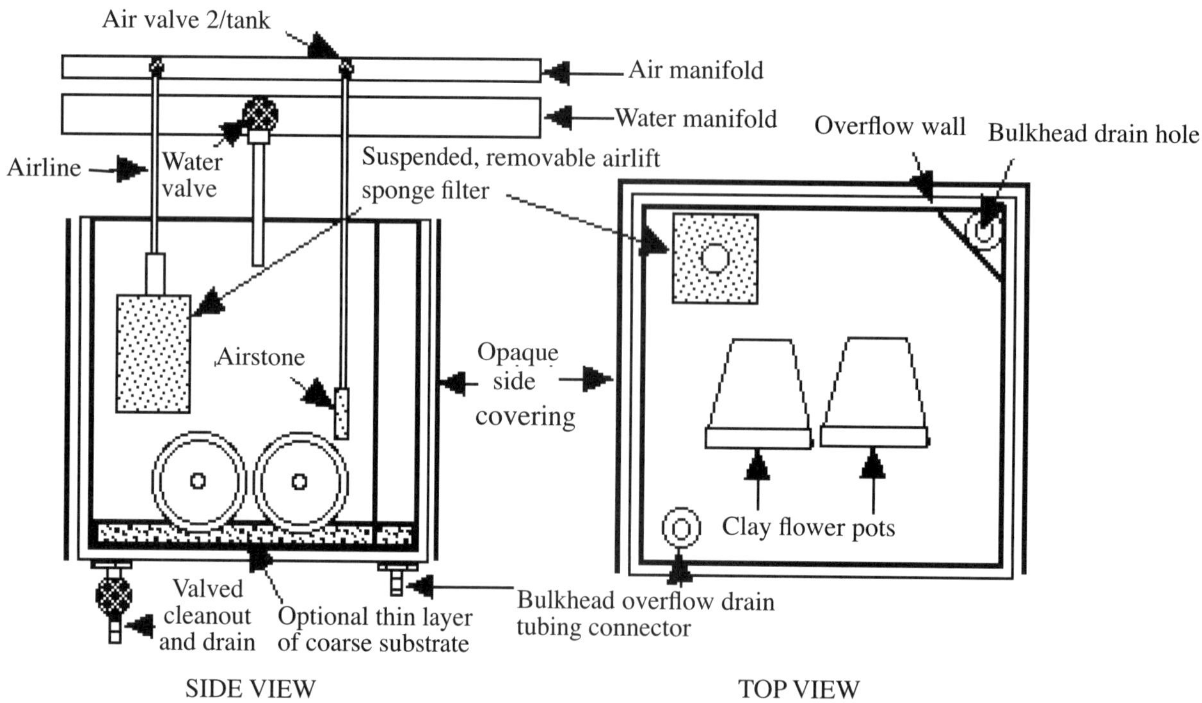

Figure 33 - Broodstock tank arrangement for benthic or substrate spawning fish.

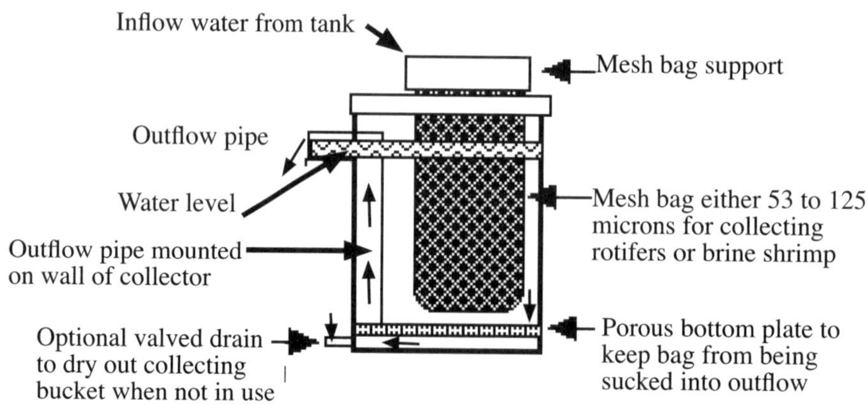

Figure 34 - Flooded collecting chamber for pelagic eggs mounted on the output line from the broodstock tank.

PAIRING BROODSTOCK

Clownfish pair for life, which is true of many benthic spawning fish. Patience is the most important factor towards obtaining a successful spawning pair. Primary criteria for determining sexuality of adult clownfish are color, size, behavior, and sex reversal. During my nine years of rearing clownfish I constantly looked for external clues to help distinguish sexually ready males and females especially when working with adult size fish.

Sexual Maturity

Age of the fish is the most important factor in determining sexual maturity. When selecting or establishing a pair, I do not advocate purchasing or using full grown adult fish. Several factors make them less than ideal candidates. First, the extreme cost of obtaining adult clownfish in good condition. Secondly, the fragile characteristics of newly captured adult clownfish make them a high risk. Lastly, wild populations take 12 to 18 months to reach sexual maturity. However, purchasing non-sexually mature sub-adults takes only 3 to 6 months to reach sexual maturity. With this in mind, it seems useless to take the risk to obtain adult fish when you realize how quickly sub-adults can reach sexual maturity once in captivity.

My experience, after spending thousands of dollars on broodstock, is that the worst buy on the market are the larger adult fish. Basically these fish have the hardest time adapting to change. It is not unreasonable to understand this since it is probably not much different than trying to move your grandmother into a nursing home after she has lived in the same home for 30 years. In either situation, they are mentally stressed more by the move than younger less settled individuals. Likewise, handling and shipping is more stressful on larger fish than smaller individuals, especially females. Paired female, most important individual, are more fragile, usually the oldest individual, and expend the most energy. Even if you are fortunate in obtaining a healthy, mature, adult pair, they will probably still take from 2 to 4 months to re-initiate spawning.

When selecting possible pairs or purchasing fish for pairing the best buy are sub-adults. These fish are around six to eight months old, show most characteristics of adults, are less dominating, less fragile, and have better survival rates. Sub-adults do not form immediate pairs or fully display adult coloration or sexual dimorphic differences. Yet they quickly adjust and mature under good environmental conditions.

Sex Reversal And Pairing

I refer to this pairing technique as, "hope there is a pair in the crowd method". A distinct advantage when pairing clownfish is their ability to change sex. Sex reversal, changing from male to female (protanderous hermaphrodite), has been reported in most clownfish (Table 8). Although not the normal mode of sex development within animal communities it is fairly common in many species of fish. Sex reversal in clownfish is always from a male to a female and cannot be reversed. This ability to change sex makes pairing of clownfish a lot easier especially when dealing with juveniles and sub-adults.

The first, easiest, and probably best approach in pairing clownfish is purchasing three or four young fish about equal in size, 1" to 2" (2.5 to 5 cm) total length. Or, if you can obtain sub-adults, 2" to 3" (5 to 7.6 cm), you might want to select two larger and two smaller individuals. Certain species like *A. ocellaris* or *P. biaculeatus* are often not equal in size when sexually mature (Table 8) . Put all the fish in one established aquarium with no other fish. Provide a simple uncluttered habitat consisting of four new, clean, clay flower pots about 4" to 6" (10 to 15 cm) in diameter depending on size of species being paired. Since sex reversal is prevalent in clownfish, they simply decide who will become the male and who will become the female. From this point, routine observations are very important

Table 8 - Distinguishing characteristics for pairing clownfish based on data from Instant Ocean Hatcheries.

	Unequal Size	About Equal	Color Differences	Sex Reversal
A. akallopisos	U	N	U	Y
P. biaculeatus	U	N	U	?
A. clarkii	N	U	S	?
A. ephippium	N	U	U	Y?
A. frenatus	S	U	N	Y?
A. melanopus	N	U	U	Y
A. ocellaris	U	N	N	Y
A. percula	U	N	N	Y
A. polymnus	U	N	U	?

INDEX

U = Usually the case
N = Not usually the case
S = Sometimes the case

Y = Yes, known to occur
Y? = Suspected to occur
? = Unknown

Eventually two fish will tend to stay together, chase others from specific areas and eventually attempt to destroy the others. If you find one facing head-up in a corner or with extreme tail and fin damage, it is time to remove that fish before it is annihilated. At times a pair will accept and allow one smaller individual to remain in a reserve situation. In aquarium situations, this is usually the exception rather then the rule. Utilization of reserve fish is a unique adaptation in clownfish. In the wild it assures an immediate replacement if one of the pair should die. Usually this spare individual will not grow as fast when compared to those not living with a pair. Reserves remain as males until the pair situation changes. When the female of a pair dies, the original male becomes the female and the reserve fish becomes the new male. Within a pair the female is usually the first to die. The reason she dies first is based on the fact that she is usually the oldest individual in a pair and expends the most energy within the pair. In nature, this is a unique adaptation for continued survival of a pair. Yet, in commercial operations, I do not feel it is good practice to leave an extra fish in the tank since it interferes with normal hatchery practices. At IOH we documented numerous cases of sex reversal within clownfish pairs, especially *A. ocellaris* . When females died, usually smaller, younger, individuals were introduced to the surviving male. Within two to three months the original male would change to a functional female. When known mature females were introduced to surviving males, the males remained functional.

Pairing By Coloration

When working with mature adult fish there are several methods of sexing them. Coloration can be used fairly well for most species of clownfish (Table 8). In order to utilize this technique it is important that the fish is not stressed, healthy, relaxed, and well adjusted to its surroundings. Although sexual color differences have not been widely reported on wild fish, I found those held in captivity often displayed striking differences.

Premnas biaculeatus (maroon clown) females are generally a dark, drab, maroon color whereas the males are a brighter red to red orange in color. *Amphiprion polymnus* (saddleback clown) females have a bright yellow colored face whereas most males have a dull brownish colored face. *Amphiprion melanopus* (cinnamon clown) females have a dull whitish red face while the males have bright reddish face and fins. *Amphiprion ephippium* (red saddle clown) females have an overall dull red body color and whitish red faces like the cinnamon clown, while the males have bright red bodies and faces. *Amphiprion akallopisos* (orange skunk) females are usually a brighter yellow orange than the males. *Amphiprion perideraion* (pink skunk) males have a narrow orange trim on the edge of the soft dorsal and upper and lower margin of the caudal fin. Due to the wide dispersion of *Amphiprion frenatus* (tomato clown) and *A. clarkii* (sebae clown) these exhibit wide differences in coloration and make them virtually impossible to distinguish. Yet in some pairs of *A. clarkii* males will have bright yellow upper and lower caudal rays on the tail.

When observing color differences, keep in mind that they are more evident and intense when fish are feeding or about to be fed. Color differences are only found on sexually mature adults. Fish have a tendency to "fade" in color over time in closed aquaria, especially if not properly fed to bring out color and/or maintained in poor water quality.

In addition, geographic location of the same species may have significant color variants. When you purchase fish there is no way to determine where they came from. Moyer (1976) studied *A. clarkii* from Bonin Islands, Yaeyama Islands, Okinawa, and Miyake-jima, Japan. The Yaeyama populations showed the greatest degree of color variation, but sexual dichromatism was not always pronounced. In Okinawan populations, adult males can be consistently recognized by orange borders on the caudal fin, compared to uniformly yellow caudal fins in adult females. Populations from Kyushu to Miyake-jima show subdominant fish of both sexes displaying orange caudal borders, similar in appearance to Okinawan males. Dominant males have solid orange or nearly solid orange caudal fins, whereas dominant females show pale yellow caudal fins. He further states that variations in color patterns appear to be related to selective pressures imposed by differences in habitat. Bonin Island individuals have a melanistic form yet sexual dichromatism is not present in populations. Melanistic individuals from the Bonin Islands inhabit *Heteractis* (previously known as *Radianthus*) anemones whereas, dark phase fish on Okinawa and Miyanke-jima inhabit *Stichodactyla haddoni* (previously known as *Stoichactis*) anemones.

Pairing By Size

Size can be another criteria in determining males and females in many species of fish including clownfish. In general, females are the largest individual in a pair. Within clownfish species I have worked with, female *A. polymnus, P. biaculeatus, A. akallopisos, A. ocellaris,* and *A. percula* are usually the largest individual in a pair. However, use of size criteria for pairing wild tomato and sebae complexes, *A. frenatus, A. ephippium,* and *A. clarkii,* is not of much value. Both members of wild pairs of these species at IOH were of equal size yet, tank reared pairs showed significant size differentials.

As opposed to many animals, fish continue to grow throughout their lives but at different rates as they grow older. Growth rates are mainly influenced by water quality, environmental conditions, and food supply. In some cases growth can be subdued by unknown factors within a group of fish such as reserve fish often associated with an active pair of clownfish.

Since many species of clownfish readily exhibit sex reversal, usually females are the oldest individual of a pair, and the first to die. Smaller younger males in a pair that change to females, will suddenly begin growing faster but will not achieve the size of the original female until years later. In older, more established pairs, size differences between a male and female are not as dramatic. This is due to the fact that as fish get older their growth rates slow proportionally to their age and eventually they are about equal in size. See page 60 for examples.

Utilization of size characteristics can be deceiving when trying to select a pair from wild stock. Since clownfish are collected throughout the Indo-Pacific, regional differences between isolated populations may be significant enough to make pairing difficult. Age/growth ratios may be completely different from region to region. Clownfish have lived under various environmental conditions and may comprise several variant complexes from remote areas. Keep in mind that wild clownfish are actually opportunistic feeders, therefore, the same size fish could be a different age or sex just based on the supply of food and population pressure within its specific environmental habitat. A small tomato clownfish from the wild could easily be a female and be four years old just based on environmental conditions. Under aquarium conditions tank-raised fish are maintained under constant environmental parameters, receive bountiful food and grow at about equal rates. However, as they reach sexual maturity (9 to 15 months) tank raised individuals normally show more dimorphic size differences.

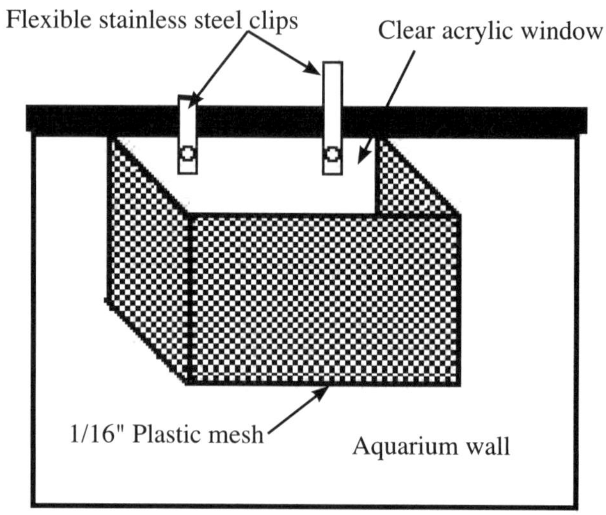

Figure 35 - Hanging plastic mesh acclimation basket (Florida Aqua Farms)

Pairing By Behavior

Behavior is another criterion for selecting males and females from adults or subadult fish. It takes time, and if not carefully controlled this method often results in injured or dead fish. If you are trying to pair adult size fish, this is usually the only method you can use. Adult fish have established a set way of life, similar to older humans, and even though you can pick out a pair by using external color or internal sexing, there is no guarantee that they will accept each other. Fish, like humans, can indeed become individuals with distinct individual attitudes and preferences. Once an adult fish becomes settled in a specific tank, it becomes the dominant individual.

The first way to use behavior is by introducing one adult after another to the tank. This should only begin in the morning and not extend overnight. Remove all distractions and other fish in the aquarium. With net in hand, lower the new introduction slowly into the aquarium and let them go at it. Stand back to the side of the aquarium so that they are not distracted by your presence. In the case of *P. biaculeatus, A. ephippium,* and *A. clarkii* it is usually over in 15 minutes or less. They will either accept each other or start to tear each other apart. Usually one will approach the anal region of the other as if trying to detect that special scent. The approached individual appears to frantically attempt to produce an acceptable scent and convince the other that he or she is compatible. Head shaking, vibrating mouths, and nipping of fins, especially the anal and caudal fins, occur. Often they mouth the fin or knock a scale or two off each other's side. A good possible pair will act basically gentle, do not engage in excessive contact, and are more relaxed. Nipping may occur, but is superficial and not overly injurious.

Even though you think you have established a possible pair, there is a strong possibility of rejection later in the day or at night. Therefore, when using the introduction pairing technique, it is best to start early in the morning, and make continuous routine observations throughout the day. During the evenings for the first two to four days, it might be wise to remove the newly introduced fish and place it in a retaining or acclimation basket within the aquarium (Figure 35). In the morning remove the fish from the basket and again begin to make observations. I prefer the extruded, ridged, plastic mesh, hanging baskets (Figure 35) made by Florida Aqua Farms or you can make one with plastic window screening material.

Established, dominant, individuals often seem to respond instinctively and without provocation, often injuring the new introduction before it has time to even signal its appeal. At times, I have found it better to put both fish in new environments. Placing both in a new environment greatly reduces dominance of established individuals and puts both fish on a more equal basis. Placing both fish on an equal basis often helps slow down initial instinctive aggressive behavior.

A second and fairly successful method of pairing adult fish or even smaller subadults, is using an acclimation basket on a continuous basis. Leave the newly introduced fish in the basket for two to three weeks or alternate them every other week. This technique is very good for pairing the most aggressive clownfish, such as *P. biaculeatus* (maroon) and *A. clarkii* (sebae) which have a tendency to destroy initially without provocation.

Determining If You Have A Real Pair

Well, do you really have a pair or not? A word of caution: two females will often accept each other. I have observed this in aggressive fish, such as *A. clarkii* (sebae) and passive fish such as *A. ocellaris* (ocellaris). These pairs will present the illusion of a true pair and even spawn, but the eggs are never fertilized. Unfertilized eggs are often common for new pairs but, for more established pairs, they are very unusual. Additional clues, besides unfertilized eggs, will include fish of about equal size and production of two nests within a very short time or at the same time.

To test for the possibility of two females within a pair, split the suspicious pair, and introduce a known male. Usually, within a very short time, you will receive fertilized eggs. Keep in mind, sex reversal is a one way street, there is no way two females can reverse their sex and form a productive pair. A good rule of thumb when dealing with clownfish is to stop trying to get the biggest fish you can find, which are probably females, and trying to make them into a productive pair.

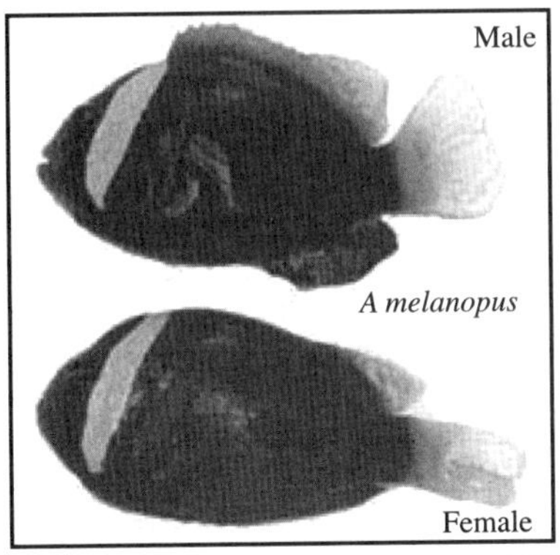

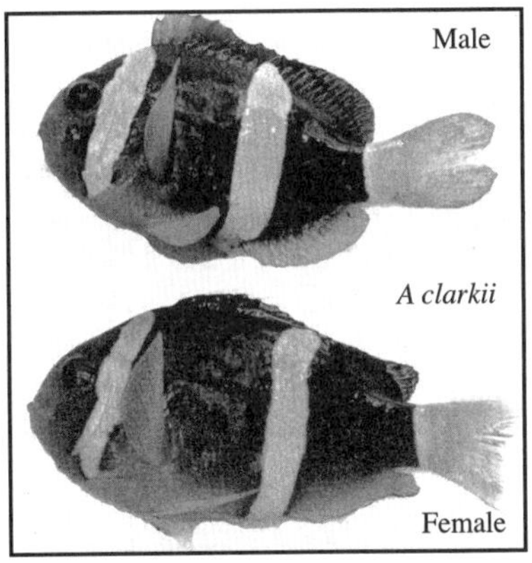

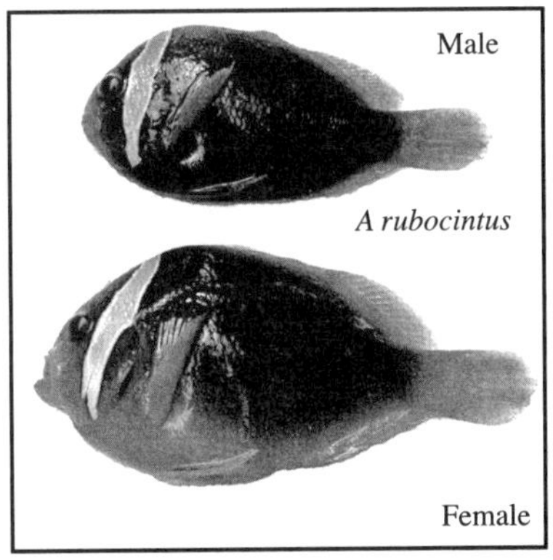

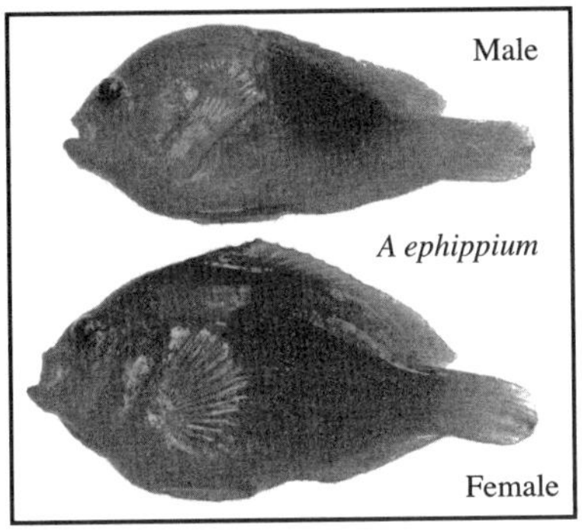

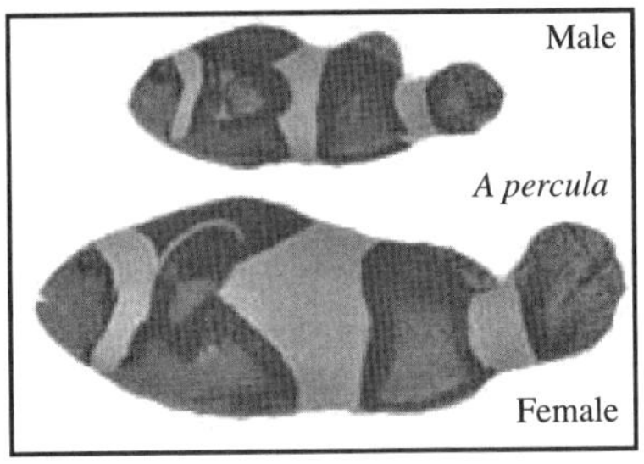

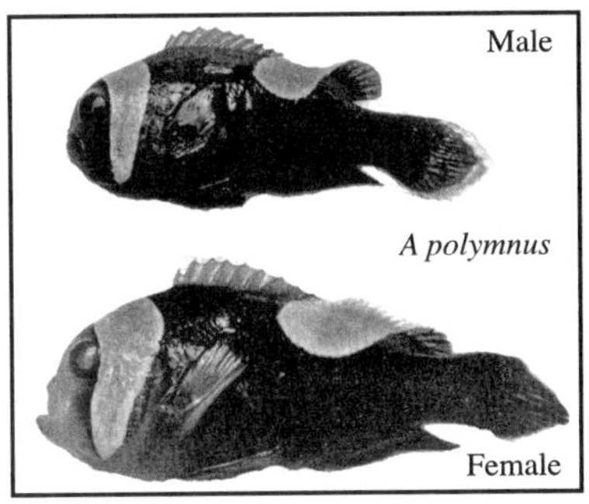

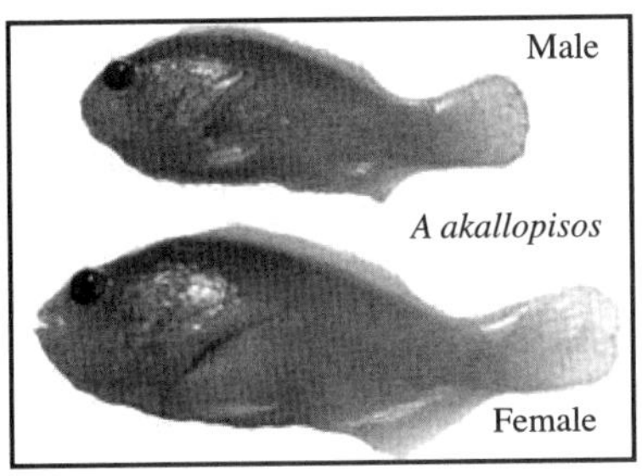

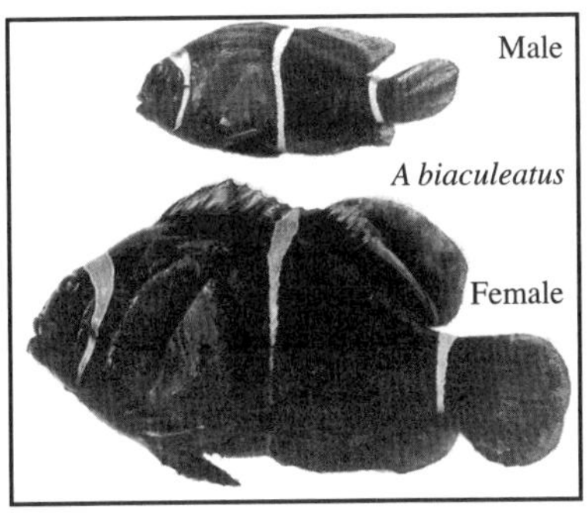

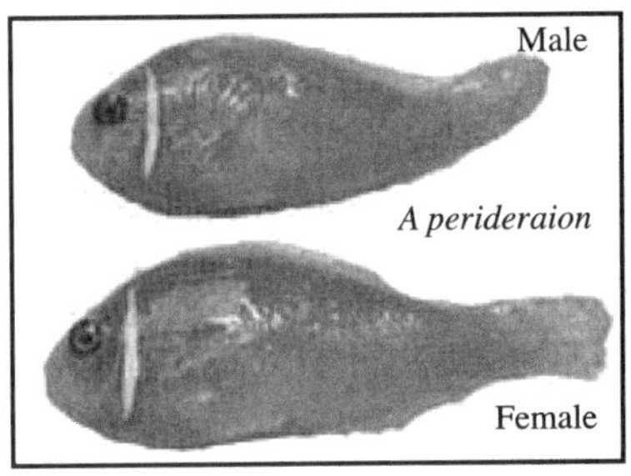

CONDITIONING FOR SPAWNING

Conditioning is a term used to describe the utilization and manipulation of a combination of environmental factors to induce gonadal maturation and, hopefully, spawning. Conditioning may include several or all of the following factors: light intensity; light duration and possibly wavelength; temperature, either constant or fluctuating; water current, direction and velocity; water quality, including oxygen, nitrogen, phosphate, ammonia, pH etc.; food including the type, size, color, amount, consistency, nutritional value, etc.; tank size and shape, aeration, habitat and many other items. All fish do not respond to the same environmental cues which trigger spawning. Finding what works best can be a challenging endeavor. Accumulating observations and data about the species you are working with and applying this information in construction of a artificial environment will enable you to induce spawning.

Inadvertent Conditioning

Conditioning, resulting in spawning, is often done inadvertently by the home aquarist or aquaculturists. A sudden spawn is usually quite unexpected and not well defined with regard to what was done or why it occurred. There have been hundreds of documented accounts of spawning fish and invertebrates in homes, public aquaria and research facilities. Yet, the prior conditions for inducing spawning are usually poorly documented or understood. In order to develop a consistent degree of control for continued routine spawning, a regimented set of conditioning parameters and records must be kept. Use of a log or diary should be mandatory by avid aquarists and aquaculturists. Proper continuous recorded documentation is essential towards continued success. I encourage everyone with a marine aquarium to maintain a log and include as many parameters as possible.

Natural Conditioning

Fish are sensitive to every day changes within their environment which includes primarily, water temperature, light period, light intensity, and water quality. Seasonal changes within their world have evolved into routine physiological and behavioral patterns called "biorhythms." Their entire body becomes a biological clock which is attuned to predictable changes. Even though, in tropical regions, these changes may be significantly less dramatic and more subtle than in temperate regions. Light and temperature are usually the "triggers" that spark gonadal development in temperate regions but in the tropics "wet" and "dry" seasons may play a more significant role. Changes within the environment are perceived by sensory organs such as lateral lines, eyes, and possibly pineal glands then passed to the brain for evaluation and direction. Like most other animals, fish are much more sensitive to environmental changes than humans.

Controlled Conditioning

Conditioning of captive fish out of their normal spawning season requires temperature- and light-controlled rooms or aquaria equipped with timers, heaters, coolers, and lights. Temperate species usually have short spawning seasons (short duration spawners) which are "triggered" by rising or lowering of temperature and/or light. Forcing gonadal development during their off-season requires condensing a year into a specific period of time. Usually the procedure consists of lowering temperature and light to a point where fish gonads enter into a resting state. This sets the physiological clock at zero. During this period, the fish is maintained on a moderate diet. After two to three weeks, light, temperature and feeding are increased. Internally, fish start to store excess fats within their body tissues which will later be incorporated into developing eggs.

Further increases in the photo-thermal regime induce maturing egg development to a point where all parameters are optimal for spawning.

Figures 36 and 37 illustrate condensed regimes utilized on a semi-protracted spawner, pompano (*Tracinotus carolinus*) and a short duration spawner, black seabass (*Centropristes melanus*)

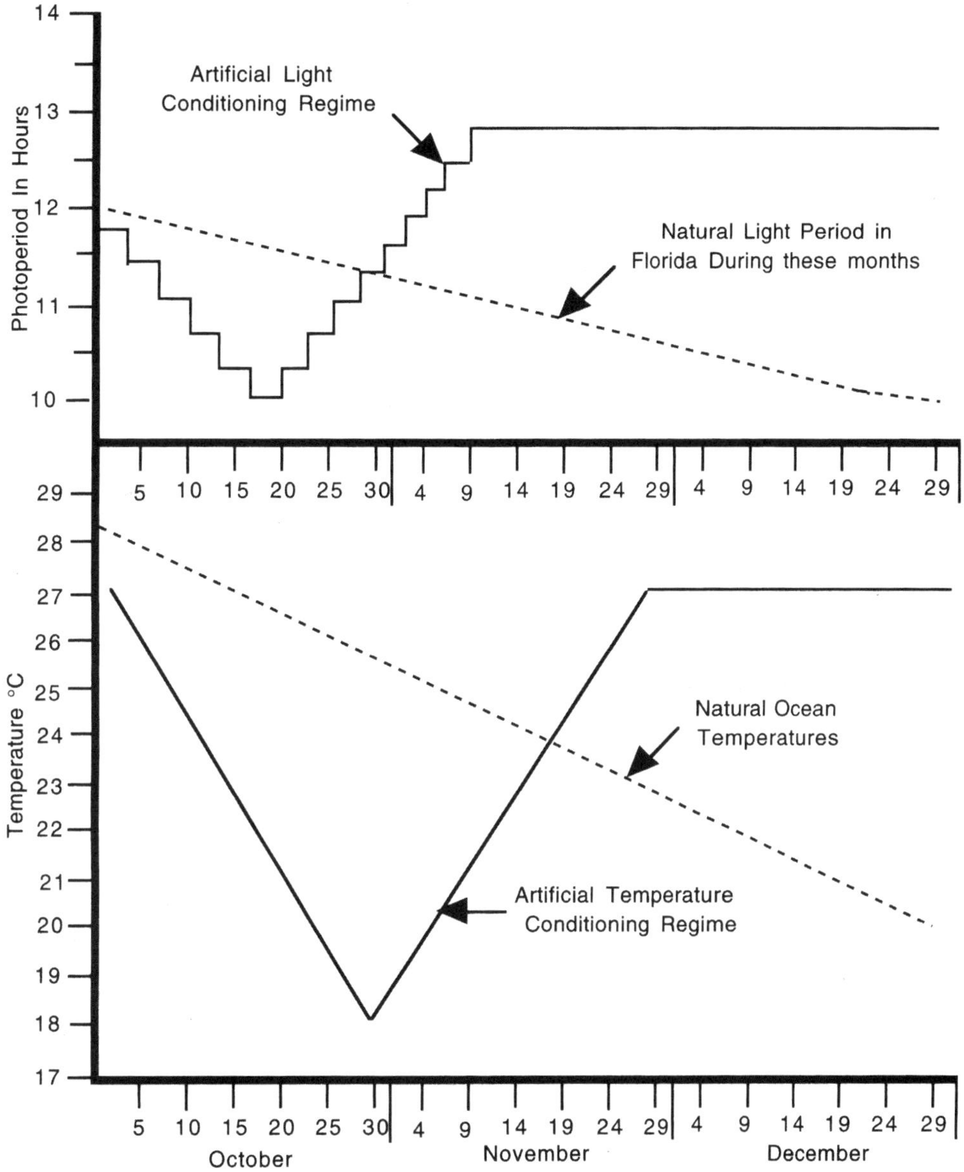

Figure 36 - Out of phase light and temperature step conditioning regime used on a summer protracted spawner, *T carolinus* (pompano), Hoff 1978a.

by Hoff (1978). To determine the parameters for such a regime, it is imperative to seek as much data as possible about the fishes' normal environment and habitats. Weather bureau and coast guard literature will provide a wealth of routine data on seasonal water temperatures. Daily light periods are available for various regions from the weather bureau or various governmental agencies. The objective of the illustrated regimes was to force pompano, which normally spawn in the spring and summer (May through September), to spawn in winter, and black seabass, which normally spawn from March through April, to spawn in the spring and summer. Yearly data

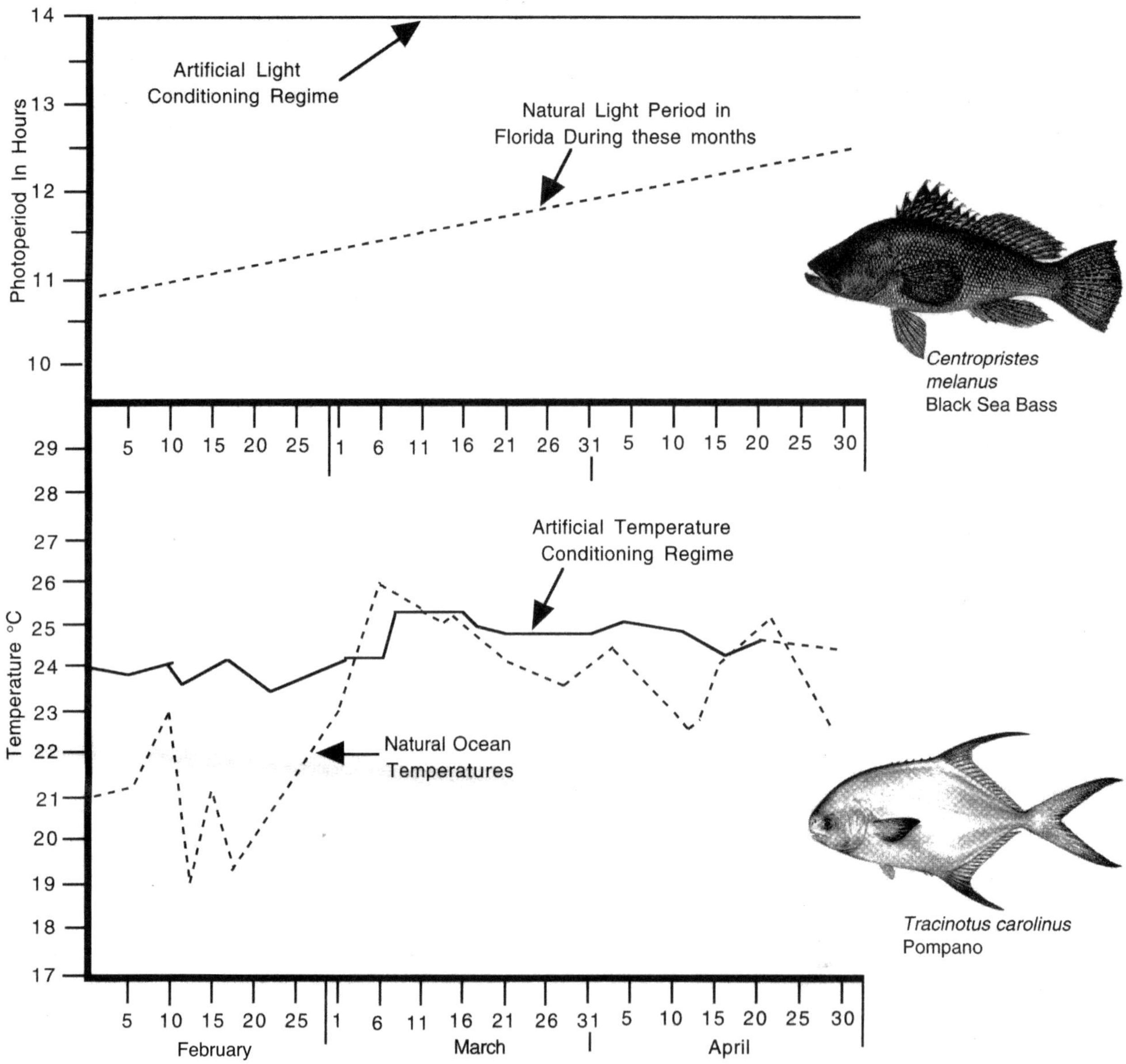

Figure 37 - Constant photo period and temperature used on T. carolinus (pompano) and Centropristes melanus (black seabass), Hoff 1978a

was condensed into a three-month period to stimulate gonadal development during the winter. If you were working with a winter spawner like black seabass, the regime would be reversed. In other words, higher temperatures and higher light levels induce a resting state, while, in a spring and summer spawning fish, the opposite would be true. Later experiments indicated that just maintaining a constant temperature and light period after conditioning caused black seabass to repeatedly spawn over several weeks until all nutritional resources needed for egg development were exhausted (Hoff 1974).

Roberts (1987) presented a very innovative illustrated time table for conditioning maturation and spawning conditions for the red drum, *Sciaenops ocellatus*, (Figures 38A & B). This is a pelagic spawner which can spawn every day for months at a spawning regime of 10HL:23°C (HL=Hours light). At times the spawning stock need to be restimulated, which is accomplished by raising the temperature to 28°C and then back, or they are put into an intermediate spawning state by lowering the temperature 1 or 2°C for a period of time (see Gonadal Cycling and Egg Development).

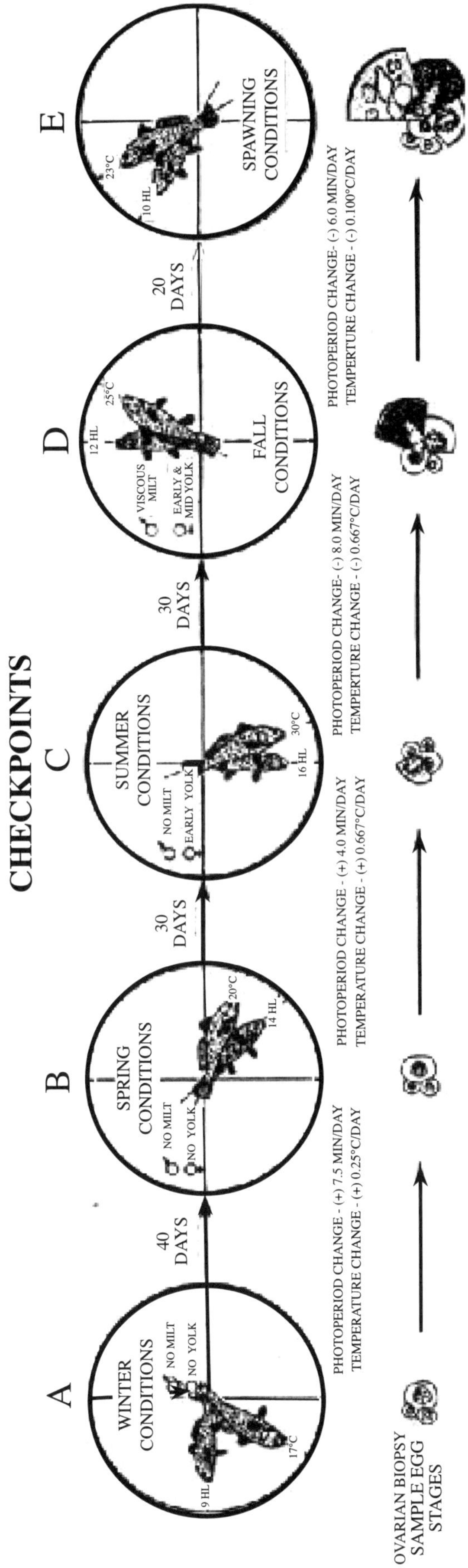

Figure 38A (Left) - Timetable for red drum maturation. Demonstration of photoperiod and temperature changes necessary to induce reproductive maturation or red drum from winter conditions (checkpoint A) to prespawning conditions (checkpoint E), and diagrammatic sample of oocyte stages for each checkpoint from in-vivo ovarian biopsies of females.

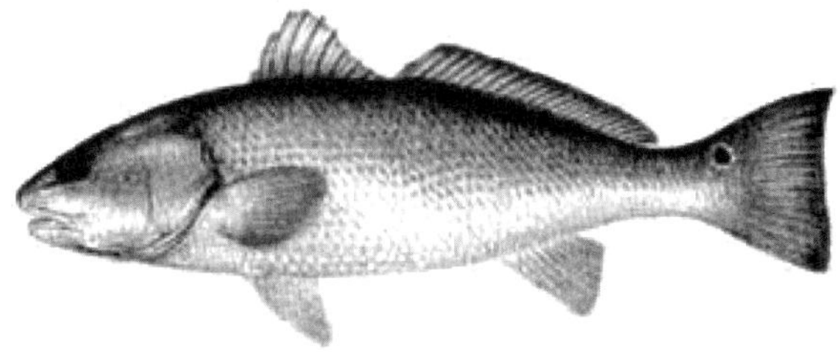

Red Drum (Sciaenops ocellatus)

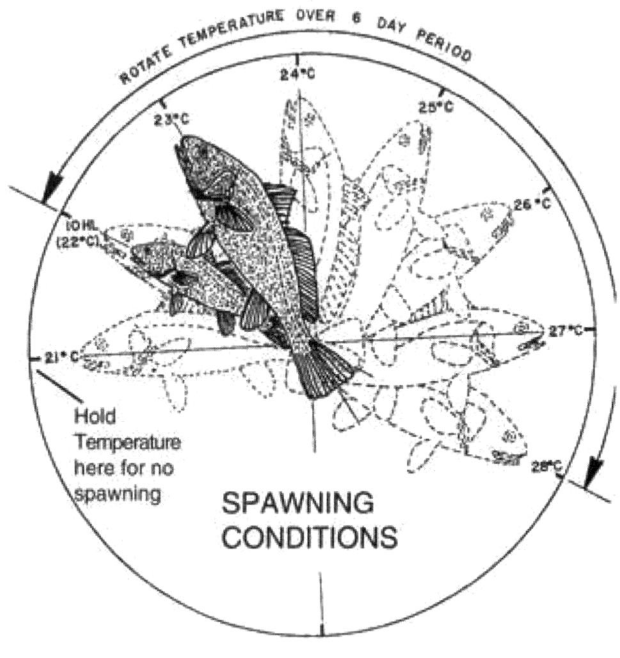

Figure 38B (Above) - Temperature fluctuations to induce spawning in mature red drum broodstock. Photoperiod remains constant at 10 HL. (Roberts 1987).

The procedure for conditioning for maturation (Figure 38A) consists of acclimating broodstock to a winter photothermal condition of 17°C:9 hours light (Checkpoint A). Maintain stable water conditions and avoid sudden temperature changes that could adversely effect the biological filter. Feed gradually, about 3% body weight per day about days a week. Fish can remain at this checkpoint for extended periods and eggs remain in a resting state. Proceed to checkpoint B over a 40 day period during which time photoperiod change is +7.5 minutes per day and temperature change is +0.27°C per day (about a quarter of a degree/day). Checkpoint B is 28°C:14HL. From checkpoint B photoperiod time is changed +4.0 minutes per day and temperature is +0.067 (about 1/2 a degree) over a 30 day period. Checkpoint C is actually the artificial summer months at 30°C:16HL. At this point the eggs are in an intermediate spawning state and can be held here for extended periods (see Figure 12 and Photo 5). When ready, proceed to checkpoint D over a 30 day period. Light duration is lowered 10 minutes per day and temperature is lowered 0.17°C per day until a 25°C:12HL is reached. Eggs in this stage are in a late stage intermediate state. From checkpoint D to E is another 20 days of declining conditions where temperature drops 0.1°C/day and light duration at a rate of 6 minutes/day until a 23°C:10HL is accomplished (checkpoint E). This is the point where eggs are in a prespawning state and are ripening for a pending spawn. Wait an additional 10 days after reaching checkpoint E to see if spawning will commence without further environmental enhancement. If they start spawning, allow them to spawn for 3-4 days if they spawn every day. Stop spawning by rapidly lowering temperature to 21°C at a rate of -0.08 to -0.20°C/hour until 21°C is reached, allow a rest period and then go back to 23°C. Note light duration remains constant at 10 HL. If the fish do not spawn at checkpoint E, gently raise/lower the water temperature through the range from 22°C to 28°C over a period of 6 days (1.0°C/day) until spawning occurs. Drumming, darkening of males from midline up, and presence of red, swollen vents in females are indicative of spawning. Once spawning occurs, allow it to occur for 3-4 days, induce rest and reinstate spawning. Spawning termination is accomplished by lowering temperature and light duration to 17°C and photoperiod 9HL over a 30 day period.

Unpublished comments from Roberts indicate that the length of the conditioning regime (checkpoint A to E) can be dramatically condensed when using high quality fish. One very large red fish has been reported to have spawned daily for over two years. However, assessment of fecundity, fertilization and egg quality are in question. His experience has shown that, if a fish is forced to spawn for over a week, there is a decline in egg quantity and quality. Yet, allowing the fish to rest at 21°C for about a week restimulates quantity and quality. He clearly points out the controlling exogenous (outside) factors for maturation and spawning are day-length and water temperature coupled with food quality and abundance.

Clownfish Conditioning

When dealing with long duration spawners (protracted spawners) such as clownfish, we utilize a different set of conditioning environments. Our goal in this case is to promote continuous spawning with a built-in natural rest period between spawns. Most reef dwellers live in the tropics around the equatorial region. In this area, a day consists of a 12:12 split between light and dark. Light duration and intensity from the sun play minor roles in triggering spawning. However, the moon may have more influence based on natural spawning data. Under natural conditions, wild clownfish spawn most of the year but usually not more than one spawn per month. Under controlled conditioning regimes and proper feeding they can be induced to spawn an average of 2.2 spawns per month. The difference is primarily due to three factors: constant water temperature, water quality and a quality year-round diet. Allen (1972) and Verwey (1930) reported on two different species. They stated that spawning was not evident in natural environments between September and December. This same period coincides with the wet season in the tropics and probably plays an important role in changes in water quality and food availability and, to some degree, water temperature. Seasonal climatic changes no doubt impart a change in the flora and fauna which make up the clownfish diets. Factors not often mentioned are seasonal changes in wind direction and current patterns in the tropics. These changes can bring cooler waters from deeper depths and/or changes in food types and abundance.

In other regions, light and temperature apparently play a roll in spawning. Populations of *A. clarkii* around the island of Miyako-jima, off South Japan, clearly show the effects of temperature and perhaps light (Moyer, 1980). This is a warm-temperate region at 34° N. latitude where temperatures range from 55° to 81°F (13°C to 27.5°C). Within this population, spawning occurs from May to October or 6 weeks after the Spring Equinox and ends nearly 3 weeks after the Autumn Equinox. Spawning commences and stops around 72°F (22°C). Yet, in the fall, the water temperature is higher than 72°F (22°C), but the light period diminishes and spawning stops.

Conditioning Temperature

Within a captive environment, consistency of temperature within a specific range is what you want to achieve. Clownfish will spawn within a fairly wide range of temperatures from 71° to 88°F (22° to 31°C) . Although spawning occurs within this wide temperature range, the "quality" of nests and hatch is usually affected. Quality eggs produced within the extremes of this range will be low. Likewise, quality of hatch and larvae produced at the extremes of this range are at risk.

In lower temperatures, metabolic rate is slowed and hatch is extended 2 to 4 days longer than optimal. Extended hatch usually results in weaker larvae and sometimes a lower hatch rate. At elevated temperatures, metabolic rate is accelerated above optimum, larvae may hatch early but are often inferior quality.

Ideal incubation temperatures for clownfish eggs are 79° to 83°F (26° to 28°C) . Diurnal daily temperature fluctuations, +/- 2°F (1°C), do not really effect the results and may indeed enhance quality. For most pelagic or benthic spawning tropical marine fish, these temperatures are suitable.

Conditioning Light

Even though the tropics have a 12:12 light and night period, I prefer a 14-hour light period with a faint (15 to 20 watt bulb) indirect, constant, moon light at night. Goldstein (1989) and others feel that a 24-hour light period is necessary to achieve spawning. He further states that this promotes continuous spawning about every three weeks. Yet, we easily achieved spawns every two weeks or less with a 14:10 light/dark period. It seems unreasonable to subject fish to 24 hours light. Physiologically this practice has to place undue stress on the fish and is obviously not necessary. What lasting effects this practice has with regard to how long a pair spawns, spawn quality, and their life span is not known.

Lights should not be located directly on top of the tank but should be elevated 12" or more above to provide easy access. The required light intensity is not clearly outlined. I have documented clownfish spawns in high and low intensities. There is some speculation that high intensity lighting may have a detrimental effect on nest quality and hatch although I do not think this has been experimentally substantiated. Goldstein (1990) stated that "tomato clownfish eggs and newly hatched fry are sensitive to light. Eggs should be hatched in a dimly lit room. All artificial light should be eliminated prior to and up to 72 hours after hatch."

Broodstock at IOH were maintained in a greenhouse covered with opaque yellowish double layered, fiberglass, roof panels. Natural light intensities within the building were measured at a high of 1500 foot candles in July to a low of 800 foot candles in December. No artificial light was provided directly to the tanks. Normally light received in the tanks was not perpendicular but at angles or diffused since the tanks had 20" (50 cm) high opaque walls and only a 12" (30 cm) square window. Limited, supplementary fluorescent lights for extending the light period to 14 hours and for use on exceptionally cloudy days were located in the attic area, 3' (1 m) above the top of the tanks. They were automatically controlled by a timer and photoelectric cell.

Effects of various wave lengths do not seem to be overly important. Yet, I prefer a bright white daylight type bulb perhaps mixed with a light-actinic type bulb. There are indications that use of near ultra violet (actinic) type bulbs may aid healing and have a therapeutic effect on vision problems. This is based on unpublished experiments I conducted on vision impaired red grouper.

Water Quality Conditions

Typically you would expect that a salinity of 34 to 35 ppt (parts per thousand) or a specific gravity of 1.0220 to 1.0228 at 80.6°F (27°C) to be necessary for spawning clownfish and many other oceanic marine fish. At IOH we typically maintained all our systems including the broodstock at 28 ppt (1.0175 Specific Gravity). Lower salinities were maintained for several reasons. It was theorized that lower salinity helped reduce osmotic stress and may reduce disease problems associated with parasites that demand higher salinities to survive, such as parasitic copepods, etc. Secondly, it allowed a large variance in salinity due to evaporation which can be considerable in a large enclosed hatchery. During warm, dry weather it is not unusual to lose 45 to 55 gallons (per 7,000 gallons) a day to evaporation. Thirdly, rotifers reproduce and survive better in low salinity. Since this is the primary larval food, rotifers should not be subject to high salinity stress. Lastly, in a hatchery containing 110,000 gallons of artificial saltwater, the cost between maintaining 28 ppt against 34 ppt is significant! Therefore, high salinity does not appear to be a significant factor in maintaining and conditioning broodstock and is not recommended.

Water quality (ammonia, nitrite, nitrate, phosphate, pH) at IOH was recorded weekly and measured on a spectrophotometer. Salinity was measured with a refractometer. Of the parameters measured and maintained, nitrate is the hardest to control in closed systems. Normally broodstock and larvae were maintained at 20 to 30 ppm (parts per million) nitrate-N. Nitrite and ammonia were insignificant and were usually less than 0.1 ppm. pH was maintained around 8.0 to 8.3 by routine additions of sodium bicarbonate. Calcium levels and redox potential were not considered at IOH but may play a significant role based on recent information about mini-reef aquaria. Trace metal and vitamin supplements (Macro Algae Grow, Florida Aqua Farms) were routinely added to the entire hatchery.

Conditioning Habitat

Habitat plays a role in conditioning but is not well defined. Experiments we conducted in 1972 with black seabass (bottom dweller) and pompano (water column dweller) gave mixed results (Hoff 1978). During their conditioning regimes some tanks were provided with habitat consisting of sand bottoms, artificial sea grass made of green plastic ribbon, random size sections of pipe, and some common red bricks, while other tanks remained bare. During these experiments we assessed the feeding rates, weight gain, and sexual development by in-vivo oocyte sampling (removal of live eggs). All fish were fed to satiation once or twice daily. Diet consisted of a gelatin-based mix. Results indicated that habitat did not encourage gonadal growth, feeding, or weight gain. Use of habitat appeared to retard but not inhibit maturation (gonadal development). Larger individuals in either case, with or without habitat, matured more rapidly. Weight gains were lower in tanks with habitat and food consumption was generally lower. Removal of habitat during the latter part of the experiment resulted in higher food consumption. Although both tanks consumed about the same amount of food, conversion was more efficient in tanks without habitat. Yet, this experiment did not fully address all the parameters or needs of fish.

Experience with other fish has indicated that habitat is more important with regard to attitude or reduction of stress. Yet, habitat in broodstock tanks should be strictly limited or modified to the point that only the basics are provided. For example, pompano are a pelagic, constantly swimming fish whereas seabass are a sedentary bottom-hugging fish. Bottom habitat for a constant pelagic swimmer is probably only important with regard to food gathering whereas in a benthic dweller it becomes a protective habitat, hiding place, and food gathering place. When food is constantly being fed to fish, the importance of habitat as a food gathering spot diminishes. Within several weeks of consistent feeding, they normally come directly to the surface to feed.

Habitat For Clownfish

With regard to broodstock clownfish, there are several points to consider. Normally these fish utilize a live anemone as their protective habitat and the hard surface beneath the anemone as their spawning substrate. Utilization of anemones results in a very difficult situation for commercial spawning of clownfish. By creating an artificial environment in which the spawning pair are comfortable and feel protected, the anemone can easily be eliminated. Experiments conducted at IOH did not show significant benefits in allowing an anemone to remain in the tank with regard to spawning. However, utilization of anemones with nervous species such as *A. polymnus* (saddleback clown) or a specific touchy pair was beneficial with regard to stress control. Since security is the primary function of the anemone, habitat selected for broodstock tanks must reflect this security. Clownfish are territorial and will not accept interference from other clownfish or most other fish and invertebrates. For best results, the pair should be kept in individual tanks without influence from outside. This tank should have a window in front and opaque sides. A moderate shade of aqua blue coloration is normally used on the opaque sides. Lighting should be moderate and located above the tank.

Clownfish are basically bottom dwellers preferring some sort of habitat to hide in. Since the anemone can easily be eliminated, a suitable multifunctional spawning substrate is required. We found red clay flower pots very suitable. They provide a measure of security for each individual and a desirable spawning substrate. In addition, they can easily be removed for hatching and cleaning, and replaced quickly. Substrates such as PVC pipes etc. can be used but keep in mind that whatever is selected must be: (1) large enough to house (cover) one or two adults at a time, (2) large enough to allow the two adults to spawn inside, (3) easy to remove and replace without disrupting the tank or upsetting the pair.

When habitat is removed for cleaning or hatching nests, a new clean piece should be quickly replaced. I prefer two pieces of equal size diameter habitat per tank. This allows each adult a place to reside and hide. I also prefer habitat with a closed or partially closed back end since clownfish generally orientate themselves with their heads facing towards the front of the tank. The closed back of the flowerpot allows an extra measure of security they desire.

Location of the "security-spawning" habitat is very important with regard to finding a spawn and recovering the nest for hatching. Pots or pipes should be placed side by side in the center with the open end facing towards the viewing glass (Figure 33). Habitat should not touch the sides of the tank.

Cleaning the security-spawning substrate should be routine. I don't mean sparkling clean but it should not be overgrown with heavy mat or slime algae. Clean substrate seems to stimulate the pair to lay eggs. Monthly replacement or cleaning of habitat is recommended. Don't switch types of habitat during replacement. Everything should be routine and remain the same each time. Don't make big changes or move too quickly when removing or replacing habitat. Clownfish will bite, but I have found that if you wear gloves or a shiny bracelet, watch, or ring they will attack that first before biting your hand. Quick movements often result in injured fish so move slowly when reaching in or going by the tank.

Conditioning Diets

Adverse effects of poor nutrition or feeding practices on reproduction in livestock, poultry and laboratory animals are well known. Conversely, little research has been done on the effects of diets with regard to fish reproduction and egg quality. Broodstock diets are virtually the main keys to success.

Eggs contain considerable lipids (fats and oils) which are high and long-lasting energy resources needed for the protracted development of embryos within the eggs. Examination of intestines and body cavities of black seabass (short duration spawner) about a month or two before spawning, reveals accumulated fat deposits. Following a period of heavy feeding, female seabass actually slow or stop feeding for a month or longer prior to spawning. During spawning, which lasts about 6 to 8 weeks, the fat deposits diminish drastically. These deposits are actually reserves to be incorporated within the eggs and provide energy to the female during her fasting period. When working with a short duration spawner it is important to recognize and realize when to feed heavily. Suitable diets and enough food must be fed to prime them with enough reserves to produce large quantities of quality eggs within a short time frame.

Lovell (1979) conducted a two-year nutritional study at Auburn University on broodstock catfish. Seven different diets were tested against a natural pond forage diet. Results showed that growth and reproductive characteristics (number of eggs, size of eggs, number of spawns, spawning date, hatching rate, and quality of fry) of females fed nutritionally balanced high protein and energy diets alone were superior or equal to that of fish on any other treatment. Females fed vitamin-deficient, low protein and energy diet alone spawned later and had fewer eggs and smaller egg size than females fed nutritionally adequate diets. Moe (1991) stated that if there is not enough chitin or shrimp tails in clownfish diets, the eggs did not adhere to the nest wall very well.

When working with a protracted long duration spawner (clownfish) it is very important that diet is consistently administered and of the highest quality you can provide. Quality is the key word, nothing else will suffice. Experiments we conducted on clownfish have shown direct correlations with diet consistency, how often it is fed, how much is fed, and how much lipid is incorporated in the diet. Fatty acids (lipids) are numerous and found in most organic based foods. To produce 2.2 spawns per month on a year-round basis for 5 years, the amount of lipids or lipid-producing nutrients is important. Females are constantly expending tremendous amounts of energy to produce 500 to 3,000 eggs per nest. If brood fish are not properly fed, results are directly reflected in the number of eggs laid, fertilization rate, hatch rate, and the quality of hatched larvae. Poor quality eggs develop slowly, hatch late, and often result in significant early larval mortalities. At times a more direct approach may be taken to increase the proper lipid content by feeding female fish roe. To date, there are foods available which are marginally adequate but are basically maintenance diets and not designed as a conditioning diet. Usually the aquarist is obligated to seek a variety of foods rather then one specific diet.

When dealing with pelagic spawning fish, egg buoyancy, clarity, and oil globule formation are good indicators of egg quality. In red seabream culture, specific diets consisting of krill, mysis shrimp, and crab wastes are administered prior to spawning to enhance egg coloration, fertilization, hatch, and larval quality. A study of five diets fed over five months prior to spawning clearly indicated that egg buoyancy, egg quality, percent hatch, and percent of early survival are directly related to nutritional quality. Broodstock (control diet) were fed white fish meal (45% protein) had an average of 80.9% buoyancy over 60 days of spawning, 69.4% hatch, 37.6% abnormal larvae and a post larvae production of 24.3% of the eggs produced. Those fed a low protein and low phosphorus diets (2 and 3) showed an average buoyancy of 54.4% and 62.1%, total hatch of 23.6% and 26.3%, abnormal larvae of 96.2% and 93.8%, and post larval survival of 0.1% and 0.3% of the eggs produced respectively. Those fed cuttlefish diets (4 and 5), with one being deficient of essential fatty acids, showed marked contrast in that the egg buoyancy was 23.9% and 88.5%, total hatch of 0.9% and 93.9%, abnormal larvae 100%, and 2.4%, post larval production of 0.00% and 78.9% respectively (Shepherd & Bromage 1988). This experiment clearly demonstrates the importance of a well-balanced broodstock diet and the overall effect diet has on production.

Conditioning foods should be administered routinely to broodstock clownfish three times a day during the week and once or twice a day on weekends. When used, supplemental diets, such as roe, live *Artemia*, etc., should constitute a fourth feeding. When utilizing roe, feed the whole eggs, do not break them down.

Gelatin Diets

Since 1971, I have made frozen gelatin-based diets to feed broodstock fish and juveniles of various species. As a pioneer in the development and utilization of gelatin-mixed diets, I feel they offer the best solution in providing a well-balanced fresh diet for broodstock and juvenile rearing. Mixed gelatin foods offer a variety of pluses including the natural desire of fish to consume moist soft diets and the ease of adding specific vitamins, color enhancers, minerals, lipids, protein, fibre, algae, carbohydrates, and drugs. Diets utilized at IOH were developed specifically for clownfish based on nutritional information we could find on specific food items listed by Allen (1972) (See Feeding Habits). Since many items could not be easily obtained in quantities, substitutes were made which we felt were adequate or similar in nutritional values. Although not a perfect diet, it was used for over eight years with good results.

This diet was made and marketed by IOH under the brand name of "Sea Food™" as well as a companion diet, Sea Salad™" from the late 1970's until 1985. Analysis of Sea Food™ was 56.04% moisture, 26% crude protein (nitrogen x 6.25), crude fat 4.44% and crude fiber 2.64%.

Table 9 is a receipt for a general diet suitable for clownfish broodstock and juveniles. Several items such as vitamins, mineral mix, kelp meal, krill meal, gelatin, and special fish meals can be obtained from Florida Aqua Farms.

Table 9 - Ingredients for conditioning and growout diet for clownfish based in part on formulas used at IOH.

Ingredient	Amount	% of Total
Beef heart or Chicken heart	10 oz.	12.2%
Fish roe or Whole eggs	5 oz.	6.1%
Fish flesh	5 oz.	6.1%
Shrimp tails	5 oz.	6.1%
Clam meats	5 oz.	6.1%
Kelp meal	5 oz.	6.1%
Krill meal	5 oz.	6.1%
Dry fish meal	5 oz.	6.1%
Gelatin	4 oz.	4.9%
Water (fresh)	38 oz.	46.3%
Vitamins	<1 oz.	0.1%
Minerals	<1 oz.	0.1%

Procedure For Making Gelatin Diets

1. Heat all water to almost boiling.
2. Trim off fat and liquefy beef or chicken hearts in blender with the smallest amount of hot water possible. (blender will bind if not enough water is added) Pour into mixing bowl.
3. Blend, do not liquefy, the fish roe, attempting to retain the whole eggs with the smallest amount of hot water possible. Pour into bowl.
4. Liquefy shrimp tails, with shell but no heads, along with the clam meats. Note: if using canned clam meats, the liquid portion can be substituted for an equal portion of the hot water. Pour into bowl.
5. Mix all dry ingredients together (kelp meal, krill meal, dry fish meal, vitamins, and minerals) together with the gelatin and add together into the mixing bowl containing

the liquefied ingredients. Place under mixer and blend while adding the remaining hot water.

6. The mixture should be thick so that a spoon will stand upright yet liquid enough so it can be poured. Add more hot water if needed, being careful not to add too much. If too liquid, add more dry ingredients.
7. Desired temperature of mixture should be around 85°F (30° C). Just hot enough so that light steam comes off the mix and the granular gelatin melts into the mix. It is important to realize that you are not cooking the mix, but just binding it. Test for the gel-factor by removing a spoon full, place in freezer for several minutes until it cools and see how well it stays together. It should feel "rubberlike" and not a sticky liquid consistency. Add more gelatin if necessary.
8. Pour desired quantities into zip-lock bags or plastic refrigerator containers and store in freezer.

The consistency you want to achieve is a moderately soft, "rubberlike" food that will remain together underwater for 12 hours. To achieve this, the gelatin content can be manipulated from 5% to 7%. Certain ingredients will not work with gelatin, including shrimp heads, excess of *Spirulina* etc. These contain oils and enzymes which are not compatible.

Gelatin mixtures should be removed from the freezer when needed and defrosted in the refrigerator overnight. When feeding mix, crumble between the fingers into particles about 1/8"^{2} (3 mm^2) cubed. Drop or dip food into the central surface of the tank. Amount of food and size should be adjusted to the size of the fish being fed. *A. ocellaris* or *A. akallopisos* need only a couple of 1/8" pieces while large fish like *A. frenatus, A. clarkii,* or *P biaculeatus* may need three to four pieces per feeding. Avoid over feeding, give only what the fish will take. Leaving food on the bottom, with the intention that they will eat later, is not recommended. If they are not hungry at the first feeding perhaps they will feed on the second or third. Excess uneaten food should be removed the following morning. A turkey baster or small siphon is ideal for routine waste removal.

Live Feeds

There is a misconception that *Artemia* are "the food" for conditioning and spawning tropical fish. Use of *Artemia* for conditioning should only be supplemental and not a steady diet. Live foods do indeed impart certain attributes not found in nonliving diets regardless of how they are processed or stored. But, a steady diet of these foods usually leads to nutritional deficiencies and disorders. Live foods are like "candy to a child or dog". Fish, children, or dogs will consume "the candy" first above everything else regardless of the nutritional value. If brine shrimp or other live foods are included in the conditioning diet they should be offered only as a supplement and only in the evening after several feedings of the routine conditioning diet have been administered. Other considerations for live foods include amphipods, copepods, worms, and wild zooplankton.

Live foods are excellent carriers of nutritionally specific fats, oils, and amino acids if they are programed prior to feeding. A starving live animal is not a nutritionally balanced food. Therefore, it is imperative that the prey (live food) is fed prior to feeding it to the predator (fish). Essential fatty acids (lipids), microalgae etc. can be administered to adult brine shrimp prior to feeding to boost their food value. More information on live food enhancement can be found in our companion text "Plankton Culture Manual" (Hoff & Snell, 1993). However, it is important to realize that "enriched zooplankton" must be utilized within 3 hours to realize the full potential of nutritional enhancement. Rotifers can loose their nutritional enhancement value in less than six hours.

BROODSTOCK MAINTENANCE

It is important to establish routine maintenance habits for all phases of your garage or full fledged hatchery operation. Regardless of whether or not you have 1 or 200 broodstock tanks, in order to produce repeatable results you must establish routine maintenance programs. All working conditions and tanks are treated equally unless you are running an experiment. Table 10 provides a checklist for adult broodstock tanks. See the section on data taking.

Disruption

Broodstock tanks take minimal maintenance. Minimal disturbance of pairs results in more consistent spawning and fecundity. When dealing with any fish, it is important to establish exactly when the fish spawns. For example pelagic spawning fish such as pompano and black seabass spawn at dawn whereas clownfish generally spawn in mid-afternoon. Maintenance and feeding schedules should not be implemented during normal spawning periods. Disruption during the spawning of clownfish often results in scattered, not fully fertilized nests (see Broodstock Spawning). Sight-seeing, just walking through broodstock areas, can have an effect on more sensitive species. Any movement can cause the pair to temporarily stop spawning and move away from the new nest. There are indications that wearing bright colors also disrupts pairs. If you are serious about spawning and rearing clownfish, every effort should be made keep the pair isolated from external distractions and other clownfish pairs. Tank mates should be very limited. Often other fish and invertebrates such as crabs will distract and devour newly laid eggs.

Feeding Schedule

Once the spawning time has been determined, your feeding and cleaning schedule should be established. For spawning pairs of clownfish I prefer three feedings per day and on weekends only one per day is needed. Feedings should be early in the morning, noon, and around five. At IOH, access to the broodstock was virtually stopped from 1:30 pm to 4:30 pm. See Conditioning For Spawning section for more details on the type and size of food for broodstock.

Tank Maintenance

Work on broodstock tanks should start about one-half hour after the morning feeding. Routine daily procedures include feeding, checking for new spawns from the previous afternoon, checking the general health of the pairs and nests, and adjusting air and water flows. Daily checks on water and air flows is very important since the heart of a health-care program depends on good water quality.

Weekly routines would include cleaning windows and spot siphoning of detritus buildup and uneaten food. Bi-weekly maintenance includes removal of dirty spawning substrates (flower pots) and replacing them immediately with clean ones and general bottom siphoning if needed. Monthly or less, depending on bioload in the tank and food being fed, the sponge filter should be removed, rinsed in cool freshwater and squeezed several times to remove heavy detritus. While still damp the filter should be replaced immediately after cleaning. A bucket of freshwater would suffice for cleaning about 12 filters or a hose could be used.

Every 12 to 18 months the entire tank should be broken down and thoroughly cleaned, gravel removed and new or reclaimed gravel replaced. Disruption by moving a mated pair usually results in no spawns for several months. However, we found that if a pair is transferred when a nest was present, spawning resumed much sooner. Average recovery back to a spawning state was 32.5 days (N=10 range from 6 days to 60 days) compared to 60+ days for pairs that were moved without nests. When cleaning broodstock tanks, remove the pair in conjunction with their nest and place them into the same bucket and provide aeration. Their tank should be quickly cleaned and the pair replaced.

Broodstock Technician Checklist

Table 10 is a checklist used at Instant Ocean Hatcheries for Broodstock Technicians. Each person was assigned specific areas and given a list of exactly what was expected from him or her. Assigning specific duties in writing is important so that each knows their responsibility. Time was listed in a 24 hour clock form to help eliminate mistakes of using AM and PM. I have modified these instructions to eliminate the use of subgravel bottom filters by replacing them with removable sponge filters. For hobbyists, the standard gravel bottom airlift filter is fine for use in broodstock tanks.

Table 10

BROODSTOCK TECHNICIAN DUTIES

ORIGINAL: 7/3/79

REVISION: 9/12/82

1. Feed all broodstock

A. Older Established Broodstock
MIX: 0800 hrs, 1100 hrs, and 1630 hrs
ARTEMIA or ROE: 1300 hrs

B. New Unestablished Broodstock
MIX: 0800 hrs, and 1630 hrs.
ARTEMIA: 1000 hrs, and 1300 hrs

2. Daily maintain all adult tanks (new and old)

A. Check water flow daily.
B. Check air lifts (top and bottom filters).
C. Check spawns, mortalities and sickness in morning hours after first feeding.
D. Record mortalities and sickness in early morning and report to biologist.
E. Check and remove nests for hatching in evening (1600 hrs).
F. Maintain inventory tags on tanks and on hatch inventory board.
G. Replace spawning pots after removing nest for hatch or bi-weekly cleaning.
H. Clean windows weekly.
I. Rake bottoms of tanks weekly.
J. Cover exposed bottom filters weekly (if subgravel filters are used).
K. Replace gravel when necessary.
L. Clean empty tanks.
M. Clean all tanks yearly.
N. Clean sponge filters monthly.
O. Collect "Spawn Tags" from larval rearing area and replace on tanks.

3. Maintain new broodstock in isolation area

A. Keep mortality records.
B. Maintain tanks for new shipments of broodstock.
C. Assist in mating new pairs.
D. Separate poorly matched fish.

4. Administer treatments and keep records as instructed.

5. Maintain Adult Record Form and Adult Inventory Record Form.

6. Assist when requested in any capacity on the farm.

Nest Consumption

Periodically adult clownfish consume their eggs only hours after laying them. If this practice continues, it indicates a specific problem that must be corrected. First understand that nest consumption is not natural for clownfish or any other fish. Why would they go through the effort of spawning and then eat the nest? This is especially true of benthic spawners which lay a limited number of eggs and display parental care. Then why does it happen? There are several possibilities including: being a newly formed pair; the pair was scared or nervous, too many tank inhabitants, diet deficiencies, parasitic destruction of the eggs, poor water quality resulting in poor fertilization, or the male is nonfunctional.

Of the possibilities, diet is usually the problem. At times two females will reside together and spawn, obviously resulting in non-fertilized nests which are usually consumed. Poor water quality, to a lesser degree, can cause poor quality or unfertilized eggs. Experiments I conducted on pompano indicated that low pH reduced sperm activity thus resulting in poor fertilization. Frequently, new spawning pairs consume initial nests but usually this practice stops after three to six spawns.

Others have found that specific components in diets may have a direct reflection on egg quality. Moe (1991) found that a diet heavy in whole shrimp (with shell) is important. He found that not enough chitin or shrimp shell in the diet caused the eggs to not adhere properly to their substrate. Lindner (1986) noted that, when he reduced the amount of frozen brine shrimp fed to the pair, they resorted to consuming their eggs. Although I do not advocate overuse of brine shrimp, I feel this is an interesting observation.

As we have established, lipids or fatty acids constitute a major portion of the composition of an egg. Theoretically if a female cannot obtain and/or store enough reserve lipids, then she might indeed resort to eating eggs. Or perhaps clownfish consume eggs because they have abilities to detect inferior composition.

Based on the lack of stored lipids theory, we found that feeding pairs whole roe (fish eggs) decreased nest consumption. An added benefit in feeding roe was an increase in egg production, higher hatch rates, and higher initial larva survival. Use only "yellow roe" (fish eggs) of about 1 mm in diameter. Do not blend or break the eggs but feed them whole. We preferred "mullet roe" but other types will no doubt work. Availability of roe is seasonal so purchase when available. Wrap the entire gonad in airtight, quality plastic bags in desired portions and freeze while fresh. Purchase only what you can use in 18 months. Warning: do not assume that roe, like live foods, is a panacea and feed it exclusively. Roe, like any singular food, is detrimental if fed in excess. No singular food item on the market beats a balanced, frozen, blended diet for primary food.

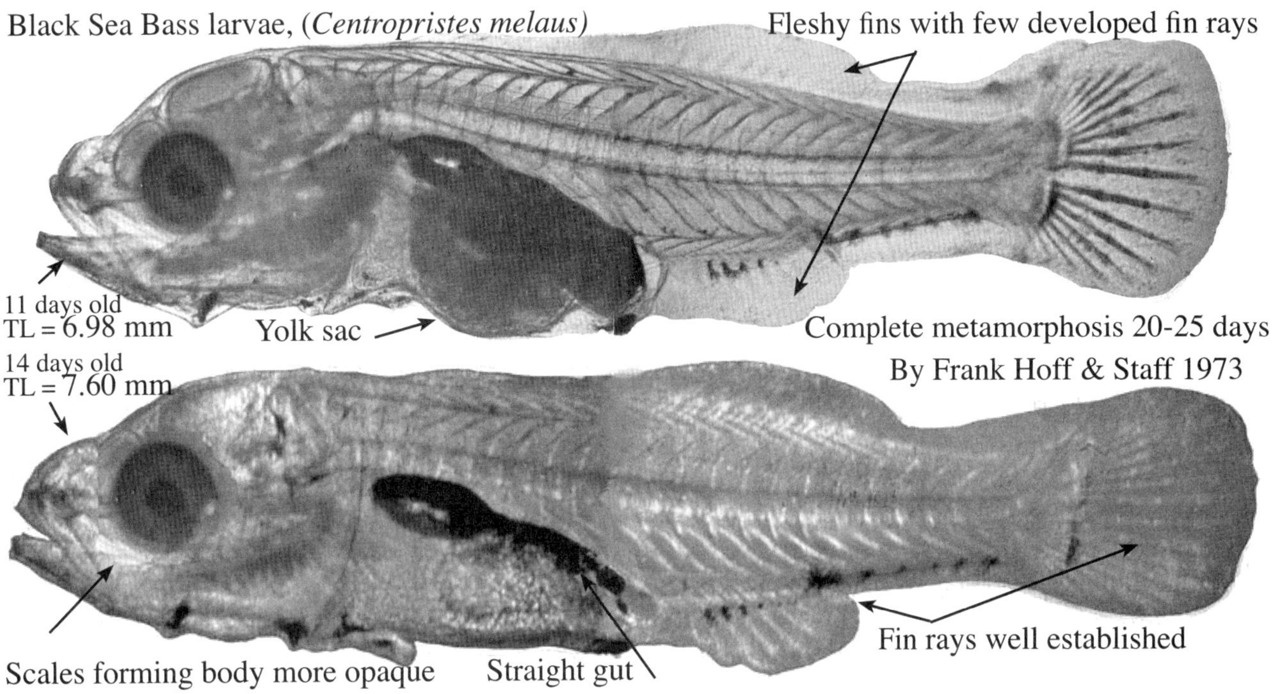

RECORD AND DATA KEEPING

Few points within the operation of a hatchery are more important than record keeping. How often we say to ourselves, "I will remember and don't need to write it down". Yet, we turn around, pile more data into the same memory spot in our brain and previous data slowly slips out of sight. Initially data should be recorded for virtually everything since it is usually hard to determine what is important with regard to your specific conditions and fish. Usually after 12 to 18 months you have formed a good picture of your specific conditions and can start narrowing the amount of data you need to record.

Record keeping should be designed to be quick and efficient. It should consist of simply making check marks, recording numbers, circling or marking in codes, or symbols. Messages or additional observations should be short and to the point. Perfection and generation of a data sheet takes time to develop and may depend more on your specific needs or biological background. Only when things become routine and results are predictable will data keeping become simplified. For example, if you are using an open system, semi-open system or closed system the parameters of data needed will be different. Where your facility is housed, such as a garage, greenhouse, or open structure will require different information such as lighting intensity, heating and cooling problems, etc.

Historical Records

After making many mistakes, I now feel that perhaps the most important records kept are historical records. Historical records are logs which detail everything not normally recorded during routine daily record keeping. These records include odd items such as changing to a new salt or even a new batch of salt, putting a new filter on line, installing a new UV bulb into a system UV unit, changing a food ingredient or source, adding something to the water that is new, changing the lights or light regimes, etc. Many times I have consulted with people who are alarmed that this or that is not responding the same as a month or two ago. My immediate response is to ask what they have changed. "Nothing" is the most common answer. After tremendous prodding it always amazes me how many things they have changed but did not think it needed to be recorded or would have an effect. Often very simple things can cause a drastic change. An important habit to develop, is to avoid making multiple changes at one time. Combinations of several changes may influence your system or biological responses and are harder to sort out. If several changes are recorded on the same date, you should switch back to what worked previously and by process of elimination make one change at a time. Usually allow two weeks to observe any change and effect.

Historical records must contain enough detail so that you know the product used, the dosage administered, and all other important details. These are the most detailed records you will maintain since they have the most overall influence on your entire operation. Table 11 is an example of typical historical records at IOH.

For quick referencing, the log can be divided into organized categories including chemical changes, food changes, mechanical changes, and miscellaneous changes.

Broodstock Records

Record keeping begins where the hatchery begins, specifically the adult broodstock area. Adult Record (Table 12), Pair Inventory Record (Table 13), Monthly Pair Summary (Table 14) forms were maintained at Instant Ocean Hatcheries for seven years. It is important to understand how we recognized a pair before we discuss using the forms. A pair that commenced spawning was immediately assigned a pair number. Each species had a set of pair numbers.

Table 11 - Typical examples of a historical log.

HISTORICAL RECORDS
Chemical Changes
Adult Area

7/15/80	Added 12 case of Instant Ocean Salts to broodstock system A.
8/24/80	Divided 3 gallons of GP-1 trace metals to systems A, B, & C.
8/30/80	Divided 1 liter dry sodium bicarbonate to systems A, B, & C.
9/10/80	Made a 30% water exchange on systems A & B. Replaced water with recovered water from outside pond RA.
10/3/80	Water turning yellow, turned on carbon filtration for three days.
11/28/80	Added 5 cases from new lot of Instant Ocean Salts #26578.
12/12/80	Divided 1 pound of chelated iron from new lot # 3335689.

HISTORICAL RECORDS
Food Changes
Adult Area

10/4/80	All systems being fed extra feeding of fish roe.
10/25/80	Ran out of mix, fed only flake food for the weekend.
11/6/80	Fish being fed mix #10 with new vitamin mixture from Vita-Fil.
11/25/80	From today on we are feeding the adults only 3 times a day Monday through Friday and only once a day on the weekend. Roe will be a supplemental feeding when needed.
1/7/81	Adults fed live artemia for three days as a supplemental feeding.
3/21/81	Mix #15 contains new pigment enhancer from Rowell Enterprises.
3/25/81	Roe feeding stopped.
4/18/81	Now feeding mix 3X per day but at 8:00 am, 12:00 pm and 4:30 pm. Roe, if fed is at 10:00 am.

HISTORICAL RECORDS
Mechanical Changes
Adult Area

5/25/80	New pump on system A.
7/26/80	Replace all UV bulbs on systems A, B, & C.
7/28/80	UV unit on C system not working for the last 3 days.
9/10/80	All tanks cleaned in C system. Fish removed and replaced. in one day.
9/15/80	Trickle filter on system B cleaned.
10/23/80	New DE filter put on line in system B.

For example: *A. frenatus* Pair 1, 2, 3 etc. and *A. ocellaris,* Pair 1, 2, 3, etc. Each pair had a separate data sheet (Adult Record Form). Pairs were further segregated into four categories:

1. Possible Pair = PP = A pair that have accepted each other but not spawned. A "Pair Record" form (Table 11) was started on each possible pair but included only the tank number, date they were set up, and source of the male and female. Source means where the fish came from. For example: F-1, F-2, F-3 meant they were tank raised and the numbers indicated what generation. WP meant the pair consisted of wild fish.
2. New Pair = NP = A possible pair that just started spawning within the month is considered a new pair and is assigned a pair number.
3. Spawning Pair = SP = After one month from the initial spawn all new pairs are elevated to a spawning pair status.
4. Inactive Pair = IP = Is a pair that had a previous spawning record but have not spawned within the last month. The pair number is not dropped but only the activity category was changed.

A pair retained the same number throughout their lives, providing the female remained alive. If the male died, a new small one was added and the pair retained the same number. Numbers were never reassigned even if both fish in the pair died. Once the female died, the male was re-mated with a smaller fish. If the old male transformed into a female and began spawning the pair was assigned a new number.

Broodstock Tank Identification Tags

Each tank containing a new possible pair was initially assigned three color-coded tags (Figure 39) which quickly indicated the activity of each pair. These hung below the widow of the tank. A green tag with no markings signified the tank contained a possible pair. Once the pair spawned the green tag was removed and a yellow tag was filled in with the species, pair number, date pair started and tank number. Tank numbers were assigned per row. We had six rows of adult tanks lettered from A to F. The tag would say A1, A2, A3 or E1, E2, E3 etc. The yellow tag was removed whenever a nest was found. The third tag was white and was a replicate of the yellow pair tag. This tag always remained on the tank as identification of the pair and a constant reminder that the tank was occupied and must be observed and maintained. Since the yellow tag had been removed, the remaining permanent white tag indicated that a spawn had occurred and the nest condition should be checked daily. In addition, each tank had a large permanent identification number mounted at the top.

Pertinent comments on pairs were recorded on a clipboard as the technicians made their morning rounds to check for spawns. Comments might include nests that were eaten, fungused or died, sick or injured fish, unusual behavior, etc. In addition, possible pairs that commenced spawning had their green tags removed and the yellow and white tags were filled in with species, pair number, and beginning spawn date. Spawning pairs with new nests had their yellow tags lifted and retained for data recording and assignment for night of hatch.

Yellow or green tags removed from the tanks were then recorded on the proper "Pair Record" form for that pair. Date of the spawn, spawn number, and estimated hatch date were recorded. Comments obtained during the "spawn check" were also recorded on the pair or possible pair data sheets.

After recording the "new pairs" and spawns, the yellow pair tags were placed on a row of 31 pegs representing a running calendar numbered from 1 to 31. Depending on when a particular species would hatch (clownfish spawn in the afternoon and hatch in early evening after seven to ten days) the tag was placed on the appropriate date peg. For example: *A. ocellaris* usually hatch

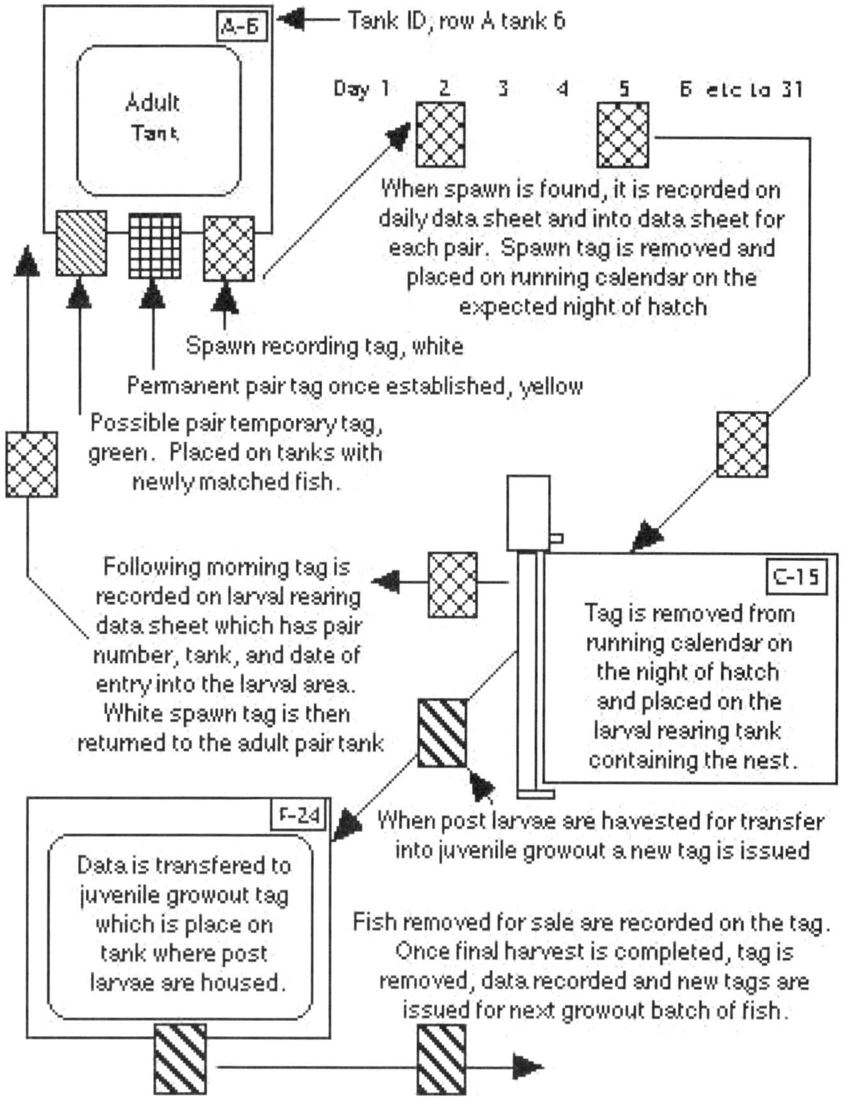

Figure 39
Adult broodstock identification tag system used at IOH to keep track of spawning and non-spawning pairs.

on the 8th or 9th evening after spawning. If the spawn was found on the following morning of November 12, the tag was placed on the November 18th peg. This is exactly eight evenings, including the night of the spawn, after the spawn actually occurred. When this date occurred, all tags on the 18th peg were removed in late afternoon and each nest was checked to determine if the eggs were ready to hatch. If ready, the nest was removed for hatching and the tags were mounted on the larval rearing tank in which the nest was placed (see Larval Rearing Records and Data).

Pair Record Form

Table 12 is a hypothetical example of data maintained on a single clownfish pair. I have attempted to demonstrate the type of events and data that might be listed on such a form. Basically the "Pair Record" is the birth certificate, diary, and death certificate of a single pair. These forms were always retained in the hatchery in a notebook maintained by the broodstock technician. As the pages filled, a separate folder was established for each pair. Keep in mind that, as long as the female of an established pair was alive, the pair data remained intact. Once she died, the pair number was terminated, not reused, and no further data was taken. Although record keeping on a pair is time consuming, it serves as an invaluable source of information for assessing the entire operation.

Table 12 - Example of a Pair Record Form

PAIR RECORD

SPECIES: A. ocellaris PAIR NO. 33 TANK NO. A-20

Male	Intro	Removed	Female	Intro	Removed
F-1	3/2/78	6/22/78	WP	3/2/78	
F-1	6/30/78				

DATE	SP. NO.	HATCH	LOST	COMMENTS
4/29/78	3-1	9		New pair
5/17	3-2	8		Spawning on side of tank
6/6	3-3	7	PH	Nest lost, hatched early
6/22	3-4	7		Male died - inactive pair
6/30				Introduced new F-1 male
8/20	3-5	8		Pair restarted
9/15	3-6	8		Cryptocaryon on female, no treatment
10/5	3-7	8		
10/22	3-8	8		Tank cleaned pair removed with nest
11/12	3-9	4	ND	
11/29	3-10	9		Female acting strangely
12/22	3-11	8		
1/7/79	3-12	?	NT	Spawn not used, not needed
1/13				Female died - pair terminated.

INDEX: PH = Pre-hatched ND = Nest Dead NT = Nest not used

Pair Inventory Form

"Pair Inventory" (Table 13) was the basic record maintained by the broodstock technician within the hatchery. This form provided a gross summary for all pairs within a row of tanks and was used as an aid in recording a "Monthly Pair Summary" form (Table 14). It provided a quick overall picture of what an entire row of adult pairs were doing during the month. First column listed tank numbers, and the second column the species in an abbreviated form such as, "fre" for *A . frenatus* , "ocell" for *A . ocellaris* "cla" for *A . clarkii* etc. The next three columns were recorded at the end of the month. Using the individual "Pair Records" the following three columns placed the pair into one of the three following categories. If a pair began spawning during the month, a check was marked in the spawning column (NP) indicating it was a new pair. If a pair was set up as a new possible pair (PP) during that month, a DP (domestic pair), WP (wild pair), or MP (mixed pair) was placed in the PP (possible pair) column. If an established pair did not spawn during the month, they became an inactive pair (IP). The next three columns were used to record the dates the pair spawned. Keep in mind, nests were checked the following morning so the date actually recorded was the day before. The next three columns are for recording lost spawns either due to an early unexpected hatch (PH = pre hatch), a nest that fungused or died (ND = Nest Died) or it might be not taken (NT = Not Taken) because of low fecundity, no need for any more nests of this species for this month's production, or simply limited space in the larval area. The last column was for specific comments about the pair for later recording on the "Pair Record" and "Pair Summary" forms.

Table 13 -Example of Pair Inventory Record form

PAIR INVENTORY RECORD

Row: A System: AB Month: Jan. Year: 1982

Tank	SP.	NP	PP	IP	Pair Num.	Spawn Date	Lost Spawns	Comments
1	fre	x			7	19	PH, PH	New pair
2	ocell				8	22	NT	
3	eph				2	15,28		
4								Empty tank
5	bia		WP					Set up 1/15/82
6	ocell				3	16,29	ND,ND	Lost nests 1/16 & 1/29 fungus
7	ocell				6	21		
8	cla	x	DP		20			Set up 1/1/82
9	cla		MP					Set up 1/7/82
10	fre			x				
11	fre				6	18,31		
12	aka	x			20			
13	poly	x			2	12		

INDEX: PP = possible pair, DP = domestic pair, MP = mixed pair, WP = wild pair, NP = new pair, IP = inactive pair, NT = nest not taken, PH = pre-hatched nest, ND = nest died.

Monthly Pair Summary Form

The "Monthly Pair Summary" form (Table 14) was compiled by the biologist from data provided by the broodstock technician. This provided an overall picture of the entire production from all six rows of adult broodstock pairs. It served as a basis for graphs, flow charts, etc. The first column showed the species. The next four columns provided information on the status of the pairs for each species. For example: how many active spawning pairs (P) did we have of *A akallopisos* (aka), how many possible pairs (PP) did we have set up, how many spawning pairs were put on the inactive pair (IP) list because they did not spawn during the month, and how many possible pairs became new pairs (NP) during the month. The next column listed the number of total spawns per species of fish. The last column listed the number of spawns lost.

Figures were tallied and summarized as to total spawning pairs for the month, how many possible pairs, how many inactive pairs and how many new pairs, total spawns, and total spawns lost. Total active pairs included all that spawned during the month.

Table 14 - Example of the Monthly Pair Summary form.

MONTHY PAIR SUMMARY

MONTH : Jan **YEAR: 1982**

SP	P	PP	IP	NP	NUM. OF SPAWNS	SPAWNS LOST	COMMENTS
Aka	2	7	1	1		6	
Bia	2	2	3	1	5	1	
Cla	9	12		3	32	7	
Eph	2	4			5		Need more pairs
EM	1				1	1	
Fre	26	3	2		50	23	20 Nests not taken, too many
FE	4				9	2	
FM	2	1	1		4		
Mel	3	2	1		4	1	
Ocel	36	6	6	1	67	11	
Perc		4	1				
Peri		1					
Poly	2	7	1		3	1	
Rubo	1				2	1	
San		3					Need more new pairs
Neon				1	2		
Total	90	52	16	7	90	49	

Total Number of Spawning Pairs 113
Average Number of Spawns Per Day.............................6.1
Average Spawns Per Spawning Pair 1.9
Comments and Changes: Good increase in new pairs this month. Looks like more are about ready to spawn.

SPECIAL NOTES:

Heater malfunctioned on Jan 20 - water temp dropped to 23°C over a 12 hour period.
Second feeding changed to Zeigler salmon pellets on Jan 5.

INDEX: SP= Species, P= Pairs Spawning This Month PP= Possible Pairs
IP = Inactive Pairs NP = New Pairs

Ostracion sebae mexicanum

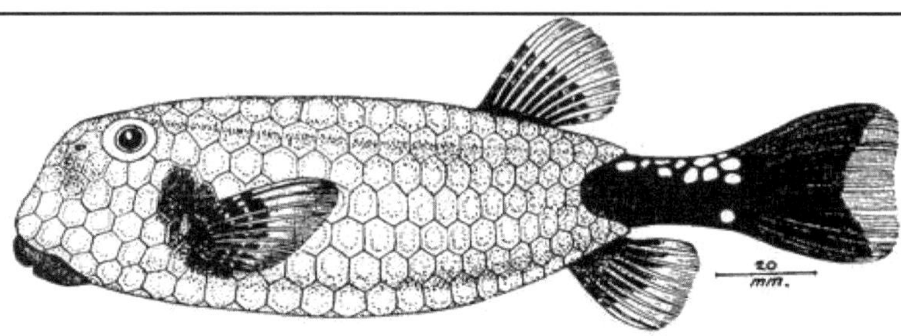

Drawing By H. W. F. del.
Results of the Fifth Vanderbilt Expedition
By Fowler, 1944

SPAWNING AND CULLING OF BROODSTOCK

Maturation Time

From larvae to sexual maturity and eventual spawning takes from 9 to 15 months (normally 12) for clownfish. Some species like *A. ocellaris* reach maturity earlier than *P. biaculeatus* and others. Subadults take about 6 months to reach sexual maturity. Full size adults obtained from the wild usually take two to three months to begin spawning again. If a consistent spawning pair is severely disrupted, such a tank change, illness, treatment, etc., they usually take one to three months to resume spawning.

Spawning Activity

Clownfish display several common characteristics that indicate a spawn or false spawn is pending. Often the female takes the initial lead role in forming a new pair and in spawning. She will clean several hard surface areas and often pushes the male on his side or belly as if teaching him what to do. At times she appears aggravated, frustrated, and nervous, often shaking her head. Eventually she may actually push the male toward the spawning substrate. Loss of several scales on the side of the male during these early courtships is not unusual. Head shaking, standing on their heads, nipping, and chasing are common prior to spawning, especially in new pairs. In conjunction with this display, excessive digging in the bottom substrate or cleaning of many areas usually occurs. As the pair matures, the male becomes less docile and more aggressive. Eventually he assumes most of the nest cleaning and tending of the eggs. Actually this pattern might be expected when we consider sex reversal and the fact that females are often the oldest, most mature, individual of the pair. Initial dominance and teaching by the female has been noted by Moyer and Bell (1976) in wild populations of *A. clarkii*. They found this to be typical of females with young, new male partners. As a pair matures, both become active in site selection and cleaning which occurs 3 to 5 days before spawning. In addition, head shaking, nudging by the female, substrate biting, and belly touching by the male increases, chasing and head stands are all common postures prior to spawning.

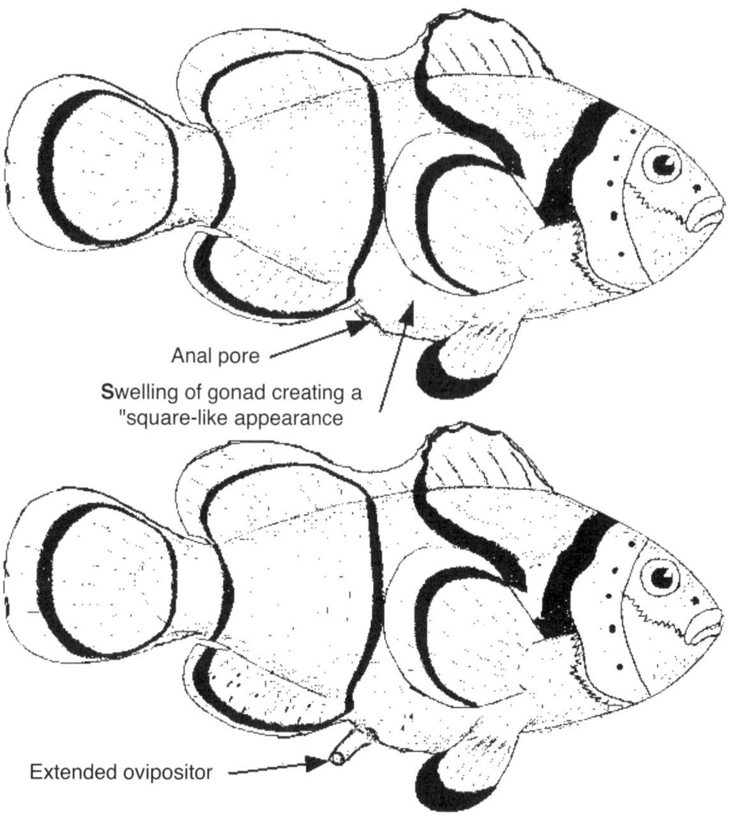

Figure 40 - External indications of oocyte development and pending spawn in clownfish.

As females become sexually mature their physical appearance changes. They begin to display a "squaring" of the abdominal area (Figure 40) which is actually internal maturing of eggs into a pre-spawning state within the ovary. At this stage they will often dig holes in the gravel below and around the spawning substrate.

Clownfish normally spawn in mid-afternoon, before dark. Pending spawns that day are evident by morning observation of a partially or fully extended 1/4" long tapered cone ovipositor on the female (Figure 40). Partial extension may be evident several days prior to the spawn date and may be retracted daily. Before spawning, several dry runs or passes will be taken by the female over the selected area. Although several areas may be cleaned she selects a single site. Eventually the male follows her by passing over the same area. Once spawning commences the, female presses her body towards the substrate, quivers slightly, and slowly moves in a rowing fashion using her pectoral fins. She moves in a circular path, depositing a continuous spiral of eggs from the center outward. The male swims behind the female, releasing sperm over the newly deposited eggs. Often the males goes back over areas and releases more sperm after the female has moved aside.

After spawning, the male starts to assume a more dominant role. Although both tend the nest, the male eventually becomes the real caretaker. He intermittently fans the nest with his caudal or pectoral fins. He also cleans the eggs by gently mouthing them without removing them. Dead and fungus infected eggs are routinely removed and eaten. Substrate around the nest is also often cleaned.

While a nest is present, males do not feed aggressively but females eat heavily. As pending spawns becomes closer, females feed less and males increase feeding.

Nest Care

Parental care of nests is mainly the function of males. Allen (1991) stated that his observations on natural populations of *A. chrysopterus* indicated that initially, after a nest is laid, males spend less then 30% of the daylight hours tending nests. During the last two days prior to hatch, males may spend as much as 87% of the daylight hours tending nests. When nests are present, the average male spends 30% to 60% of his daylight time tending nests. Night maintenance is minimal from observations made at IOH.

Fanning eggs is frequent on the day after spawning and diminishes considerably about midway in the incubation period. On the day of hatch, fanning increases again. Fanning is usually accomplished by utilizing pectoral fins and occasionally caudal fins.

"Mouthing" of eggs by both fish is common. They literally place their mouths around the eggs. This is considered a cleaning process for removal of attached algae, fungus etc. It has been speculated, that during the process of "mouthing" fish secret an anti-bacterial/fungal slime on the eggs. Often they remove and digest eggs that are fungus infected and dying. At times, entire nests may be eaten within the first few days. This is not natural and should not be considered a part of nest care unless the entire nest is dying. If this habit continues to persist then other factors such as nutrition, water quality, or a pair consisting of two females might be the problem. Of the three options nutrition is usually the key problem (see Nest Consumption).

Nest Location

Nests will be located on any hard surface. Ideally you want to train a pair to spawn on a portable substrate that can be removed for hatching and/or cleaning. Placing spawning substrates centrally helps minimize spawning on the side of the tank.

When using red clay flower pots as spawning substrates, nests will normally be located on the upper inner surface. Usually nests are deposited in the pot where the male resides. Using a bright flashlight is helpful in locating the nests. Since clay flower pots are orange and newly laid eggs are yellow orange to red, nests are often hard to distinguish without direct lighting. When locating nests, observe the rims and holes in the back of the pots. At times pairs will spawn on the back or side of a pot and a few eggs are often deposited around the edges.

Excessively dirty spawning substrates should be avoided since it makes nest location harder to detect. As mentioned, routine replacement of spawning substrates is recommend regardless of whether a pair has spawned or not.

Initial spawns from new pairs are often hard to locate since they consist of few eggs and are often scattered. Don't be surprised to only find a dozen or so eggs since this is especially true of young new pairs. Older pairs establish patterns which are typical for that pair; you know what to expect and where to locate the nest. Observe pair attitudes; does the male dart back to a specific spot or the female charge toward the glass? Carefully watch where they go as you approach the tank. Their instinctive reaction is to swim towards the nest to protect it. When nests are present, the pair is generally more aggressive towards outside intrusion. They charge the front glass as you approach or bite your hand when placed in the tank. They chase other fish or physically remove invertebrates, gravel, or plants.

Nest Coloration, Fertilization, and Hatch

A key factor in regard to success and failure of the spawn is nest pigmentation, and this should not be treated lightly. Certainly pale nests will be deposited and hatch, but there are direct relationships between nest coloration and hatch, initial survival, and juvenile growth.

After consulting with several freshwater farms I have found they have a tendency to feed and treat adult pairs the same as juveniles in grow-out. New breeder size fish from outside ponds are robust, have good coloration, and produce large, brightly colored, tightly packed nests, for the first 3 to 6 months. Nests then become pale, eggs are loosely packed, and hatch rate is drastically reduced. This does not mean the pair is spent or finished in most cases. What pairs are fed and how they are treated directly influences productivity, nest size, color, hatching success, initial larval survival. These advantages are no doubt is passed on to overall juvenile survival. Broodstock diets are extremely important with regards to quality production (see Conditioning For Spawning).

Egg pigmentation of benthic spawners, either phytophilic or lithophilic, (eggs deposited on plants or hard substrates), is very common. Pigmentation has a direct relationship to the success of hatch and initial larval survival. Healthy, well-fed, clownfish pairs produce orange to bright red nests depending on the species. Orange colored nests are produced by *A. ocellaris, A. percula, A. clarkii, A. akallopisos, A. polymnus* and *A. sandaracinos.* Red nests are produced by red colored fish such as; *A. frenatus, A. melanopus, A. ephippium,* and *P. biaculeatus.*

The degree or intensity of color is directly proportional to the amount of pigments within the diet. Highly colored nests hatch better than pale whitish color nests (Table 15). There is some evidence that pigments may enhance the conversion of certain vitamins within the eggs. Certainly a whitish clear egg and a brightly pigmented egg exposed to light must have different physical and chemical makeups. Chickens raised on the ground produce thick, sticky, bright yellow-orange colored yolks while those in cages normally produce thin, yellow yolks. I have often wondered the role of pigmented yolk, located within the egg, with regard to a developing chicken. Egg yolk of caged chickens can be drastically changed from a very pale yellow to a bright red by simply adjusting what they eat. The question is why? Certainly natural coloration must be best.

Invariably the ovaries and eggs of most dissected wild fish are pigmented. Pigmentation in spawned eggs may act as a form of camouflage but, this can be discounted due to the fact that: (1) why wouldn't the eggs be transparent? (2) why would camouflage be necessary when the parents tend the eggs constantly? (3) clownfish eggs change from orange or red to gray, to black and then black and silver therefore, camouflage would not make sense, 4) the color is not there simply because it is pretty since Nature does not work this way.

Experiments conducted at IOH in 1980 (Hoff's Theory) showed that addition of the pigment astaxanthin to the gelatin-based mix resulted in bright colored nests within two weeks. In addition, nests became tighter, more compact, nests and eggs were larger, eggs were less susceptible to low oxygen or poor circulation, hatches became regular and not delayed, initial larval mortalities dropped significantly, larval activity increased, larval coloration was brighter and delayed metamorphosis decreased drastically (Table 15). With all these plus factors there is no doubt that pigmentation of eggs is essential, and reflects directly on larval development and no doubt future juvenile growth and survival.

Changes in adult diets were noted within just weeks in the larval rearing area. Suddenly newly hatched larvae were clear rather than black, which is normally considered an indicator of stress. Postlarval survival did not clearly reflect positive results but the vigor of the fish was much improved. Although not documented, there is no doubt that egg coloration had a direct correlation of juvenile survival and growth.

Nikolsky (1963) stated that development under more or less favorable conditions of oxygen supply is connected with the greater or lesser development of respiratory pigments in the egg, usually carotenoids, pigments of yellowish or reddish tones. Deeply pigmented eggs develop in low-oxygen environments. This statement may be true of wild populations, but in closed culture environments the type of feed also influences pigmentation intensity and / or the ability to produce pigmented eggs.

Table 15 - Effect of Egg Pigmentation on Hatch and Survival of Post Larvae.

1981	**Produced**	**Hatched**	**% Hatch**	**% Survival**
Jan	43,455	24,911	57	34
Feb	38,819	27,887	72	56
Mar	57,814	42,166	73	55
Apr	55,678	37,466	67	49
May	62,067	41,781	67	49
Jun	55,243	37,808	68	56
Jul	58,401	41,642	71	58
Aug	57,298	40,631	71	45
Sep	64,439	40,656	63	43
Oct	67,903	40,222	59	44
Average before pigment			**68.75%**	**46.42%**
Pigmentation added in October and reflected in November				
Nov	48,179	37,304	77	39
Dec	64,559	51,525	80	29
1982	**Produced**	**Hatched**	**% Hatch**	**% Survival**
Jan	47,144	35,743	76	31
Feb	52,166	42,478	81	53
Mar	47,484	41,193	87	54
Apr	57,900	50,536	87	54
May	61,988	55,178	89	62
Jun	60,351	54,181	90	50
Jul	54,525	43,665	80	51
Aug	53,976	48,578	90	36
Sep	34,824	28,903	83	28
Oct	60,480	53,827	89	41
Average with pigment			**84.08%**	**44.00%**

He stated that the need and intensity of pigmentation is directly related to the conditions of oxygen supply under which development takes place. He presented an example of the sockeye salmon, *O. nerka,* which has a deep crimson egg which is buried in the lakes close to where oxygen-poor spring water emerges. Intense roe coloration in salmonid fishes especially the genus *Onchorynchus* was found to more effectively resist various environmental conditions such as low oxygen content, with a subsequent reduced hatching mortality rate (Meyers and Huei-Mei, 1983). Carotenoids have been shown to function in respiratory processes in which oxygenized carotenes (xanthophylls) can store oxygen for periods when the supply in the environment is exhausted (Czeczuga, 1979a).

Fish are considered to assimilate and modify specific carotenoids prior to maturation. Carotenoids in muscle tissues, but not in skin (chromatophores), were found to be mobilized by ripening ovaries and deposited into the egg cells in trout during sexual maturation (Steven 1949). This pigment transfer was reversed in sockeye salmon that were gonadectomized (ovaries removed) even when fed pigment-rich marine zooplankton over an 8 week period (Donaldson & Fagerlund, 1970). It was noted that 65% of the carotenoids present were metabolized during spawning, during which retained pigments, originally concentrated in the muscle, were transported and deposited at different sites, mainly in the skin of males and the eggs of females (Crozier, 1970). This statement alone explains why fish brought into a facility from the wild will initially deposit brightly colored nests but, if not properly fed, subsequent nests will become pale to clear white over a period of one or two months. In addition to its occurrence in eggs, astaxanthin has been reported in the sperm of *Coregonus albula* and *Salmo gairdneri* (Czeczuga, 1975).

Hartmann et al. (1947) demonstrated that astaxanthin present in eggs can enhance sperm mobility in rainbow trout and possibly could serve as a reproductive hormone. It has been long suspected fish sperm must have a way of sensing eggs that have been released in billions of gallons of seawater. Recent studies have shown that sperm are attracted to eggs by scent since molecules related to scent have been found on eggs and sperm. Pigmentation may play an important link in the development of specific scents. Studies on finding human male contraceptives has revealed molecules found in the nose and used to pick up cologne scents has been found on sperm. This suggests, a drug that blocks the sperm's ability to sense that enticing aroma could be used as a male contraceptive (Lefkowitz, 1995). This statement may have great implications with regard\ to explaining why fertilization rates vary. Experiments we conducted on pompano and seabass sperm showed that mobility was greatly hampered by low pH (<7.9).

Pigmentation is also important in pelagic eggs. Mikulin and Soin (1975) found an intensive expenditure of carotenoid pigments, with a suggested relationship between the degree of egg pigmentation and mortality during embryonic development. Shepherd and Bromage (1988) reported experimentation on broodstock red seabream (pelagic spawner) which showed diet deficiencies of EFA (essential fatty acids), low phosphorus, low protein, and low pigmentation had a direct effect on egg quality, hatch, and post larval survival.

Egg quality of pelagic eggs is usually judged by the percentage of buoyant eggs and the formation of desired singular rather than multiple oil globules. In 30 to 36 ppt salinity, eggs that sink or float in midstream are considered inferior and hatch poorly. Extreme variances of 81% to 24% buoyancy, 94% to 1% hatch, and post larval survival of 80% to 0.1% clearly demonstrate the importance of a balanced diet for broodstock. Egg quality of red seabream was found to be enhanced by supplementing feed with carotenoids, such as astaxanthin, phospholipids, and vitamin E just hours prior to spawning.

Invertebrate culture has also indicated a similar correlation. Currently work being conducted on shrimp maturation indicates a relationship between egg coloration with regard to hatch and survival.

Table 16 - Color changes of maturing eggs prior to hatch.

Directly After Spawn	Clear yellow-orange with small white dot on tip of egg.
Day 1 (24 hours)	Color intensifies.
Day 2 (48 hours)	Dull grey color.
Day 3 (72 hours)	Darker grey color with slight evidence of eye spot development.
Day 4 (96 hours)	Gray color with dark strip (back bone) and more defined eye.
Day 5 (120 hours)	Gray-black with definite silver eye spots forming and dark black backbone.
Day 6 (144 hours) =	Dark reddish-brown with large eye spots, larvae has started to turn around in the egg.
Day 7 (168 hours) =	Clear reddish-brown with large (20%) bright shiny silver eye spots. Larvae are well defined and orientated with head at tip of egg. (ready for hatch)

Nest Changes Prior To Hatch

As fertilized clownfish eggs mature they change daily from the initial yellow-orange or red color to a vivid black (Table 16). Using these color changes allows you to determine when the spawn occurred and when it will hatch. Below is a typical sequence of development of eggs for *A. ocellaris* (commonly called ocellaris or percula).

During early development the larva's head is located at the attached end of the egg. When about to hatch, it rotates inside the egg so that its head is at the unattached tip. At hatch, it pushes forward with its tail, breaking the unattached tip of the egg. At times a few unfortunate individuals can be seen with the tip of the egg over their heads.

Unfertilized or diseased eggs are clear initially but turn opaque white within 24 hours. During incubation, eggs may die due to improper development, fungus, physical damage, severe water quality changes, lack of parental care, and/or parasite penetration. If an entire nest turns white after the initial 24 hours, it is probably due to excessive fungus brought about by poor water quality. Parasite penetration is also a possibility. Lastly, lack of parental care might be a problem especially with a new pair. If this continues for more than two months, break up the pair and reestablish new pairs.

Table 17 - High, low and average nest size of captive clownfish. Based on two years of IOH data.

SPECIES	LOW	AVG.	HIGH
A. akallopisos	212	301	392
A. clarkii (sebae)	435	668	981
A. ephippium (red saddle)	225	583	869
A. frenatus (tomato)	309	440	551
A. melanopus (cinnamon)	172	249	339
A. ocellaris (ocellaris)	168	236	313
A. percula (percula)	67	331	649
A. polymnus	191	526	1217
A. rubrocinctus	87	428	817
P. biaculeatus (maroon)	146	480	986

Egg Development

A series of photos depicting developmental egg stages of the benthic spawning clownfish A. ocellaris compared to pelagic eggs from black seabass C. melanus. is provided on the page 88. Keep in mind development time is mainly dependent on egg quality and pigmentation, temperature, water quality, and to some degree photoperiod.

Allen (1972) provided a fairly detailed description of internal egg development of A. chrysopterus and A.tricinctus. Development is quick. At approximately 24 hours the early nurula (earliest stage of obvious body form) is evident. At 48 hours, eye, heart, brain, and spinal cord are present and the yolk is slightly reduced and darker orange. Allen (1972) noted a heart rate of 140 beats per minute for A. chrysopterus. He further noted that during the third night, the embryo reverses its position, by aligning it's head at the unattached end of the egg. By the fourth day, darkly pigmented eyes and body parts are evident and the heart rate increases to 200 beats per minute. By day five, pectoral fins and dorsal and ventral fin folds are apparent. Major blood vessels and circulation are clearly evident. By the sixth day, semicircular canals, mouth, and well developed eyes consisting of blackened iris and silver covering are apparent. Opperculum flaps are barely discernible. Day seven is similar to day six except more refined, Early stages of gill rakers and developed gill flaps are evident. The yolk is considerably smaller.

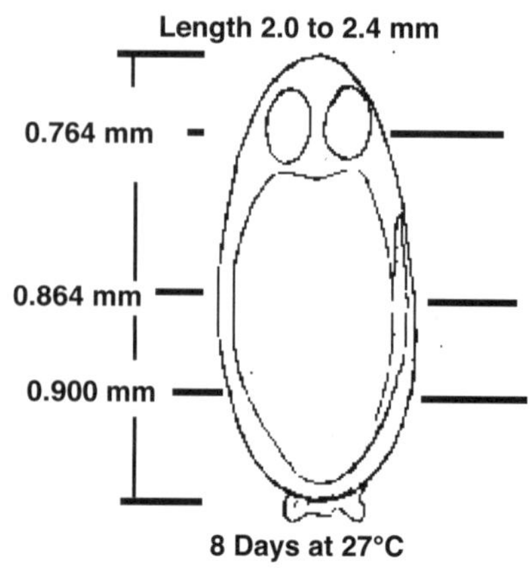

Figure 41 - Egg size , A. ocellaris, Bertschy 1979a

Photo 13 - Clownfish,
Amphiprion ocellaris

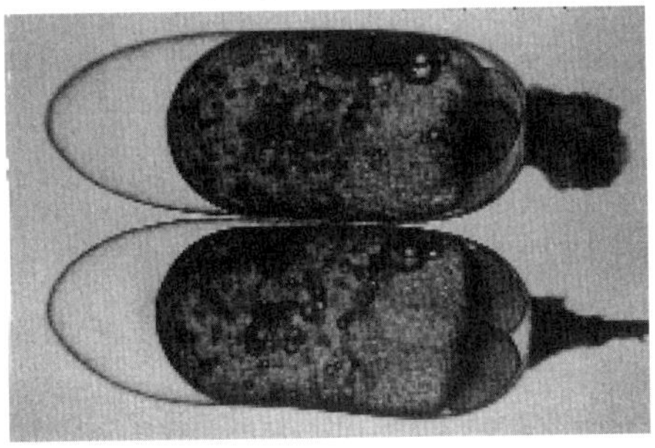

2 & 4 cell stage 1 Hr

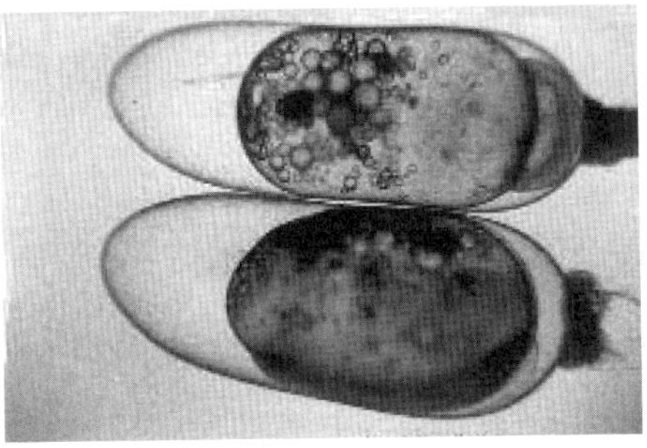

Mid to late Blastula 8 Hr

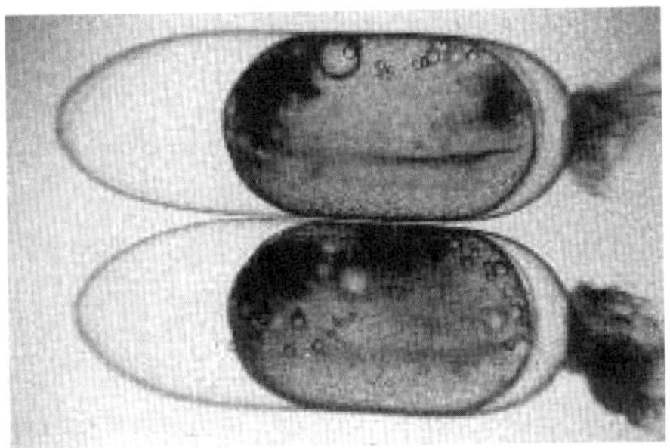

Early Neurula 24 Hr (1 day)

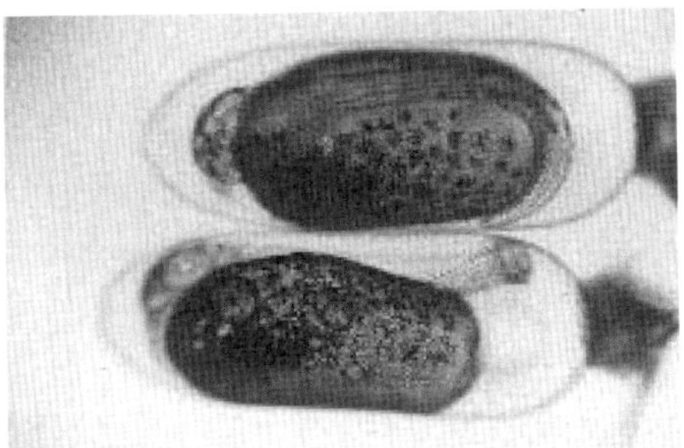

Mid Neurula 96 Hr (4 days)

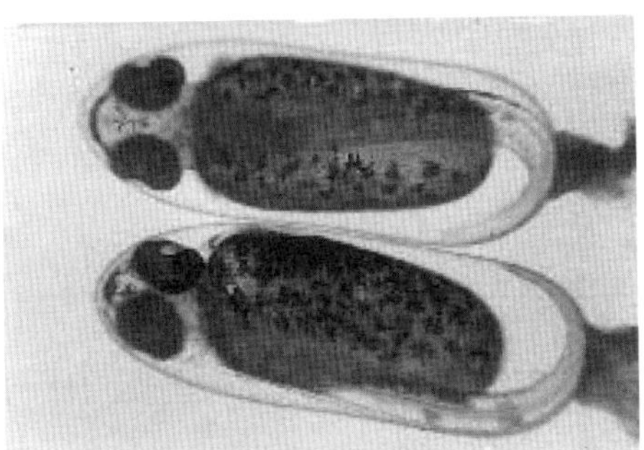

Late Neurulat

Photo 14 - Black Seabass,
Centropristes melanus

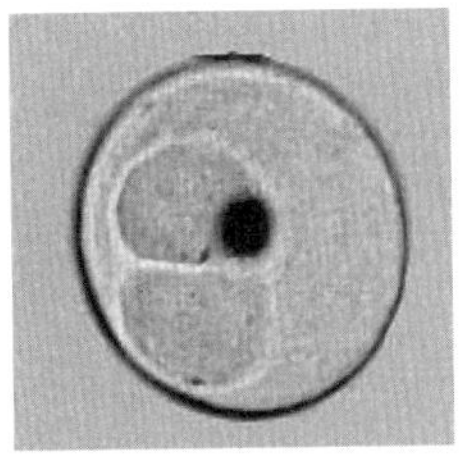

2 Cell Stage 1:20 Hr

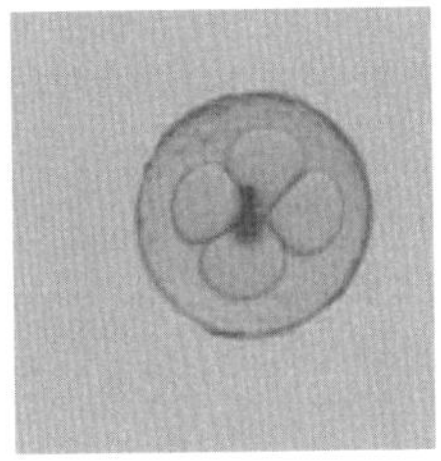

4 Cell Stage 2:20 Hr

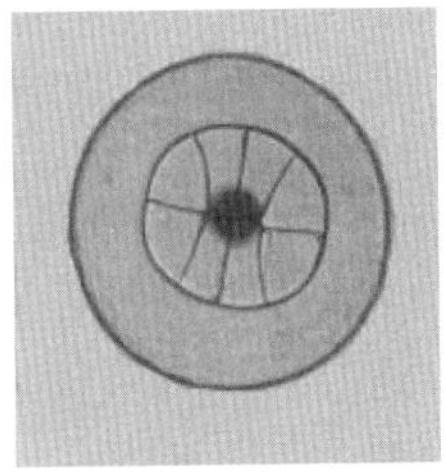

8 Cell Stage 3:00 Hr

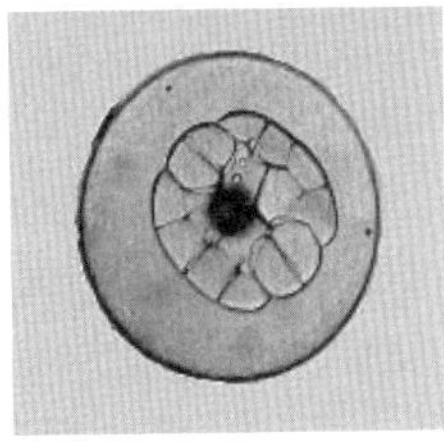

16 Cell Stage 3:35 Hr

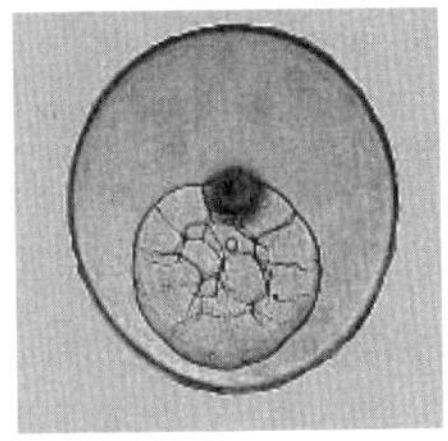

32 Cell Stage 4:15 Hr

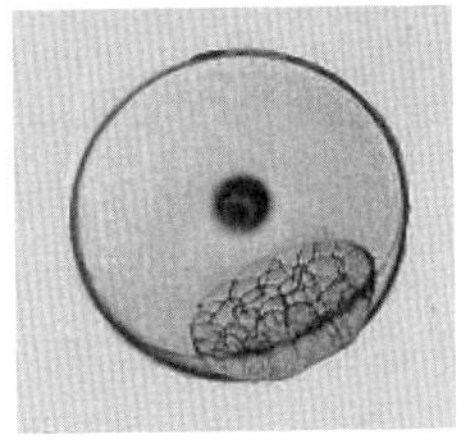

64 Cell Stage 5:05 Hr

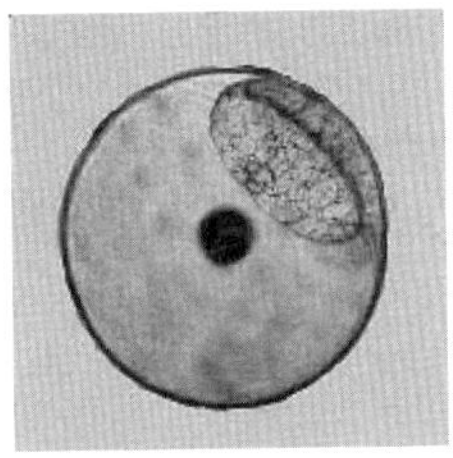

128 Cell Stage 5:45 Hr

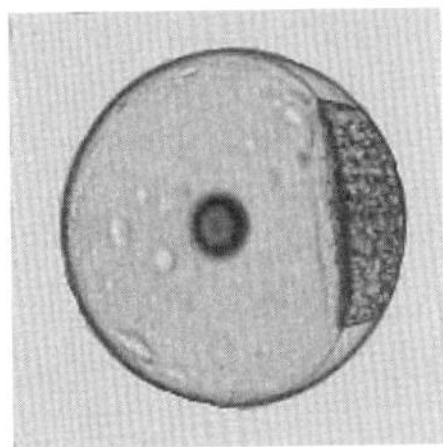

Early Blastula 7:05 Hr

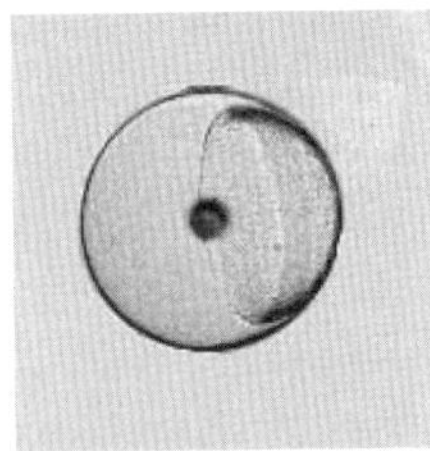

Late Blastula 26:50 Hr

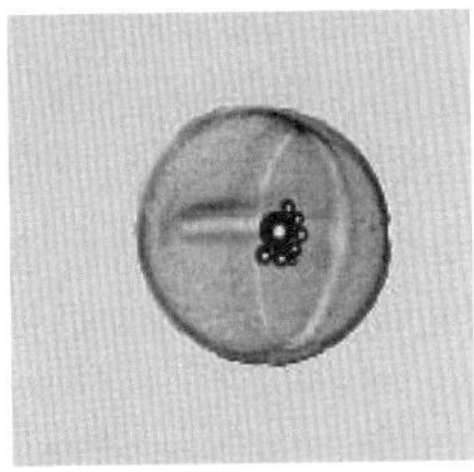

Ear;y Neurula 31:25 Hr

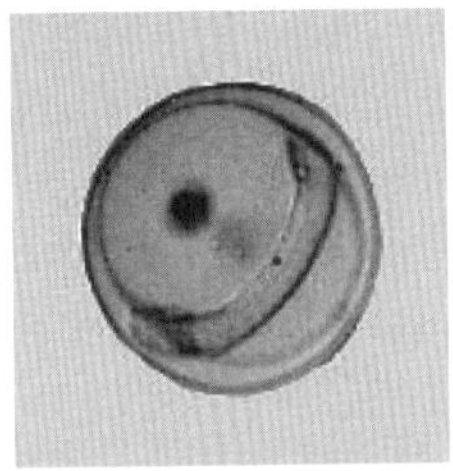

Mid Neurula 37.05 Hr

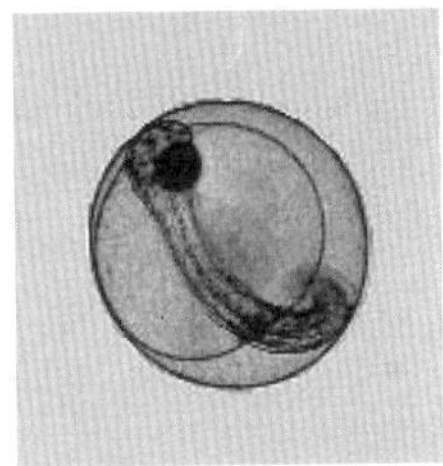

Late Neurula 61.20 Hr

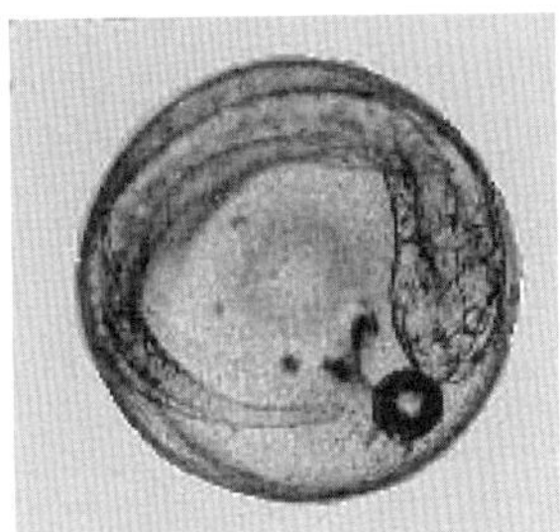

Prehatch 75:35 Hr

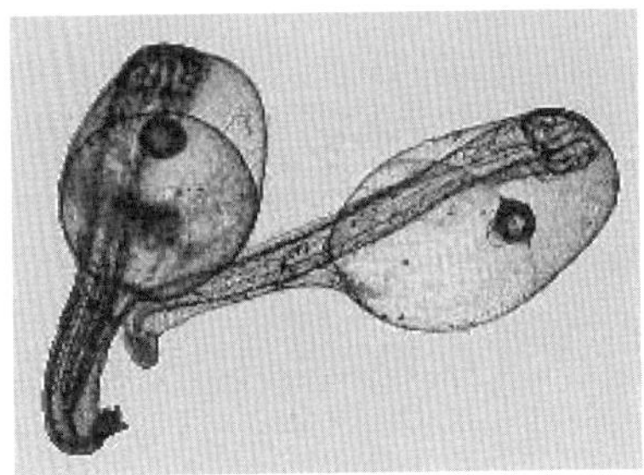

Hatch 75:35 Hr

Egg (oocyte) Development

Photos series 13 on the left are typical stages of benthic eggs from clownfish *A. ocellarus* from two cell stage to prehatch. Development at 25°C (77° F). Photos series 14 to the right are typical egg stages of pelagic eggs from seabass, *C. melanus*. Development was at 19°C (66° F).

Egg And Nest Size

Overall nest size depends more on the species, size, and age of the female. Eggs range in size from 2.0 to 2.4 mm and 0.9 mm wide (Bertschy 1979a, Allen 1972). Upon hatch larvae are slightly over 4.0 mm total length.

A good, well-formed, nest is oval in shape and tightly packed. Tightly packed nests of large individuals have about 60 eggs per square centimeter, medium size fish 50 eggs / sq. cm; and small fish 40 eggs / sq. cm. To determine the actual area of the oval nest use the formula:

$$A=\pi x R1 x R2$$

(A = 3.14xRadius 1xRadius 2).

Once you determine the area of the nest in centimeters, multiply this times the number of eggs / sq. cm. Average nest size (Table 17) is highly dependent on species. But, there may also be seasonal changes. During spring and fall, natural biorhythms come into play resulting in larger nests and the establishment of more new pairs. Within a species, nest size is directly related to the age, size of the female, how long the pair has spawned, nutrition and health. Hatch rates are highly dependent on many factors and are discussed under hatching procedures.

Nest Configuration

Another criteria used in determining good and bad nests and pairs is nest configuration. Spacing of the eggs or how well the nest was constructed indicates the quality of spawn and pair. "Quality nests" are a well defined oval shape with tightly packed eggs. A "fair nest" is one that is less dense and less defined in shape. A "poor nest" is not well defined, consisting of widely scattered eggs. Often, many eggs in poorly formed nests are not fertilized. Consistently poor quality nests are related to adult physical problems and the pairs should be terminated. However a occasional or short period of "poor nests" may indicate something is disrupting the pair during spawning. Newly established pairs often create poor quality nests and spawn sporadically for several months.

Spawning Duration

Just how long will a fish spawn? I, for one, feel fish will continue producing quality nests longer than popular breeding books have stated in the past. Typically freshwater farmers

Table 18 - Spawning records of clownfish pairs a IOH

SPECIES	TOTAL SPAWNS	YEARS	MONTHS	AVG/MONTH
A. akallopisos (orange skunk)	70	2 yr 5 mo	29	2.41
P. biaculeatus (maroon)	102	4 yr 11 mo	59	1.73
A. clarkii (sebae)	150	4 yr 4 mo	52	2.88
A. ephippium (red saddle)	138	5 yr 4 mo	64	2.16
A. frenatus (tomato)	127	4 yr 8 mo	56	2.27
A. melanopus (cinnamon)	103	4 yr 10 mo	58	1.78
A. ocellaris (ocellaris)	132	5 yr 4 mo	64	2.06
A. percula (percula)	64	4 yr 7 mo	55	1.16
A polymnus (saddleback)	52	2 yr 5 mo	29	1.79
A. frenatus x A. ephippium (FE)	156	5 yr 3 mo	63	2.48
A. frenatus x A. melanopus (FM)	98	3 yr 11 mo	47	2.09
A. ephippium x A. melanopus (EM)	102	4 yr 0 mo	48	2.13
AVERAGES	**108**	**4 yr 4 mo**	**52**	**2.08**

reestablish new pairs every year or every two years. After consulting with several farms I have found they have a tendency to treat adult pairs exactly the same as the juveniles they are growing out. Breeder size fish brought in from outside ponds are robust, have good coloration, and produce large, brightly colored, tightly packed nests for the first 3 to 6 months. After that point the nests become pale, eggs are loosely packed and hatch rate is drastically reduced. Does this mean the pair is spent or finished? In most cases I have found this is not the case. Degree of care and nutrition are directly related to the degree of continued production from a pair. What pairs are fed and how they are treated directly influences productivity, nest size, color, hatching success, initial larval survival. These advantages no doubt influence overall juvenile survival. Broodstock diets are extremely important with regard to quality egg production (see Conditioning For Spawning).

Longevity and reproduction are often extended in animals kept in captivity. This is not hard to explain when you consider the extreme hardships, temperature changes, food shortages and predation that wild populations must endure to survive. Even within humans, there are strong correlations between survival and reproduction in countries where diets are well balanced and the overall environment is temperate. Records maintained by Dr. Conde (1986) at the Aquarium in Nancy, France listed the age of several captive clownfish: *A. frenatus* (tomato) 17 years and still alive in 1986, *A. clarkii* (sebae) 14 years, *A. ocellaris* (ocellaris) 14 years, and *A. perideraion* (pink skunk) 21 years. Merrill Cohen (Aquarium Products) has a pair of *A. ocellaris* purchased from Instant Ocean Hatcheries 20 years ago. How long any remained sexually functional is not known.

Clownfish are protracted spawners and produce about one nest per month or less in the wild. Under controlled conditions and consistent ideal diets, they can easily be induced to spawn an average of 2.1 times a month. In captivity, most pairs spawn a minimum of 11 months per year regardless of the species. How long can a captive pair maintain good consistent production is still not clear but appears to be at least five years. Juhl (1992) listed a pair he had inherited with a eight year history of breeding which was still active. Pair records analyzed for the oldest

Table 19 - Comparison of gross egg production at IOH over a three year period with regards seasonal changes

	Jan	Feb	Mar	Apr	May	Jun	Jul	Aug	Sep	Oct	Nov	Dec
1979	42153	33252	42030	59509	37345	39960	58590	54868	39767	36800	52597	68442
1980	73445	58233	78942	77096	98592	86828	94452	86319	105160	97156	77585	66738
1981	43890	70700	63756	66924	71796	56925	62426	62073	64389	67903	48179	64569
Avg.	53162	54062	61576	67843	69244	61238	71822	67753	69772	67286	59454	66583
Avg Nest	458	438	432	459	454	405	443	384	387	386	367	427

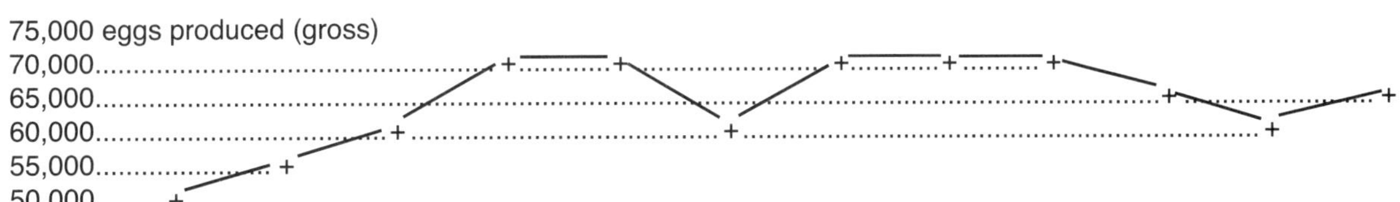

productive individuals at the closing of Instant Ocean Hatcheries are listed in Table 18. Keep in mind that these were still active, good producing pairs when the facility was closed. The last three are hybrid crosses.

Another question is whether fish that spawn for a long time continue to produce quality nests and larvae. We found that individual pairs seem to reach a typical reproductive pattern which remains fairly consistent for an extended period of time. A five year old *A. ocellaris* pair will continue producing quality nests if that is the normal pattern for that specific pair. Yet, a two year old pair may consistently produce medium quality nests and will continue to do so unless one of the pair is changed. Each pair should be considered individuals with distinct quirks, personalities

Table 20 - Seasonal comparison of nests produced (NP) verses nest that were lost (NL) and could not be used (IOH data)

	Jan	Feb	Mar	Apr	May	Jun	Jul	Aug	Sep	Oct	Nov	Dec
1978NP	59	59	88	81	94	120	100	86	50	76	82	122
1978NL	--	22	29	30	21	14	57	47	25	7	19	15
1979NP	120	136	161	171	170	161	170	178	182	144	165	202
1979NL	25	46	31	34	73	41	35	6	21	29	16	15
1980NP	205	188	195	209	239	227	254	235	251	254	214	168
1980NL	20	47	18	25	31	31	50	28	31	40	21	21
1981NP	126	212	207	202	242	219	232	202	207	--	--	--
1981NL	21	37	46	59	49	54	50	31	38	--	--	--
AVG NP	128	149	163	166	186	182	189	175	173	158	154	164
AVG NL	22	47	31	37	44	35	48	28	29	25	19	17
AVG %	17%	32%	19%	22%	24%	19%	25%	16%	17%	16%	12%	10%

NP = o = Nests produced

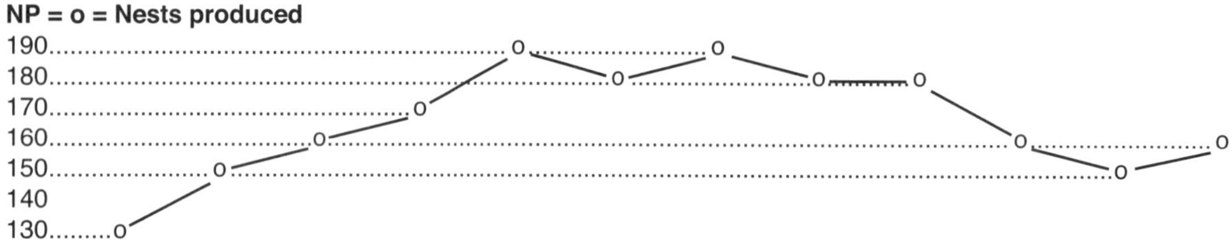

%NL = x = % Nests lost

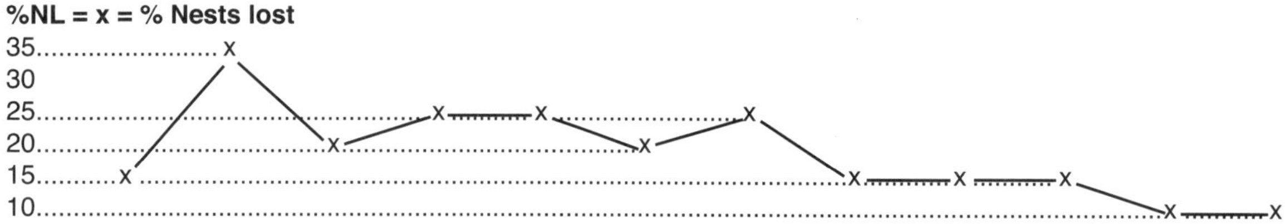

Table 21 - Comparison between spawning pairs of clownfish verses non spawning pairs over a year (IOH data).

	Jan	Feb	Mar	Apr	May	Jun	Jul	Aug	Sep	Oct	Nov	Dec
78 SP	30	33	39	38	53	58	51	33	30	37	40	48
78 NS	--	13	--	9	14	21	17	20	23	13	11	7
79 SP	52	63	68	76	77	83	83	80	81	70	79	86
79 NS	5	7	7	6	11	11	12	16	11	28	13	11
80 SP	96	92	79	83	91	93	97	94	103	105	104	98
80 NS	5	11	18	13	8	7	6	11	6	8	14	20
81 SP	86	113	107	109	108	100	101	109	105	--	--	--
81 NS	27	13	18	11	13	21	17	19	17	--	--	--
SP = Spawning Pairs												
AVG	66	75	73	77	82	84	83	79	80	71	74	77
NS = Non Spawning Pairs												
AVG	12	11	14	10	12	15	13	17	14	16	13	13
%	18%	15%	19%	13%	15%	18%	16%	22%	18%	23%	18%	17%

and patterns. They are indeed different. Old age is not usually the main criteria for producing low quality nests and larvae. Obviously there is a time when egg production and quality diminishes. This usually occurs rather quickly and happens over three to six months, eventually resulting in the death of the female. Often the old female becomes blind prior to death.

Spawning Versus Season

The following spawning data is from IOH (Instant Ocean Hatcheries). Table 19 shows a comparison of gross egg production which includes all nests produced regardless of whether they were used or lost during maturation. Basically, production followed a typical bell shaped curve with lower production during the coldest months, January and February. Light intensity and duration did not seem to have an effect since December production was constantly high even though the shortest day (December 20th) and lowest light intensity occur during this month.

When we consider the total number of nests produced and the percentage that died or were not fertile, we find another strange point. Table 20 shows that the amount of nests produced increased in the warmer months. Ironically, the percentage of poor or dead nests was less during periods of low production. Post larval production (Table 35, in Larval Rearing Section) followed the same bell-shaped curve as gross egg production with higher production in May, June, and July with a drop from mid summer into the winter months.

Table 21 compares spawning pairs with pairs that ceased spawning for one month or more over a twelve month period. Basically, there is little difference except from August through October where the percentage of non-spawning pairs increases slightly.

The spawning data and resulting larval rearing data show a general upward trend of higher production in early spring (April) through early fall (September). In some cases, the warmest months, August, September, and part of October can become detrimental with regard to the number of active spawning pairs, gross egg production, and average nest size. Careful control of water conditions especially temperature, is certainly advantageous but not absolutely necessary to obtain production providing you establish more spawning pairs than needed.

Criteria For Culling Pairs

Even though adult clownfish pair for life, this does not mean they will remain good commercial pairs throughout their lives. Commercial endeavors must establish criteria to determine when a pair is no longer productive, should be replaced, or separated for repairing. At Instant Ocean Hatcheries we established the following criteria (Table 22) for culling pairs.

Table 22 - Criteria for culling adult pairs

1. A pair that remains as a possible pair (PP) for over eight months.
2. Pairs producing less than two spawns a month over the last four months.
3. Old pairs with greater than 50% of the nest gone over the last four months.
4. Old pairs with consistently loose, scattered or small nests.
5. Pairs that continuously spawn on the side of the tank.
6. Unpopular species or overstock of a specific species.
   a. Select those that have small or moderate nests.
   b. Select those with consistently loose nests.
   c. Select those with less than four spawns over the last three months.
7. Hybrid pairs or new species of fish are the exception to these rules.

Serious breeders of marine fish quickly realize why good record keeping is a must. Marine tropical fish pairs are expensive to establish and cannot easily be changed like less expensive freshwater fish. Your goal is to optimize production and quality of each pair for as long as possible.

HATCHING PROCEDURES

How and when a nest is hatched can create success or failure. Determining methods suitable for your conditions is important. I will attempt to outline and describe several alternatives, but it is prudent to keep several points in mind. Nests can tolerate more mechanical and chemical changes than newly hatched larvae. While within the egg, larvae can adjust more easily to water changes than after hatch. The protective shell acts as an osmotic membrane which resists sudden changes by allowing slow internal adjustment. Whereas, newly hatched larvae are more vulnerable to abrupt water and chemical changes and cannot adjust as quickly. Initial failures or high mortalities are often caused by unsuitable hatching methods or adverse water imbalances between broodstock and larval rearing tanks.

Larvae cannot be reared within broodstock tanks without devising some method of containment to protect them from the parents and concentrate live food sources. In nature larvae are expelled into the surrounding area and experience a period of planktonic existence of 5 to 7 days before dropping to the bottom and seeking protection within the immediate reef habitat. Parental care stops after hatch and larva are subject to predation even by their parents. Hatching in the dark serves as a protective means for reducing initial predation.

Hatching Time

Determination of time of hatch was dependent on visual appearances (Table 16 & 23), temperature during incubation, and the species. If the hatchery experienced a cold spell during incubation, hatching was usually delayed one or two days longer than normal. Normal hatch rates were based on temperatures of 78.8° to 82.5°F (26°C to 28°C) Table 23 provides hatch rates for several species reared at IOH over five years. Water temperature within the hatchery normally fluctuated 1° to 2°F (1°C) daily.

Juhl (1992) described a direct relationship between constant temperature and hatching of *A. ocellaris*. At 78.8°F (26°C) hatch took ten nights at, 80.6°F (27°C) nine nights, and eight nights at 82.5°F (28°C). His data, based on constant tank temperature, conflicts with our five year data but does indicate variability when using temperature as the only criteria for hatch. As stated before, hatch is also highly dependent on broodstock health, quality of eggs, initial color of eggs, and water quality other than temperature and light. Yet, egg quality is not easily determined by external visual observations alone. Time to hatch, partial hatches over two or three days, and initial larval survival are strong indicators of nest quality. As you become familiar with your broodstock, there are pairs which consistently produce nests which hatch earlier or later than normal regardless of what you do. Keep in mind that late hatches are the exception and not normal. The longer the larva remains in the egg beyond normal incubation, the weaker it will be at hatch. If you are maintaining a temperature of 78.8° to 82.5°F (26° to 28°C). and continue to experience delayed hatches, then concentrate efforts towards improving adult diets and water quality.

Table 23 - Hatch Based on 26°C to 28°C

A. akallopisos (orange skunk)	7 th evening
P. biaculeatus (maroon)	6 th evening
A. clarkii (sebae)	6 th evening
A. ephippium (red saddle)	8 th evening
A. frenatus (tomato)	9 th evening
A. frenatus X A. ephippium (FE)	8 th evening
A. frenatus x A. melanopus (FM)	9 th evening
A. melanopus (cinnamon)	9 th evening
A. ocellaris (ocellaris)	8 th evening
A. percula (true percula)	8 th evening
A. polymnus (saddleback)	7 th evening

Theory of Hatching Nests

At Instant Ocean Hatcheries, two main concerns motivated our initial thinking in finding a suitable hatching method for commercial applications. Originally, like everyone else, we felt it was necessary to hatch the nest out in the same water in which it was spawned. Secondly, we felt larvae had to have the same water as in the broodstock tanks. Our initial attempts centered on meeting these parameters but our theory was basically wrong. For some reason, the primitive concept of hatching nests out in the broodstock tank and collecting the larvae is unique to marine fish culture but not freshwater fish culture. Virtually every tank-reared fish in freshwater is hatched from nests that were removed from the parent at some point prior to hatch. In addition, most freshwater hatcheries have separate broodstock and larval rearing facilities and water conditions.

A detrimental fact to commercial rearing of clownfish is that hatch normally commences from one to two hours after dark. Hatching takes from 15 to 20 minutes. If larva are allowed to remain in broodstock tanks overnight the yield is significantly lower either due to adult predation or other reasons. A commercial operation forced to employ people to work at night increases overhead and reduces normal work-hour efficiency. Therefore, scooping and siphoning hatched larvae is very impractical for commercial applications. To expedite removal, a net is sometimes utilized but this practice is prohibitive due to severe mechanical injury to the delicate larva. Whatever means are used, it is important to realize that larvae cannot tolerate being touched by a solid object like a net.

Considering low yield normally associated with allowing nests to hatch in broodstock tanks, alternative methods have been devised.

Hatching In The Broodstock Tank

Common practice is to allow the larvae to hatch within the broodstock tank which is suitable for small scale operations and hobbyists. Siphoning or scooping newly hatched larvae can be easily accomplished. Since clownfish larvae are phototrophic (swim towards light) they can be drawn to a specific spot for removal by using a flashlight. Other, more sophisticated, siphoning methods include using large diameter plastic tubes mounted above the nest area and equipped with a small light source at the top. This helps concentrate larvae more efficiently and quickly.

Several methods have been employed to expedite or delay hatching until a more convenient time. The problem is to trick larva into hatching prematurely in the afternoon or delay hatch until the next morning. During the late afternoon of the day they are expected to hatch, cover the tank with black plastic to imitate darkness. This should be done as close to the evening as possible. To delay hatch, the lights can be left on all night or the temperature can be lowered. Both methods are not foolproof, especially with regard to delaying hatch.

Since siphoning takes time, automatic mechanical methods have been devised to concentrate larvae. If your tank is equipped with an overflow or stand-pipe a special collecting tank can be made (Figure 34). This tank should have a 100 -150 micron mesh wall. It can be made to mount directly on the stand-pipe, or wall overflow. A tight flush fit is needed to prevent larvae from being trapped or flowing into the filter. The addition of a small light will aid collection. Larvae should be removed that evening and not allowed to remain in the collector over night.

If your tank is not equipped with an overflow and is self contained, another method (Figure 42) employed by Reef Propagations Inc. might be suitable. In this case two, one gallon or larger stacking, plastic aquaria are modified into a pump-powered collector. The outer aquarium is equipped with a "U" tube which is mounted on the bottom and supported on the side. The second aquaria has areas cut out of the bottom and 100 micron filter cloth is permanently mounted over the entire bottom. This aquarium is then mounted within the first in such a manner that a false bottom like a submerged biofilter is created. The rim of the second aquarium is cut to be flush or lower than the outside one, then glued in place to provide a permanent seal. To use this device, clamp the collector to the side of the aquaria so that it is about 1/4" below the water level. Insert a powerhead into the external "U" tube then connect a tube to the output which pushes water into another mesh covered box or diffuser mounted at the other end of the tank. The diffusor box could be a "Florida Aqua Farms" hanging acclimation box with a mesh glued in place. Water should flow into the collecting box, but not fall into the box so the level adjustment is critical. Locating a small light over the collector will expedite collecting. About one or two hours after the lights are off should be a good time to collect most of the larvae. Do not retain them in the small collector overnight. To remove collector, detach, and lower the water several inches in the collector. Then turn off and remove the power head, move the entire tank into the larval rearing tank and gently empty contents.

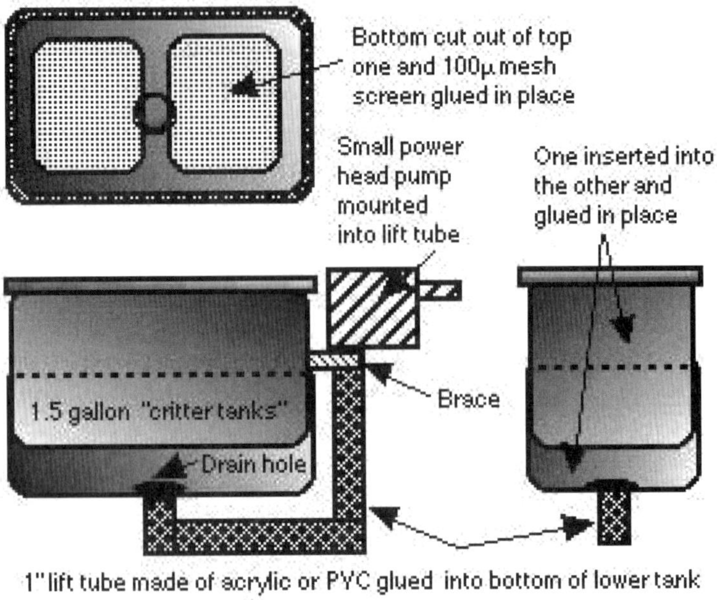

Figure 42 - Side overflow collector used by Reef Popagations Inc., courtesy of Joe Lichtenbert

Hatching Nests Remotely

Besides the factor of having to wait until the evening for the nests to hatch, there are other points which make remote hatching more advantageous especially to commercial operations. As stated in the beginning of this chapter, larvae within the eggs are more tolerant than newly hatched larvae to physical, chemical, and water quality changes. Hatching nests within broodstock tanks may yield 100% hatch but not 100% recovery of the larvae. Many larvae are consumed, drawn into the filters, entrapped, or die before being captured. This loss has motivated others to look at obtaining nest control and 100% capture of hatched larvae. Initially, attempts were made to physically scrape the eggs from the spawning substrate. In theory, this practice sounds good but has an inherent flaw. Attached eggs provide a solid surface from which larvae push themselves out of the egg shell. Free floating eggs do not offer the resistance necessary to accomplish this.

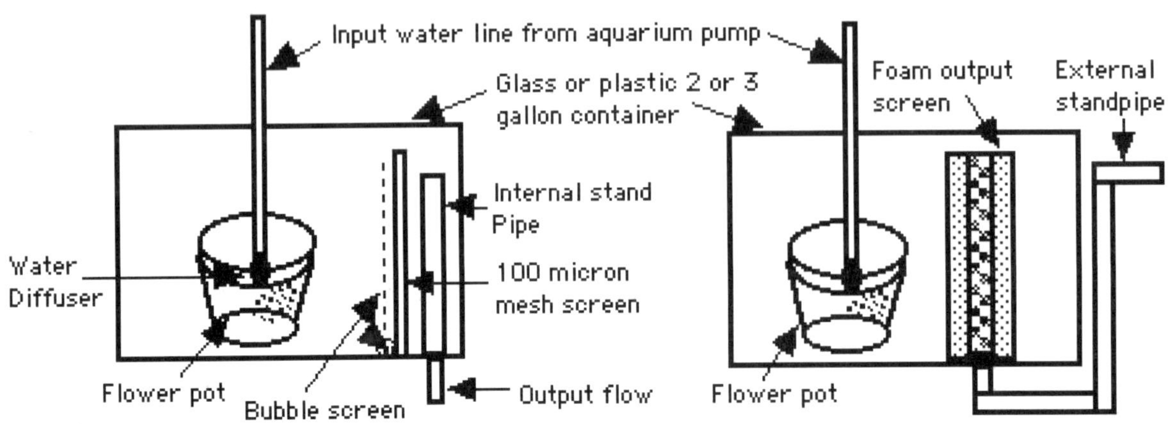

Figure 43 - Remote flow through hatchers designed and tested at IOH.

At IOH we tried to control the entire nest and hatch using several modifications of the hatcher shown in Figure 43. The concept was to remove the nest, place it into a flow-through hatcher that was connected to the flow on the adult broodstock tank. To keep the eggs moving and well aerated, they were either aerated or incoming water flow was directed onto the eggs. When hatched, the larvae were trapped in the compartment containing the spawning substrate. This concept works, but allowing larvae to remain in hatchers overnight resulted in high mortalities. These hatchers were further modified by making them dark to try and force hatching before the evening. This worked to some extent, but not consistently. Utilizing remote collectors allowed more freedom and provided complete control over the nest and hatch, but were cumbersome to use within the broodstock area. In general, using small, contained, hatching and collecting devices usually results in higher mortalities due to mechanical injury.

Based on these results, we then attempted to physically remove intact nests just prior to hatch and place them directly in larval rearing tanks. Initial attempts were filled with mixed results and cumbersome since we also transferred broodstock water into the larvae tanks prior to setting up the nest. Later experiments revealed that as long as the water was similar, water transfer from adult tanks was not necessary. After considerable work we devised a method that provided consistent results with a minimal amount of work.

Hatching Method Used At IOH

Prior to removing a nest for hatch, technicians would remove tags from the "running calendar" (Figure 39). With tags in hand, each nest was checked to determine if it was ready or worthy for hatching. Determination of hatch was dependent a visual appearances (Table 16), temperature during incubation and species (Table 23).

Nests that appeared ready to hatch were slowly removed from the broodstock tank. Wearing rubber gloves will prevent the adults from biting. The flower pot or spawning substrate was entirely cleaned of excessive attached algae and dirt then transported to the larval rearing area. To help reduce parasite infestation and fungus, the entire substrate and nest were submerged in a bucket containing a mixture of 1 ml formalin (37%) per gallon of aerated saltwater (1 ml/3.75 L) for 1 to 2 minutes. While submerged, the substrate around the nest was rubbed and cleaned until it felt rough and not slick. Use of rubber gloves is highly recommended when using formalin or any drug or chemical (liquid or dry). After treatment, excess water was shaken off and the nest was placed into the larval rearing tank. Residual formalin was not a problem but when using

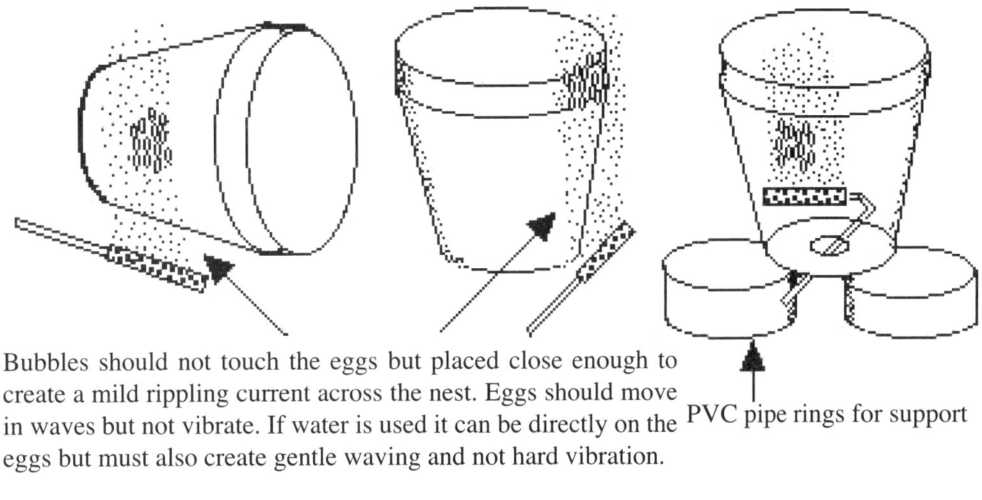

Figure 44 - Setting up for hatch using aeration as a means of creating a current.

small larval rearing tanks (5 to 15 gal) you might want to rinse the nest with a little temperature-compensated, freshwater first. Concentrations as high as 5 to 15 ml formalin per gallon have been used with no observable detrimental effects. Others use a 10% iodine solution (25 ppm), submerging of the eggs and substrate for five minutes.

Nests retained in broodstock tanks are constantly maintained and fanned by the pair. Fanning becomes more vigorous prior to hatch and no doubt aids larvae in escaping from the eggs. In order to artificially hatch a nest, efforts must be made to mimic the "fanning" activity of the adults. Ideally you want to gently move the eggs back and forth in waves across the entire nest without air bubbles touching the eggs.

Figure 44 shows various methods of using aeration to circulate water over nests. Utilizing small sections of 2" and 3" PVC pipe help support the nests in more suitable positions for optimal circulation. If absolutely necessary, air bubbles can touch the nest, but a continuous stream of bubbles will often kill a significant portion of the eggs. The portion affected is usually the area where the air bubbles are most concentrated. Coarse bubbles are recommended over fine ones. This is further supported by Juhl (1992) who stated that "very fine bubbles have a tendency to stick to the eggs and very large bubbles will harm newly hatched larvae". Maintaining bubble diameter around 1 to 4 mm is essential if you are forced to allow them to touch the nest.

Airlift pump hatchers (Figure 45) were found to be ideal. Basically they are airlift recirculating water pumps equipped with a reservoir and several delivery tubes. These simple devices are constructed from standard PVC plumbing fittings. On top is a 3" (75 mm) diameter x 6"(150 mm)

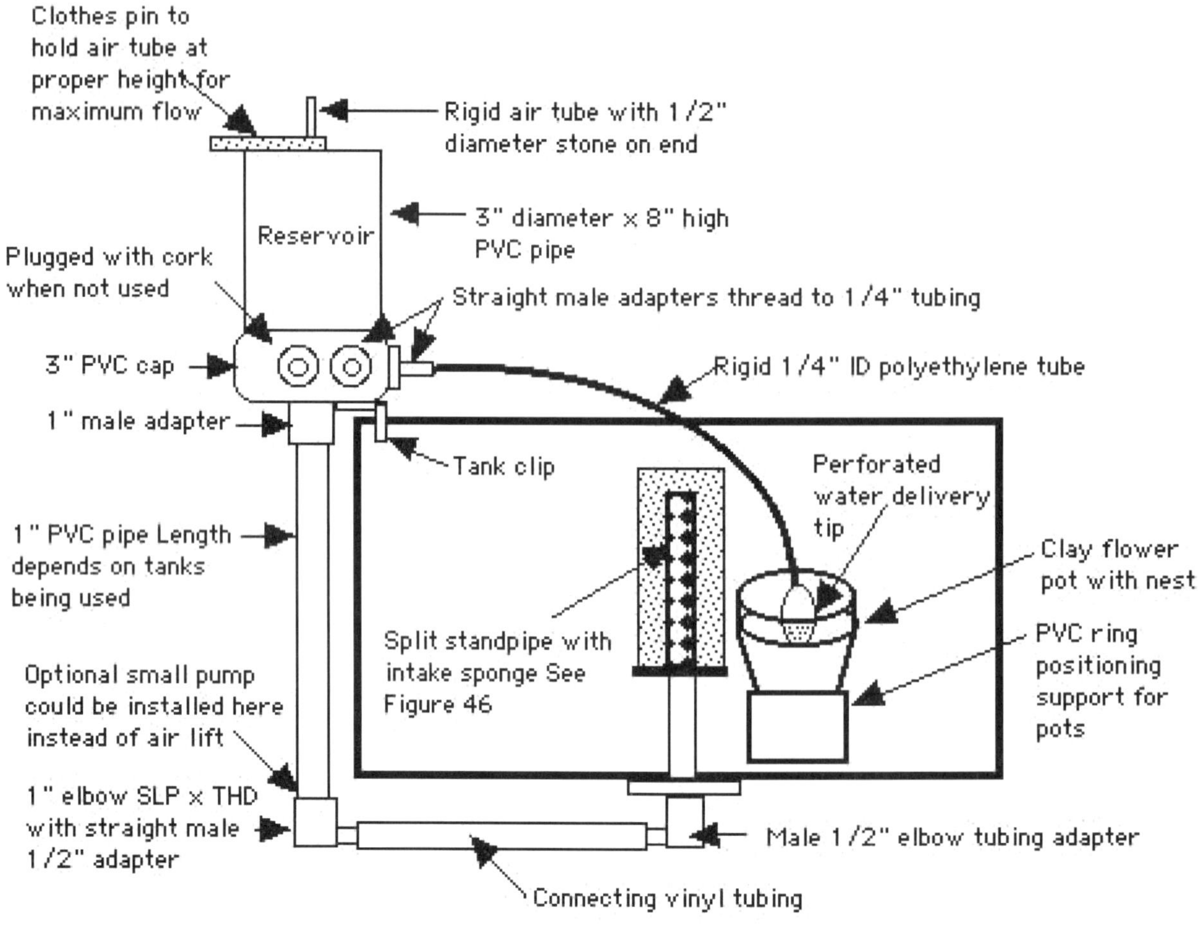

Figure 45 - Air lift water hatcher pumps used at IOH for providing circulation to nests. Up to four nests could be aerated with each hatcher.

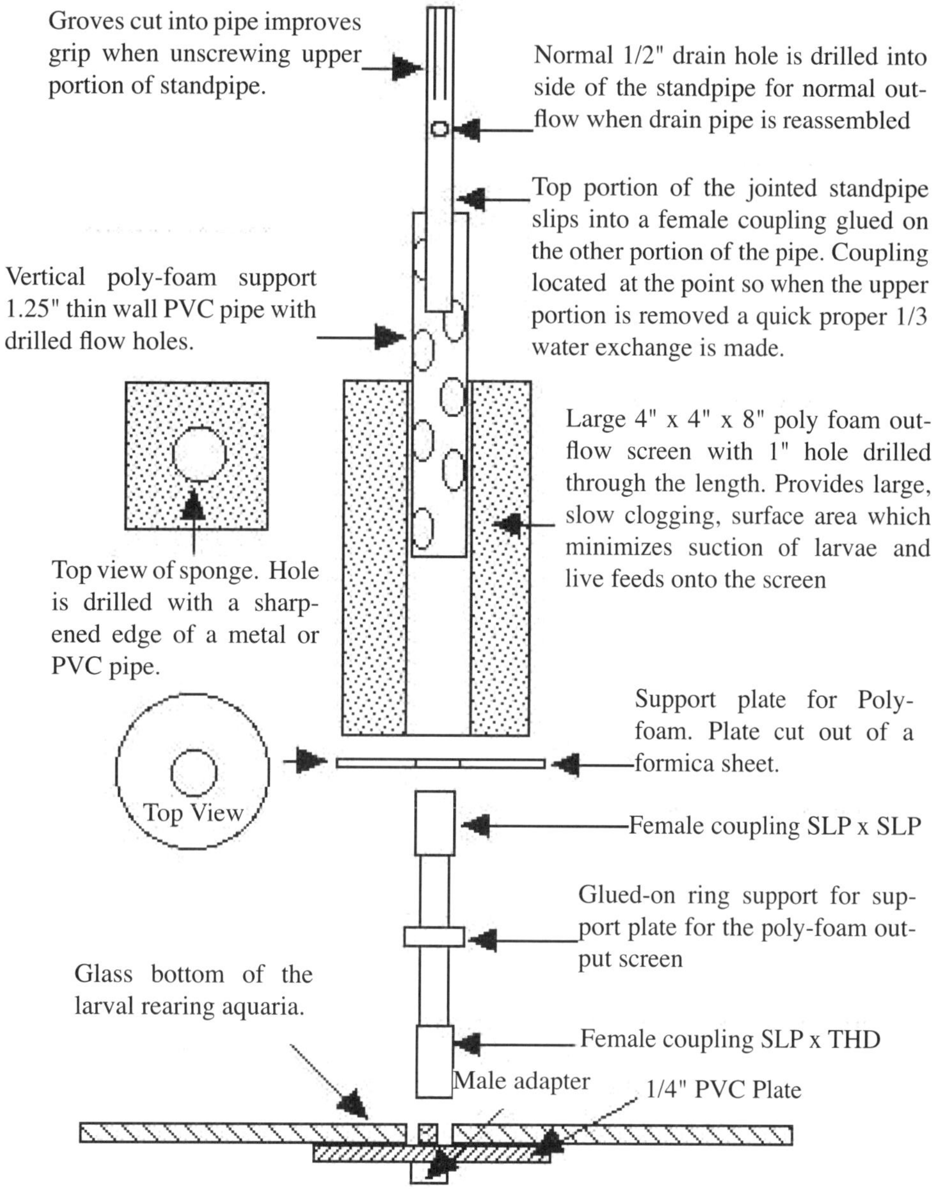

Figure 46 - Jointed stand pipe used in larval rearing tanks for quick uniform water exchanges without trapping larva.

high reservoir attached to a 1" (25 mm) diameter airlift tube. On the bottom of the airlift tube is an elbow with a tubing connector which attaches to the overflow tube on the larval rearing tank. Mounted on the bottom of the reservoir are four male adapters (1/8" NPT x 3/8" tubing) used to attach the semi-rigid nylon or polyethylene delivery tubes. On the end of each tube is a spray device to disperse water. Perforated thumb stalls, a small rubber cap used in the banking industry to count money, are very suitable. A stall is attached on the end of the delivery tube with rubber bands. Delivery tubes were constructed in various lengths to accommodate any situation. Multiple delivery tubes allowed several nests to be set up in a single tank. Inside the airlift tube was an airline made of a long 3/16" rigid tube with a 1/2" diameter airstone. By adjusting the height of the airline you can regulate water flow into the reservoir. Outlets not used were plugged with corks. In hatcheries where electricity is available to each tank, small powerhead pumps could be attached to the reservoir. To prevent new hatched larvae from being sucked into the pump, a large surface intake sponge was mounted inside the tank (Figures 45, 46, and Photo 15).

If your larval rearing tank does not have external outlets, a modified internal hatcher can be made. A low volume powerhead or airlift pump attached to a large surface area sponge filter can be used (Figure 47). This unit could be attached to the side of the tank or preferably suspended within the tank. If attached directly to the side, there would be a tendency for larvae to become entrapped between the hatcher and the wall of the tank. Caution: make sure new sponge filters are well aged in fresh or saltwater before using.

In either case, it is important that flow on the nest be optimal and not too vigorous. You are trying to create currents that appear as waves across the nest. Therefore, if you choose a powerhead make sure it is not too strong or if needed, shunt a portion of the flow from the nest.

Once nests were positioned properly, identification tags were placed along the edge or within the tank directly by or above the corresponding nest. The following morning the technician recorded the nests on the daily larval rearing log sheet. Hatched nests were removed and mortalities counted before the bottom was cleaned. Tanks with unhatched nests were given a 30% water exchange and the nests were re-dipped in a formalin solution and placed back into a tank. Mortalities were recorded daily and determination of the total nest size was calculated when the postlarvae were harvested from the larval rearing tank. Tanks containing several nests were averaged. Again, I must emphasize record keeping. It becomes especially important when sorting out problems. See larval rearing section for more details.

Photo 15 - Overflow and exchange standpipe used at IOH

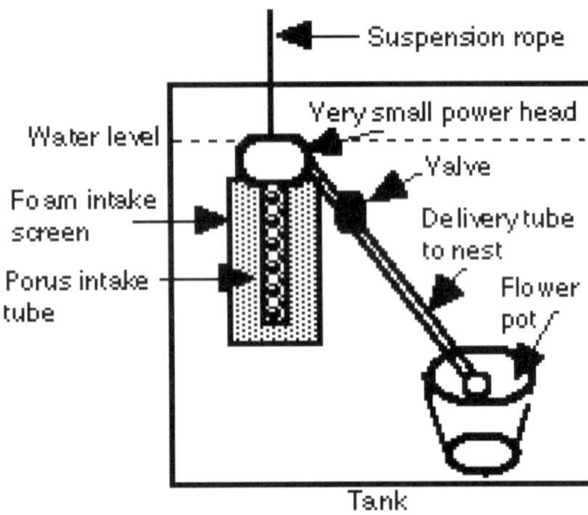

Figure 47 - Above, Internal powered hatcher utilizing a small power head pump and suspended inside the larval rearing tank

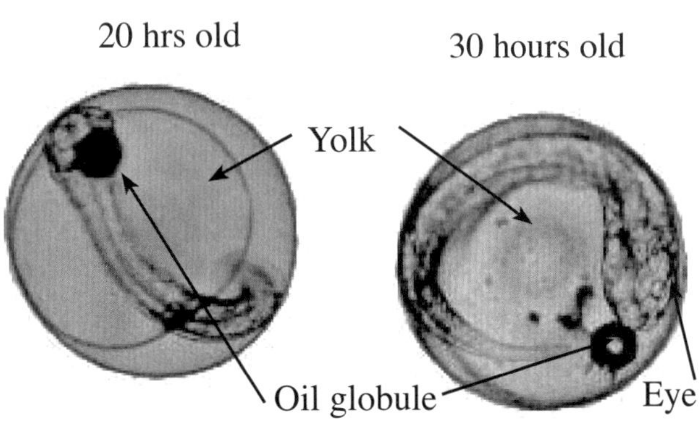

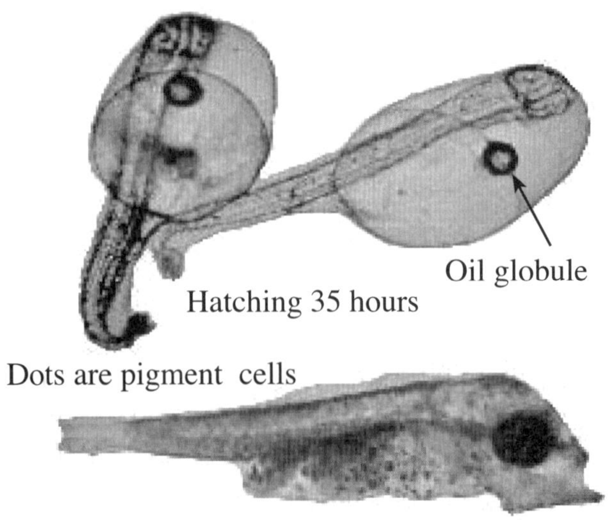

Black Sea Bass (*Centropristes malanous*)
9 days after hatch

BROODSTOCK DISEASES

Diseases in adults are not as serious as in juveniles. Adults apparently have developed immunities to common bacterial and protozoan infections. Treatment of adults should be cautious and careful since no treatment might be the better alternative. Once a pair has adjusted to their surroundings, a well balanced diet regime, and good water quality helps minimize disease and injuries. Aggressive rather than passive treatment of adults may also lead to temporary or permanent internal organ damage with regard to egg and sperm production and may affect egg quality.

Cause of Diseases

Adult diseases are more often a result of old age, prolonged environmental stress, and nutritional disorders. When environmental parameters are routinely held constant, adult fish can usually resist and overcome most diseases and infections. Regardless of what you do, harmful bacteria and protozoans are always present in a stable aquarium environment, but are just part of the ecosystem and insignificant to healthy, well-adjusted fish.

Neither humans nor any other animal, insect, etc. live in sterile environments. Pathogens are always present but our ability to resist is directly related to our mental and physical health. Breast-fed babies are known to inherit immunities from the mother and this might be true of fish since there are few recognized diseases in larval fish. During our juvenile lives we suffer many viral diseases and build up our resistance. Throughout life we build a vast immune system and as long as we maintain our mental and physical health we suffer few illnesses. As we get older our resistance starts to break down and we struggle to maintain our health until death. So why not fish?

Prolonged stress creates a suitable environment for diseases to multiply and attack weak fish. Deteriorating water quality is the most significant form of stress to fish as air pollution is to man. Other forms of stress might be alteration of physical environments by addition of fish, change in habitat, routine tank maintenance, excessive water exchanges, etc. Changes often act as the triggers or the "straw that broke the camel's back" towards a disease outbreak. Coupling sudden alterations with prolonged stressful poor water quality or poor feeding practices creates the perfect turning point. A word of caution: do not make sudden changes on fish that are already under environmental stress.

Detection of Pending Diseases

Early detection is necessary if treatment or correction is to be effective. Pairs should be routinely checked externally for skin disorders, swelling, and behavioral changes. Usually within spawning pairs of clownfish, the female is the most susceptible to disease. Within a non-spawning pair, either fish may become sick. The most susceptible adult fish are new ones, disrupted fish, and non-spawning pairs. If a pair is already routinely spawning and is quite at ease, it is usually not necessary to administer any treatment. After years of working with clownfish I found that well fed, comfortable, spawning adults can usually cure themselves of most diseases without the risk of adding chemicals.

The most common disease problems in broodstock fish are due to old age. Keep in mind that the female of a clownfish pair in most cases is the oldest individual unless you made the pair from your own tank-raised fish. Second, she is the individual who expends the most energy and is most susceptible to nutritional disorders which lead to weakness and eventual disease.

When dealing with broodstock, the age and production of a pair should be a consideration of whether or not treatment should be initiated. Once treatment is started, you are obligated to complete the entire process in order to prevent bacterial resistance and immunity.

Table 24- Initial indicators of possible disease or stress

1. Paleness or darkening of color.
2. Excessive mucus or white surface film.
3. Dry looking skin especially around head region.
4. Exposed mucus pores around head region.
5. Rapid breathing.
6. Gills flared outward.
7. Gulping air at the surface.
8. Hiding in the corners of the tank.
9. Resting on the bottom.
10. Swimming on the surface.
11. Erratic swimming or apparent loss of equilibrium.
12. Sunken abdomen, forehead, or dorsal crest.
13. Sores which do not heal quickly.
14. Frayed fins that do not heal quickly.
15. Bloated abdomen or raised scales.
16. Excessive scratching.
17. White spots on surface of fins and body.
18. Disinterest in mate, usual habitat, or feeding.
19. White strands of fecal matter hanging from anal region.

Common Seasonal Diseases

Yearly, you might expect a break-out of *Cryptocaryon irritans* (white spot or saltwater "ich") in the spring and fall. This is a ciliated protozoan. In the initial stages, the large visible trophonts (active parasitic stage) and tomonts (reproductive stage) are confined to the caudal and pectoral fins and upper head region. Normally, the tomont stage is visible without magnification. At this stage, they are as large as 250 μm X 440 μm (1/4 to 1/2 of a millimeter). In order to visually detect this disease, it must be observed in the early morning since the tomonts (white spot) drop off later in the day. These egg cases settle to the bottom, hatch out into free swimming tomites and re-infect the fish. *Cryptocaryon* is usually a minimal infection with regard to well-adjusted, settled, adults and usually does not require treatment.

This ciliate is very susceptible to hyper >55 ppt or hypo <16 ppt saline conditions. Use of high salinity dips for five to ten minutes will help remove attached tomites. During treatment the tank should be drained, cleaned, and refilled. Several dips may be required. Freshwater dips can also be used providing the pH is adjusted to 8.0 to 8.2. Lowering the salinity to less than 16 ppt in a system for several weeks may help eliminate a crisis situation. If further treatment is needed, antimalarial compounds (quinine hydrochloride, quinacrin, or chloroquine) are the most effective and least injurious to the fish. Dosage for quinine HCl alone is 2 gm per 100 liters for a treatment of 2 days. Spotte (1992) proposes a dual treatment consisting of a 20 minute dip in a hypersaline solution >45 to 48 ppt followed by a tank treatment of 3.96 mg/L quinine HCl and 15.90 mg/L chloroquine for 2 days. When using quinine products keep in mind they are light sensitive therefore, lighting should be minimal during treatment. Residual effects of using quinine products are minimal even to anemones, except for skin discoloration which eventually goes away. A second, and often less effective treatment, is a 10 -14 day treatment of copper sulfate at < 0.2 ppm free copper (stock solution 16 gm/473 ml H_2O). Administer at 2.25 ml/30 gallons (113 L) and monitor levels especially over carbonate substrates. Warning: all invertebrates including anemones must be removed unless a chelated version is used that specifies invertebrates can remain in the tank. Excessive treatment with copper is injurious to the fish. Lastly, the Shedd Aquarium treats *Cryptocaryon* by dropping the salinity to 1.010 (2.5 ppt) for 14 days and initially treating with one drop of formalin per 10 gallons for two days followed by two drops per 20 gallons for the remaining treatment period.

A second disease which might show up yearly is *Amyloodinium ocellatum* (oodinium or velvet disease) which is caused by a dinoflagellate and usually occurs in the hot summer months. In adult fish, it manifests itself in high concentrations within the gill cavities and gill filaments. In later stages the fish characteristically have a dull velvet film on the skin with very small sparkling dots. Other symptoms include excessive scratching and labored breathing. Development of this parasite is a three-stage cycle like *Cryptocaryon* consisting of an attached parasitic stage (trophonts) which form reproductive stages (tomonts) which hatch out into free swimming dinospores (tomites). Like *Cryptocaryon*, it is during the free-swimming stage that treatment is the most effective.

Again we usually did not treat adults unless the disease progressed to an advanced state. If treatment was needed we found quinine hydrochloride at 15 ppm (1.5 gm /27 gallons = 100 L) to be very effective as a quick treatment followed by copper sulfate in a long term treatment. We have also had good results using a DE filter, or reducing the salinity to 10 ppt (1.0069 density) over a two week period. Use of copper should be considered the last resort since it has many detrimental and injurious effects on gills and specific organs associated with reproduction.

Dinoflagellate blooms are triggered by high levels of specific trace nutrients in conjunction with increasing temperatures. Use of high grade carbon, protein skimmers in conjunction with 50%+ water exchanges may help break the reproductive cycle. With adult clownfish, both diseases can also be subdued and/or cured by increasing the water flow to the tank, adding a efficient DE filter to the system and/or adding an anemone. Constant bathing of adults in an anemone helps reduce external parasitic diseases.

Less Common Diseases

Lymphocystis is a noncontagious, viral disease, common on newer pairs and especially on *P. biaculeatus* (maroon clownfish). *Lymphocystis* is characterized by large raised nodules on the fins and in advanced cases on the skin. Unless the case is severe, no treatment is needed since the nodules disappear in normal closed aquaria conditions. Photos 16 to 18 show an adult female maroon clown with an extreme case of *Lymphocystis*. Simply remove the fish, scrape off the nodules with a scalpel and place iodine solution on the open scars. One treatment is usually all that is needed.

Another, more lethal, disease due to bacteria is what we called "dropsy or raised scale disease." Lacking for a more proper identification we will leave it at "dropsy." In this case, females were indeed the most susceptible. Their sides around the abdominal area swell to a point where scales would stick out at angles from the body surface. Often the mouth will lock open and the fish swims around lethargically. It is similar to constipation but more dramatic. Treatment works best if fish can be induced to ingest drug-laced foods up to 1% concentration of drug but this is usually not the case. The best, but rather harsh, treatment for an otherwise doomed fish, is to administer orally and/or anally, via a fine tipped wash bottle, a mixture of antibiotics and wide spectrum of vitamins mixed with cod liver oil. Antibiotics that can be used include Chloromycetin, Erythromycin, Kanamycin or Gentamycin and Furanace. We never found a suitable external treatment for this disease that did not lead to problems associated with collapse of the biofilters.

External parasites such as parasitic copepods are often found on newly introduced fish. These look like dark rods about 3 mm long hanging off the fish. They can be mechanically removed with tweezers and iodine administered to the wound. In closed systems, these do not become a continuous problem.

Flukes or trematodes often appear on the eyes and skin of fish in tanks that are overfed or basically not clean. These can be easily removed using a three to five minute freshwater dip and cleaning the tank. When using a freshwater bath, make sure to adjust the pH to around 8.00 and maintain the same temperature. By adjusting pH, fish can survive for five minutes or longer in freshwater. Formalin dips are effective at 1 ml per gallon (3.75 L) of saltwater. Usually the flukes will swell and fall off in about one to two minutes. Formalin dips are effective on many external parasites.

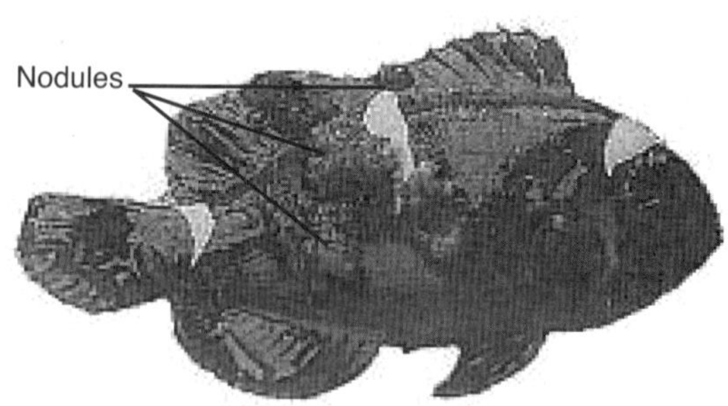

Maroon Clownfish with severe lymphocystis

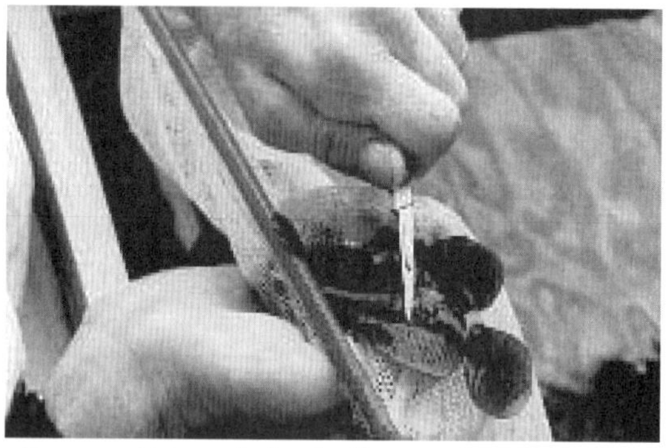

Remove fish and physically scraping off nodules

Treat sores with iodine solution.

Photos 16-18 - Lymphocystis treatment

"Popeye" And Mechanical Injuries

"Popeye" is either a disease or non-disease problem. With regard to adults, we found that in most cases it was induced by a mechanical injury from running into tank walls or habitat. Again, it is stressed not to make quick movements when going by or reaching into tanks. Most tomato complex fish will bite when you reach in the tank. If this is the case then wear rubber gloves. Most susceptible to eye injury are nervous species such as *A. polymnus* (saddleback) and *A. clarkii* (sebae). Detection of mechanically induced "popeye" is subtle but, usually only one eye is involved. Initially opaque scar tissue or scrapes along the eye are evident. In later stages of "popeye" a slight swelling or bulge of the eye is evident which exposes underlying tissue. The exposed tissue gives the appearance of a white ring, especially around the upper part of the eye. Treatment for "popeye" due to mechanical damage is simply rest, and care not to scare the fish. Over time, most swelling will go down but a slight bulge may always be evident after the injury. As a preventive measure, put nervous fish species in larger tanks with dark sided sides and a window covered with shade screen. Use of anemones helps reduce injury with regard to these species.

"Popeye" can also be caused by a bacterial infection but, in this case, both eyes usually swell evenly. This may be a secondary infection following a mechanical injury. There is no known cure and at IOH if the disease got worse, the fish was destroyed. I have experienced positive results in exposing fish, such as grouper, to extreme light levels by combining blacklight or other near UV emitting bulbs like actinic lights in conjunction with bright full spectrum daylight bulbs (1:3 ratio). Exposure time was 14 to 16 hours per day for several weeks. Often fish overcame this problem but, residual swelling remained. "Popeye" caused either by mechanical or bacterial means might result in the loss of sight or actual loss of an eye. Yet, once cured, even the loss of one eye does not distract them from spawning.

Injuries to fins and skin often occur in newly established pairs from excessive fighting or are self-inflected through mechanical damage. Initially the injury will be very red and within a day should start to turn whitish along the edges. A healthy fish in a healthy environment will heal quickly. Caution should be taken in deciding to treat an injury. If the injury remains excessively red after three days, more drastic measures can be taken. Simply remove the fish, dry the injured area with a paper towel and apply an antibacterial salve on the sore or apply iodine. Addition of any antibacterial chemicals to the water should be done judiciously since many of these drugs can kill or disrupt biological filters and may add to the problem. As with popeye, use of a near UV transmitting light such as a blacklight or actinic may help. Heavier doses of vitamins A and B_{12} may also be helpful.

Gas Bubble Disease

This basically refers to dermal blisters on the skin, large emboli in the gills or bubbles on the eye (Blasiola 1995). Several factors may be involved including improper decompression of newly captured fish, supersaturation of gasses due to sucking in air through a pump, or a physiological biochemical imbalance. Large bubbles in the eye are associated with damage to the pseudobranch (part of the gill that supplies oxygen to the eye). Once this disease has occurred it is difficult to reverse or treat. Use of an aseptic hypodermic needle may allow the air to escape. If the emboli in the gills are affected the disease usually becomes fatal. Check all connections on the suction side of the pump. Any loose fittings will allow air to be sucked and compressed within the pump.

Constipation

Constipation is characterized by a long whitish, string-like feces hanging from the anus. Often the stomach region swells but not as severely as "dropsy." Normal feces are round, dark balls that easily drop out of the anus. Diet is the major contributor to this problem and, with regard to clownfish, a diet consisting of more algae or plant material will help. In severe cases we found administering a vitamin and fish oil mixture directly down the throat and up the anus usually cured this problem. The liquid mixture was administered with a fine tipped wash bottle. Basically it was like giving the fish an enema. This treatment may have to be repeated for complete results. Constipation is very common in newly acquired fish which have not fed since they were captured in the wild. Adjustment to completely new diets is traumatic and takes considerable time for new fish. Utilization of oil enriched live foods may initially help but should not constitute an exclusive diet.

Blindness

Blindness was detected in adults and older juveniles during the late summer and fall months when water temperatures were elevated to the 86°F+ (30°C+) range. Within adult fish, 90% of the effected fish were females. Blindness was also detected (Hoff, 1979) in pompano, *T. carolinus*, subject to low temperatures of 62° to 66°F (17° to 19°C) . In both species some individuals regained partial vision after considerable time. Usually inability to obtain food led to slow starvation and eventual death before eyesight was regained. Herwig (1980) noted that captive marine fish seem more susceptible than freshwater fish. He further stated that older, larger, fish held in captivity for at least six to eight months are more susceptible.

Blazer (1979) conducted studies at IOH and the discussion provided is part of her findings. No gross ocular lesions are apparent. Blindness is characterized by iris expansion causing the eye to look entirely dark regardless of the light level. Normal eyes have a yellowish outer ring with a small dark iris in the center. Blind fish do not respond to food in a visual fashion but rather swim erratically, often in circles, seeking food by smell. They scavenge food from the bottom rather than striking food at the surface or water column. Another test is to put your hand in front of the fish and see if it moves or is scared. The only other external characteristic was pale gills. Internally the kidneys contained a white gritty material (calcium oxalate monohydride crystals). Microscopic examination showed extensive kidney tubule damage, high deposits of ceroid in viscera and muscle degeneration.

What causes the "disease" is not well known but it is believed to be associated with diet, especially fat and vitamin deficiencies, coupled with age, and prolonged exposure to extreme high or low temperatures. Rancid or oxidized fats either within the organs or in the diet are suspected to be part of the problem as indicated by the ceroid deposits found in the liver, spleen, and head kidney. Low vitamin E, which helps prevent oxidation of lipids within the diet, is a possible problem. Woodall et. al. (1964) found that Chinook salmon fed up to 5% herring oil required around 3 mg vitamin E per 100 gm of food. However, if fat sources are rancid and utilized in a mixed diet, the amount of vitamin E is quickly depleted even in a frozen state. The amount and type of fatty acid fed may also play a role. Mammal lipids should be low in the diet and higher fish and invertebrate lipids containing high omega 3 oils (essential fatty acids = EPA) should be used. Farkas et. al. (1978) stated that the first sign of EPA deficiency in rats is an increased lipogenesis (fat production). They showed there was hyperproduction and deposition of certain fatty acids in carp fed < 1% linolenic acid. This increase coupled with a decrease of essential fatty acids from the blood results in fat accumulations in the liver as well as other areas of the body.

Turk (1987) stated that green items which provide vitamins, carrotenes and other nutrients are essential in preventing nutritional blindness. He suggests utilization of fresh household greens (spinach, romaine and endive) or algae (*Caulerpa, Ulva, Enteromorpha*) in the diet. Ford (1982) indicated that use of greens should not exceed 25% of the diet.

Blasiola (1995) noted that deficiencies of vitamin A cause the retina to degenerate, eventually resulting in blindness. If the condition is detected early it can be reversed. The only way to determine conclusively that a vitamin deficiency is the possible problem is to have the food tested. He further stated that fish with no apparent corneal damage may have a transitory condition caused by too much light. This is probably true of nocturnal feeders such as squirrelfish and cardinalfish, as opposed to clownfish.

Occurrence of blindness is not always related to nutrition and/or water quality. Old, dying, female fish also became blind several weeks prior to death. Elevated temperatures seem to enhance blindness in older females.

Problems at IOH were basically related to nutritional imbalances. Maintaining a well rounded diet and controlling temperatures below 29°C (84°F) greatly reduced blindness in adult broodstock.

Disease Prevention

Diseases are prevented, not controlled. Spotte (1979) summarized the disease dilemma very well: "without the proper controls, treating diseases and at the same time trying to maintain stable environmental conditions are antagonistic processes. The culturist who is too quick with their beaker of remedies may aggravate an already serious situation by killing beneficial microbes in the filter bed. When this happens, the decline in water quality accelerates and still more factors combine to plague the animals. With this in mind, any treatment program must be approached thoughtfully and with the realization that failure is imminent in a deteriorating environment. In unskilled hands, the so called cure may be more lethal than the disease."

Disease prevention in broodstock is closely related to routine maintenance and water quality. Yet, a strong word of caution is needed. Extreme maintenance may actually be adverse to producing quality spawns. Agreed, "cleanliness is next to godliness" but within a fish hatchery this applies more to plankton culture and larval rearing and not as strongly in all other areas, especially broodstock. Too much agitation and disruption of adult pairs can impair their health and productivity. I have purposely avoided giving you a mixed bag of magic potions since I feel strongly that manipulation of diet and the environment will, in most cases, cure adult fish.

You have available to you a wide array of fine mechanical, biological, UV, and ozone equipment that enables the aquarist to virtually sterilize water. If applied in a judicious manner there should be no reason for water quality to become a primary broodstock problem.

In commercial situations, culturists should not spend a considerable amount of time trying to cure broodstock diseases unless the entire stock is in jeopardy. Isolation, separation, and destruction of diseased individuals within a system is recommended. In most cases, it is probably best to treat the sick fish in its own tank unless the disease has become very infectious. In extreme cases, if the "disease" cannot be cured in two weeks, the fish should be removed from the system and destroyed. Strip and clean the tank before it is restocked. You simply cannot take the risk of losing any more broodstock due to one sick individual.

Egg Quality Versus Treatments

Use of drugs and chemicals within adult broodstock can be directly related to the quality of the eggs they spawn. Fish, like humans, can absorb, retain and concentrate within the liver and other organs metals, chemicals, and drugs that are not easily excreted. The liver is the primary producer of yolk and lipids that are incorporated into the eggs. Within humans, this can often result in fetal deformities and mental problems. Within fish, it has been proven that adverse metals such as selenium can be passed to the eggs. Although the eggs hatch normally the post larvae died later.

No doubt the over use of drugs, copper, or other non-natural treatments on adult broodstock can have an immediate or delayed adverse effect on the eggs, larvae, post larvae,, or older juveniles. Extreme caution should be used in treating broodstock since the essence of their worth is to consistently produce quality eggs. These pathways are just now being discovered so avoidance is the best measure.

Tracinotus carolinus
Pompano

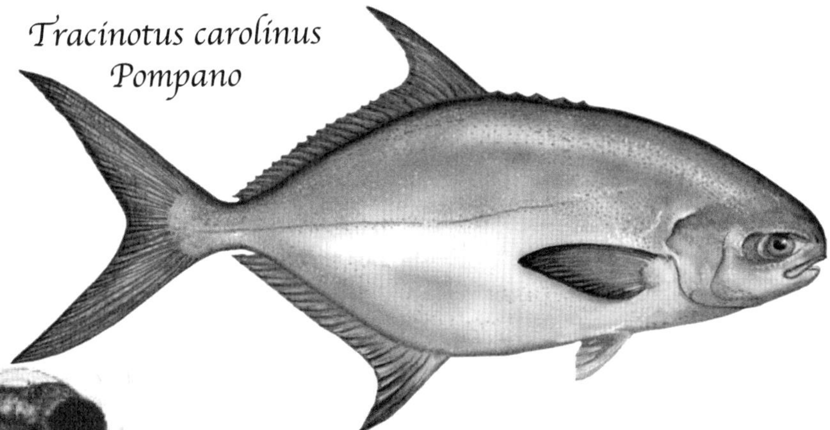

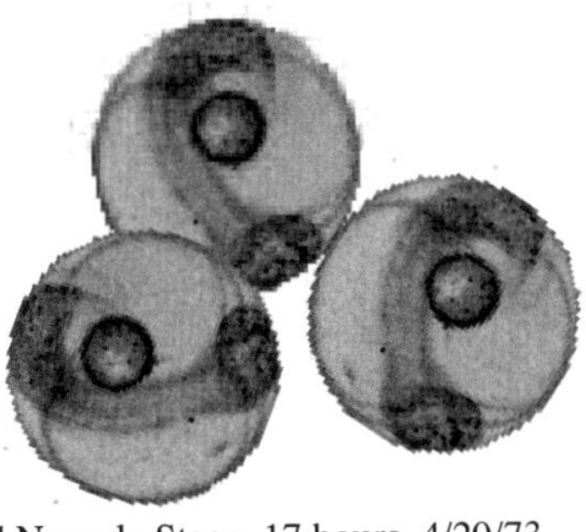

Mid Neurula Stage, 17 hours, 4/20/73

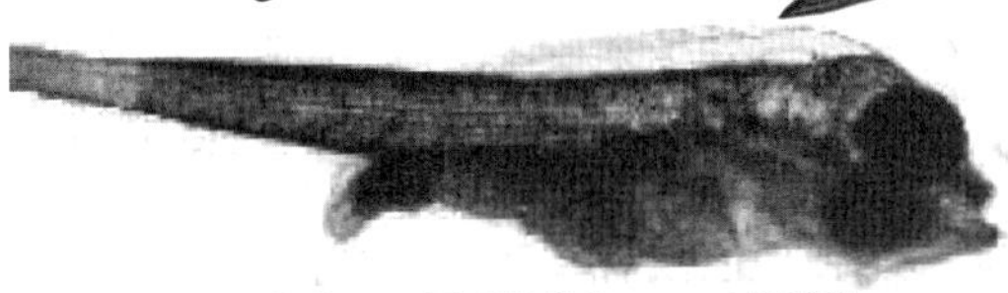

Larvae 5 days old, TL 3.2 mm, 4/26/73

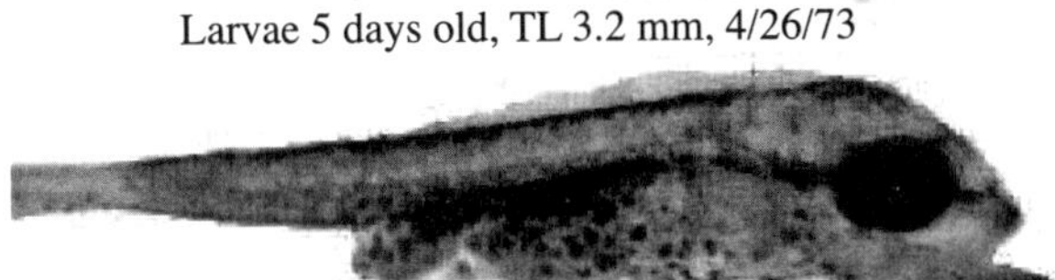

Larvae 9 days old, TL 4.0 mm, 5/15/73

Juvenile 19 days old, TL 10.1 mm, 5/15/73

LARVAL REARING

Introduction

The larval stage is defined from the point of birth to metamorphosis, the point where external characteristics and internal organs appear like adults but only smaller. Larval fish can be divided into three groups according to alimentary tract morphology and the enzymes secreted into the gut (Dabrowski, 1984a). The first group includes salmonids and channel catfish which appear to have a functional stomach before changing from endogenous to external feed. The second group includes striped bass and many marine species, including clownfish, that have rudimentary digestive tracts with no functional stomach or gastric glands and which undergo complex metamorphosis of the digestive system. The third group includes carp which develop functional digestive tracts but remain stomachless through life. Species that are like group one and three, that have structurally and functionally differentiated alimentary tracts, pose less of a problem with initial feeding, whereas those with immature systems are more difficult to feed and usually require live feeds as an initial food.

Prior to metamorphosis a certain point is reached where dry feed can be substituted for live feed. Striped bass, a typical pelagic larvae, metamorphosis in 21 to 42 days (depending on temperature) and start feeding around day 5 yet they cannot accept dry feeds until day 15 (Baragi & Lovell, 1986). Poor ability of larval fish to utilize prepared diets initially may be related to the low affinity of proteolytic enzymes in immature digestive tracts for proteins in prepared diets (Dabrowski, 1984b). Several fish species have been shown to be unable to digest protein from prepared diets (Bremer, 1980). Relatively high consumption rates common to larvae just increase passage rates and results in low digestive efficiency (Kaushik & Dabrowski, 1983). Larvae consume more food per unit body weight (50% to 300%) than subadult fish (2% to 10%) being raised to market size (Bryant & Matty, 1980). The inability of some larval fish to immediately use prepared diets may be due to absence of enzymes, hormones or their regulators, or growth factors that are provided in live feeds (Lauff & Hofer, 1984; Baragi & Lovell, 1986). Other dietary factors in live feed might inhibit or stimulate hormone action in larvae. Thyroid hormones play an important role in metamorphosis and growth (Collie & Stevens, 1985), and are believed to be influenced by dietary components of live feeds. This introduction is in part from Lovell et. al. 1993.

Larval rearing is the most critical, time-consuming phase of fish culture and probably the most complicated to adequately describe. Variables are encountered on a daily basis and the sensitivity towards fish culture becomes a fine line between success and failure. Yet, a careful contentious person can easily rear some clownfish. There are several things you are trying to achieve:

1. Provide a simple, adequate stable environment that can be easily manipulated and maintained.
2. Provide adequate, quality foods, on a consistent basis.
3. Provide strict maintenance procedures on a daily basis.
4.,Develop a good wet thumb, be observant, and keep records

Adequate Environments

Simplistic environments are key to simple procedures and eventual success. Every effort you make should be directed toward how to make each process more efficient, yet still provide the essential basic needs of the larvae. Habitat, gravel bottoms, filters, etc. are not essential to success and are usually detrimental and/or labor intensive interferences. Water quality is a key environmental factor but can be easily controlled with simple water exchanges. Typically the bioload (total weight of fish and food) in larval rearing tanks is insignificant when compared to broodstock or juvenile rearing tanks. Therefore, deterioration of the water environment is easier to control. Everything should be simple and easily arranged so it can be quickly serviced one to three times daily

Foods Versus Larvae

You have larvae and now you want to find live rotifers? Stop right here! It just doesn't work that way. First and foremost you absolutely must be ready to rear larval fish before you have larval fish. You cannot put the "cart in front of the horse" and expect to get anywhere. Success or failure in larval rearing is closely keyed to available quantities and quality of live foods and how they are administered. If you have not mastered the art of culturing microalgae and rotifers then it is time to back up. If you have not dedicated a specific area and time towards to production of a constant supply of microalgae and rotifers, then it is time to back up. Without a ready, plentiful, nutritious, live food source your larval rearing attempts will be futile, erratic, and very discouraging.

To rear clownfish larvae you will need 300 to 600 rotifers per larva per day for a period of 5 to 10 days. In addition, the rotifers must be completely nutritious and balanced with essential fatty acids and microalgae.

During my 40 years experience in rearing marine fish I have consistently seen the same mistake made by doctoral researchers, professional aquaculturists or avid hobbyists. They simply dedicate far less time in learning to culture live food organisms than they do in trying to induce spawning. Why spawn fish and hatch larvae when you can't even provide suitable foods? Just one or two days without suitable foods in sufficient quantities, spells death to larvae. Without adequate quality food how can a researcher justify any results they receive? Too many scientific papers on rearing larvae are lacking, or do not adequately address or support, basic nutrient needs. Failures and poor results in larval rearing are primarily controlled by two factors: poor quality larvae and/or poor quality food.

To save space and avoid redundancy I have purposely left out the live food culture section. This does not in any manner or form diminish the importance of being able to master the ability to produce live foods on command. Since 1993 to the present (Hoff and Snell 2008) have written and updated a complete, detailed, companion manual, "Plankton Culture Manual", published by Florida Aqua Farms Inc. (FAF). This manual clearly outlines what is needed to culture microalgae, rotifers, daphnia, artemia, etc. In addition, FAF has further expanded their product lines into providing "off-the-shelf" starter cultures, special fertilizers, and culture equipment to make your job easier. For the hobbyist trying to rear one nest at a time, two 5 gallon tanks for the rotifers, 8 or 9 two liter clear plastic cola bottles for the microalgae, a light, air pump, fertilizer and starter cultures is all that is needed.

In order to provide a balanced diet, and to develop adequate transition from live rotifers to live *Artemia*, a series of dry micro particle foods must be used throughout the rearing process. Rotifers, *Artemia* cysts, microalgae, and micro foods are all available from Florida Aqua Farms Inc.

Maintenance Procedures

Larval rearing is not simply putting baby fish in the tank, feeding them food and watching them grow. They are delicate babies, needing attention and a clean environment on a daily basis It will be necessary to clean the tank daily of detritus and uneaten food. Water exchanges and airflows must be watched. The main thing is to be consistent, and avoid losing fish due to poor water quality or improper handling.

Record Keeping

Again I cannot over emphasize the importance of record keeping. The information that forms the basis of this book is coming from cases of records kept at Instant Ocean Hatcheries over a 10-year period. If you want to learn from past mistakes, then learn to keep records. As outlined in the broodstock section of this book, several types of records should be taken in the larval rearing processes including historical data, water quality, and observations. Results and trends are obtained by combining your recorded data into graphs and tables. Eventually you begin to see where most of the mistakes are being made, where most of the larvae die, when they metamorphose into true juvenile fish, how they feed, etc. Trying to retain observations and records in your head is just not good enough. Our mind is simply not designed to be a file drawer full of little details.

Oocyte (egg) Development Stages and Information

Live Samples

Prepared Samples

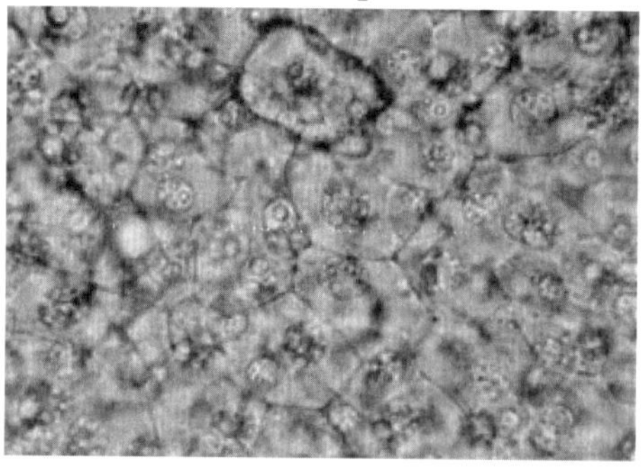

Live Stage II Oocyte sizes

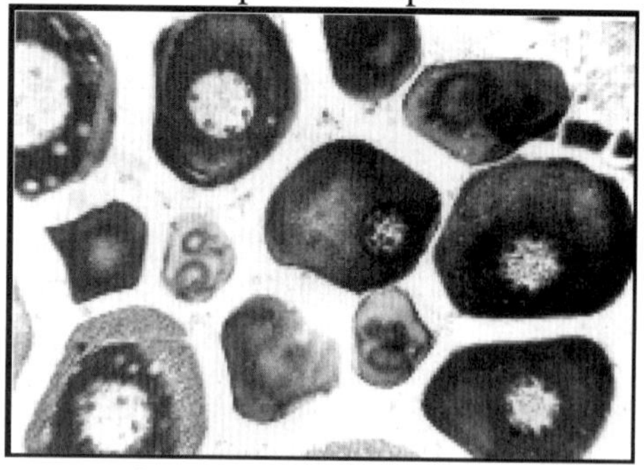

Prepared Stage II Oocyte sizes range 30 to 120

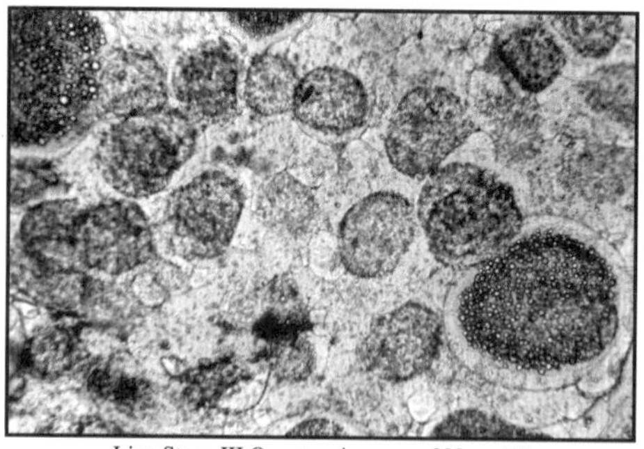

Live Stage III Oocytes size range 200μ - 400μ

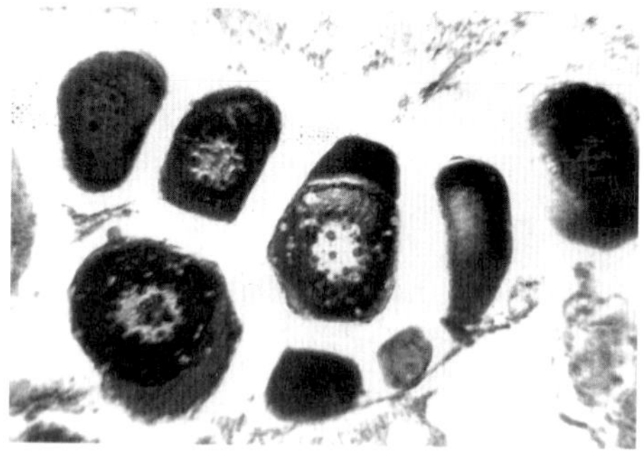

Prepared Early Stage III
Prepared Average 90μ = Live average 200μ

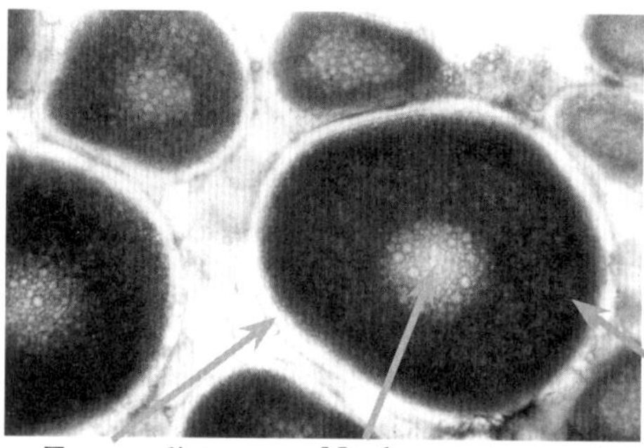

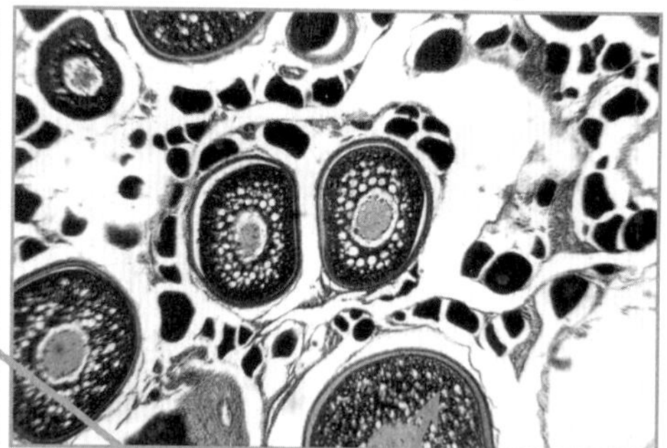

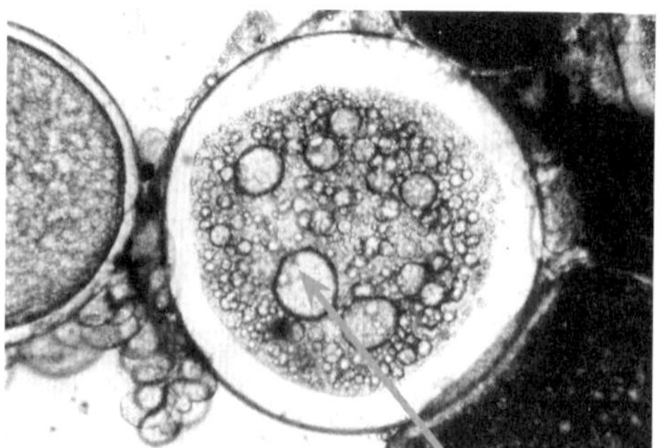

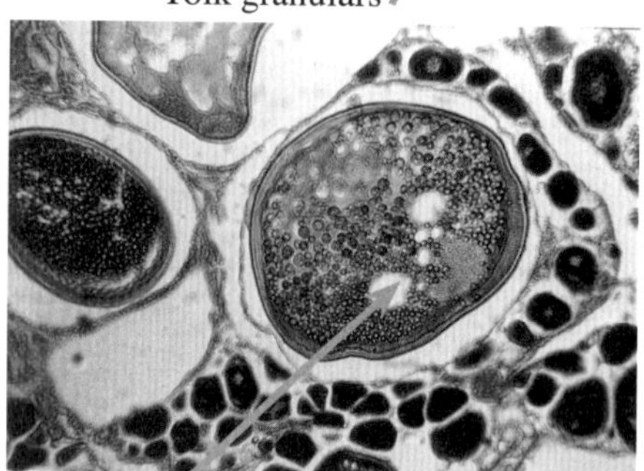

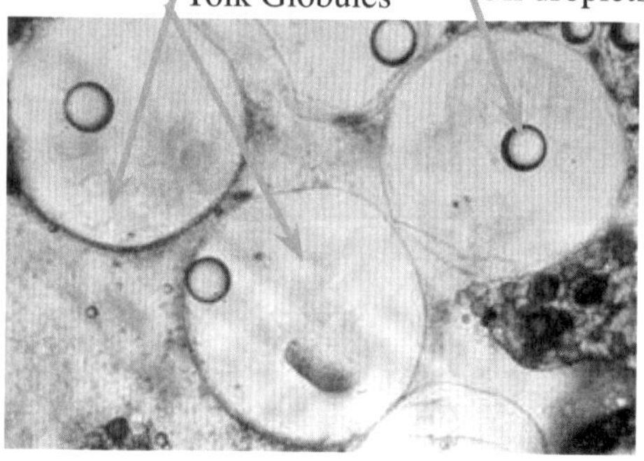

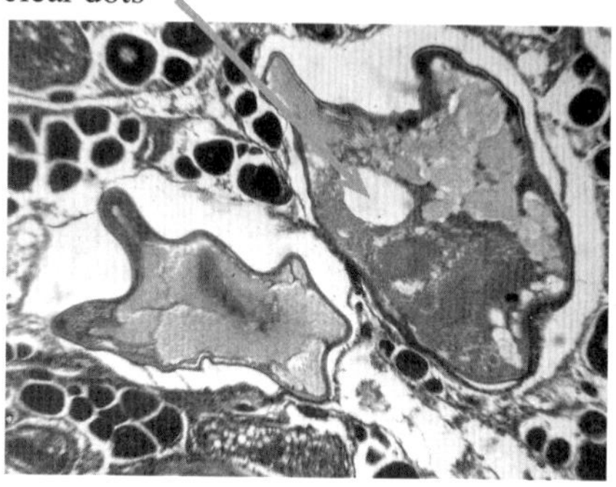

Stage II Oocytes

V shaped, often elongated, attenuated and sharply angular (prepared samples packed, predominant in resting gonads. Nucleus clearly defined; several nucleoli around periphery.

Stage III Oocytes

Stage III oocytes range in size from 120μ to 400μ for live eggs and 80μ to 250μ for prepared eggs.

Early Stage III Oocytes - Generally round or oval, tightly packed, predominantly found in early developing gonads. Nucleus diameter increasing; several nucleoli around periphery (not easily seen in live oocytes). Prepared cytoplasm strongly basophilic: small oil droplets (clear vacuoles) forming around nucleus followed by small, slightly acidophilic yolk globules forming broad band around nucleus. Oocyte grayish when alive.

Mid Stage III Oocytes

Round to oval. Nucleus, prepared- slightly acidophilic; membrane obscure in live material; several nucleoli around periphery. Nucleus surrounded by band of cytoplasm followed by band of oil globules (prepared show clear vacuoles) followed by yolk globules (prepared are acidophilic) around periphery. Zona radiata, (prepared slightly acidophilic) and is less than 10μ thick in live materials. Live oocyte is yellowish with distinct dark band around the nucleus

Stage V Oocytes

Stage V oocytes range in size from 670μ to 1150μ for live eggs and 340μ to 670μ for prepared eggs.

Early Stage V

Spherical. Live - nuclear area not clearly defined, prepared - lightly acidophilic. Prepared - cytoplasm band lightly basophilic, thicker along periphery. Yolk globules coalescing; oil droplets dispersed throughout cytoplasm and starting to also coalece. Zona radiata noticeably thinner. Live oocyte opaque, clearing along periphery, grayish in color.

Stage VI Oocyte

Stage VI

Stage VI oocytes range in size from 700μ to 1500μ for live eggs and 330μ to 730μ for prepared eggs.

Early Stage VI

Oblong or spherical. Nucleus near periphery (prepared - becoming basophilic). Globular yolk vesicles (prepared lightly acidophilic) throughout cytoplasm; no granu

LARVAL REARING TANK

Larvae cannot be successfully reared in adult tanks unless some form of isolation is utilized. A separate tank or some sort of contained facility for retaining live foods and larva is necessary. I have reared marine fish larva in small containers ranging from finger bowls and one gallon glass bottles to 536 gallon (2010 L) aluminum cylinders 6' deep and 4' wide (187 cm x 117 cm). Bertschy (1979a) successfully reared his first clownfish larvae in a simple small (1 liter) hanging rectangular clear plastic box equipped with a fine, nylon-stocking mesh, screen bottom and an

Table 25 - Larval rearing aquaria used by various investigators.

1956 - Morris	17 species	10 gal.	semi-closed
1960 - Kashahara et al	black porgy	2 & 20 gal.	??
1962 - Blaxter	herring	12 & 500 gal.	open
1964 -Delmonte	goby	85 to 150 gal.	??
1966 - Joseph and Saksena	killifish	1 gal.	??
1968 - Fahey	herring, killifish	1 gal.	open
1968 - Bardach	mackeral	500 gal	??
1968 - Blaxter	herring	50 & 350 gal.	open & semi-closed
1969a - Blaxter	pilchard	6 gal	semi-closed
1969 - Richards & Palko	herring	10 gal.	semi-closed
1969 - Schumann	30 species	500 gal	closed
1970 - Houde & Palko	herring	20 gal.	semi-closed
1970 - Kramer & Zweifer	anchovy	95 gal.	closed & semi-closed
1970 - Lasker et al.	anchovy	2.5 & 125 gal.	??
1977 - Bryan & Madraisau	rabbitfish	2000 gal.	closed
1972 - Hoff, et al	pompano, seabass	536 gal.	closed
1974 - Houde	anchovy, bream, sole	20 gal.	semi-closed
1975 - Valenti	puffer	43 gal.	semi-closed
1981 - Hoff	clownfish	40 gal.	closed
1981 - Moe	clownfish	60 to 250 gal.	semi-closed
1981 - Turk	clownfish	60 gal.	semi-closed
1984 - Young	clownfish	50 to 75 gal	semi-closed

air lift pump (Figure 52). This basket was suspended in the adult aquarium. Still others (Siddall 1979) used elaborate 100 gallon (375 L) round cylinders equipped with flat or cone bottoms and black sides, while others have used a one gallon rectangular clear aquaria. Table 25 is modified from an article by Siddal (1979) in FAMA magazine which lists numerous methods and tanks used to rear larvae.

Many say that rectangular or square tanks cannot be used to rear larvae and that a large round flat or cone bottom tank is best. Yet, I contend that when dealing with benthic spawning fish and larvae, tank shape is far less important than water quality, and to a lesser degree the species you are trying to rear. Shape is more important with regard to desired circulation patterns than anything else. Healthy, good quality, benthic larvae do not normally get trapped in corners or run into walls but, poor quality larvae do. Size and shape of the larval rearing tank is an ongoing debate. The final outcome may indeed be a reflection of the species you are trying to rear, your ability to rear larvae, or simply the tank you prefer to work with.

Selecting Larval Rearing Tanks

Criteria may indeed include species selection and spawning strategy. For example, pelagic spawning fish release eggs directly into the water column. Normally, pelagic eggs hatch in 24 to 36 hours. Initially these larvae are poorly developed and drift in the open ocean where no hard surface contact is possible. Whereas, benthic spawning fish attach or lay eggs on a hard surface and more advanced stage larvae hatch in five to seven days. Benthic larva generally remain in or live around solid bottom habitat and are probably subject to some surface contact.

Initially pelagic larvae cannot see, have limited swimming ability, and cannot easily swim backwards. Therefore, they may indeed become trapped in corners. Round tanks in this case would tend to minimize trapping. Benthic larvae normally hatch with well developed functional eyes and fair swimming ability. In this case a rectangular or square tank will suffice.

Depth can be important. Generally all larvae feed on food that is directly in front of them or within the water column. Benthic larva may feed off vertical surface areas but not normally off the bottom. Dry or nonliving foods utilized in shallow tanks settle quickly without sufficient time for larvae to orientate and strike food particles within the water column. Therefore, deeper tanks are usually preferred.

One extremely important criteria for selecting tank size is knowing the quantities of live food (rotifers and *Artemia*) you can provide on a daily basis and how many larvae you intend to rear in each tank. Larvae do not normally actively seek food but tend to be opportunistic feeders, patiently waiting for a food particle to come within striking distance. Placing 300 hundred larvae in a 100 gallon (375 L) tank taxes your ability to adequately provide proper densities of dry foods, rotifers, and *Artemia* nauplii to each larvae without polluting the water. Opportunistic contact between food and larva diminishes drastically when few are reared in a large tank. Forcing larva to swim considerable distances to seek food tends to drain their potential power supply quicker than it is replenished and results in eventual death or slow growth. Concentrating early stage larvae and food supplies minimizes production costs to provide sufficient live food organisms per unit area. By assuring more adequate food particles per unit area, energy expenditure of larvae to find food is minimized. This ensures faster growth rate and higher survival.

Water quality is a direct linear relationship with tank size and in some cases shape. For example a tank with a restricted surface area such as a bottle, will present problems in gas exchange. A small volume tank will build up dissolved, detrimental, metabolites quicker than a larger volume tank. The simple solution is water exchanges. From years of experience I have found that early stage larvae do not do well when continuous water exchanges are used. One to three day old larvae may not tolerate any exchange.

Density can vary with the species you are rearing. After the research and development stages at IOH we were achieving an average of 55% survival over a 25 to 30 day larval rearing period. Average post larval production was 288 fish per 30 gallons (113 L) of water (9.6/gal. or 2.56/L). Highest post larval production was 2000 post larvae in 30 gallons (66.7/gal. or 17.79/L). Average stocking density was around 420 larvae per 30 gallons or a density of 14 larvae per gallon (3.73/L). Our system consisted of 80 larval rearing tanks connected to a 10,000 gallon (37.5 M^3) closed system. Average production was around 24,000 post larvae per month.

Tank Coloration

Tank coloration can play an important role in the survival, coloration, and attitude of all stages of fish culture. In 1975 we conducted two sets of experiments to determine what color would be best for rearing clownfish. Prior to this work most people reared marine larvae in all black, cone shaped tanks or clear, all glass aquaria Eight, 15 gallon (56 L) tanks were set up and painted different colors. One tank was lined on the outside with plain brown cardboard. Each tank received 23 larvae. Feeding, water exchanges, and cleaning were the same for each tank. Temperature ranged between 77° and 81°F (25.5° and 27°C). The experiment was repeated once. Table 26 presents the results of this series of experiments with *A. ocellaris* (ocellaris).

It is interesting that yellow, green, and black painted and cardboard lined tanks are all closer to natural environmental colors and showed better survival than the darker red, and brown colored tanks. Yet, the latter two tanks showed good growth. Usually light colored tanks are considered detrimental in larval rearing since live-foods are less visible against pale colored backgrounds. In white colored tanks, larvae actually died early in the experiment and probably did not initiate feeding.

Table 26 - Tank coloration versus survival and size

Tank	Color	Survival	Growth
1	Painted white	0%	---
2	Painted brown	12%	medium to large
3	Painted red	16%	large
4	Painted yellow	18%	small
5	Painted dark blue	18%	medium to large
6	Painted green	20%	medium
7	Cardboard lined	27%	medium to large
8	Painted black	31%	very large

Yet, the light colored tan cardboard had the second highest survival. The extremes of a light tan cardboard lined and black painted tanks had the highest survival. Commercial larval rearing tanks are normally black or turquoise blue. Literature also supports evidence that matt finishes might be better then gloss since they reduce reflection.

All glass aquaria without background appears as a mirror to fish. Place a glass or diving mask into a tank and look at the sides. Larvae tend to be drawn towards the mirror surface and just "stick to the glass". This reduces feeding efficiency and is detrimental. Likewise, they are often attracted to the bottom since it is transparent and reflects whatever color surface the tank is placed on. Ideally you want to reduce the mirror attraction and force the larvae to orientate in the water column.

At Instant Ocean Hatcheries we settled into the use of all glass aquaria with black plastic attached to the outsides. The front piece was a flap that could be lifted to observe the larva horizontally. The bottom was painted a flat white or placed on a white styrofoam sheet.

Lighting

Lighting is closely in line with tank coloration and construction. Larvae are basically phototrophic (attracted towards light). Some aquarists actually design lighting arrangements to draw larvae away from the sides of the tank. Placing an opaque cover over the tank with a central hole and directed light source can help draw larvae away from sides (Ehrlich 1989).

Goldstein (1990) speculates that the unusual sensitivity of clownfish eggs and fry to light might, under natural conditions, be lethal. He states that direct exposure of the eggs to sunlight effects the viability of the fry but not the hatch. He further states that eggs should be artificially incubated in dimly lit rooms with no more than a normal daylight photoperiod. Following hatch he states that all artificial light should be eliminated prior to, and subsequent to hatching for a least 72 hours. Bear in mind that IOH was in a greenhouse with intense natural light and we could not find evidence to support his claims.

However, high light intensities or sudden lighting changes to stress fish. The light intensity I prefer should be sufficient to see the live food and larvae but not overly bright. Light spectrum has not been closely looked at but with the advent of bulbs that are more closely attuned to natural sea life conditions, I would select a double fixture consisting of a full spectrum white light like Bio-Lume or Vita-light and a near UV blue range light like Prime Tinic (Rainbow) or similar bulbs.

Utilization of a timer and "moon light" is recommended. Lighting can be arranged so that half goes off at a time. The "moon light" is constant and is located along the floor or some other obscure area to provide a suitable glow that will allow you easily walk through the area but low enough so you cannot easily observe fish. Usually a 15 to 20 watt bulb is all that is needed.

Light duration is another variable within the literature. There is supportive evidence that fish need darkness. Many feel that providing 24 hour lighting makes the larvae feed constantly, makes them grow faster and results in higher survival. Realistically, it does not make sense to subject larvae to 24 hour lighting. Physiologically, I doubt if this is doing the larvae any good. Certainly the acts of feeding and fasting must have an effect on proper development. I don't think we can change a biorhythm that has taken millions of years to evolve and expect spectacular results. Granted, 24 hour bright lighting does not occur in nature, so we really don't know what would happen if this was suddenly the case. Qasim (1955) working with *Blennius pholis* under continuous light and 8 light/16 dark day cycles, clearly showed that those fed *Artemia* under continuous light survived to 48 days while those with alternate lighting lived to 70+ days. He noted that after a period of darkness and inactivity, larvae exposed to light became particularly active compared to those under continuous light which often become lethargic.

I prefer a 14 to 16 hour light period. Under commercial situations lighting should be arranged around the common work day. At IOH, the larval area was housed under a double layered fiberglass greenhouse roof. In addition, we lined the ceiling with 50% shade cloth. Lights were set to come on about 7:00 AM and go off at 7:00 PM. Since natural light influenced the intensity in this building, we did not use a timed sequential lighting system except with regard to the "moon light." Ideally you want the lighting arrangement to allow you to equally space 3 to 5 feedings and provide about 2 hours light after the last feeding. In the winter months, especially mid November through mid March the light period was extended to 16 hours since natural intensities were low.

Lights should be mounted well above the tanks so they do not interfere with normal day to day maintenance. At IOH, lighting consisted of two bulb fluorescent fixtures mounted on the ceiling about 3' above and along the entire length of the tanks. All the lighting was tied into a single timer.

Commercial Tank Configuration

At IOH we used a 1/4" plate glass cube 24"L x 24"W x 22"H with maximum capacity 35 gallons (131 L). Normal operating depth was 17" (43 cm). Sides of the tank were wrapped with black plastic and the bottom was painted flat white. A 1 1/4" (3 cm) diameter hole was drilled in the center of the tank. A jointed central standpipe (Figures 46 & 49) allowed us to make 30% water exchanges daily by simply removing the upper portion of standpipe. The standpipe screwed onto a male adapter mounted into a tapped PVC plate that was glued to the outside bottom. This design allowed for complete drainage of the tank. Underneath the plate we mounted a 1/2" (1.25 cm) elbow tubing adapter and flexible drain tube. On the end of the drain tube a 1/2" straight tubing adapter was mounted for connecting air lift hatchers when needed. Normally this drain tube was stuck into a 3" (7.5 cm) diameter central drain manifold which flowed into the primary particle filter (Figure 48). The stand pipe was adapted with a large foam sponge outflow which provided a higher surface area and a lower flow resistance per unit area than screened standpipes. This large surface overflow sponge prevented larvae from being sucked onto the drain surface when water exchanges were made. Cloth covered drain arrangements were tried but easily clogged and required more maintenance when compared to foam. Nothing else was provided except an airstone in upper left hand corner. No filters or habitat were used.

Each tank had water and air manifold systems mounted behind and above the tank. The air manifold was made of 1" (2.5 cm), SCH 40, PVC pipe. The water manifold was made of 1 1/2" (3.75 cm), SCH 40, PVC pipe. Plastic air and water valves were threaded into the pipe above each tank. Heavy walled PVC pipe can be easily drilled and tapped to accommodate threaded valves up to 1" THD (threaded). Two types of valves were used. Water manifolds to each tank consisted

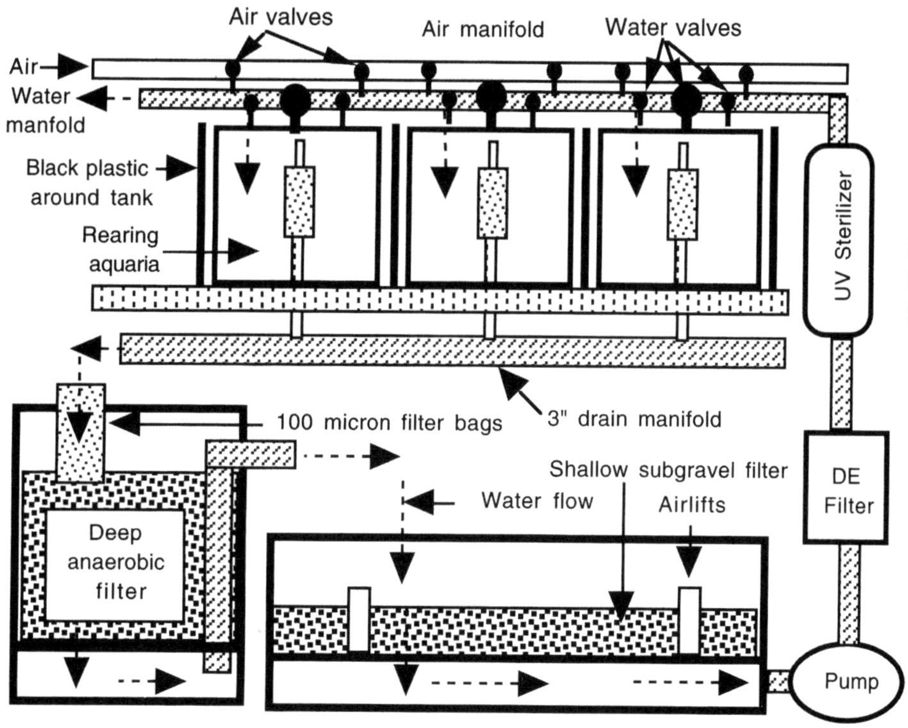

Figure 48 - Larval rearing recirculating system used to culture clownfish at IOH

Figure 49 - Details of an all glass larval rearing tank used at Instant Ocean Hatcheries. Note for details on the standpipe see Figure 45 page s 97 and 98.

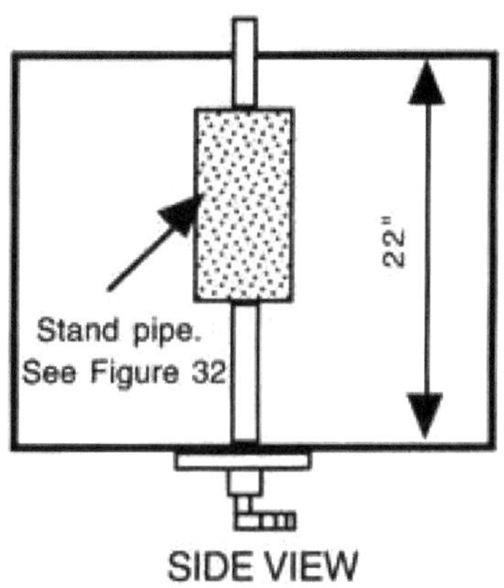

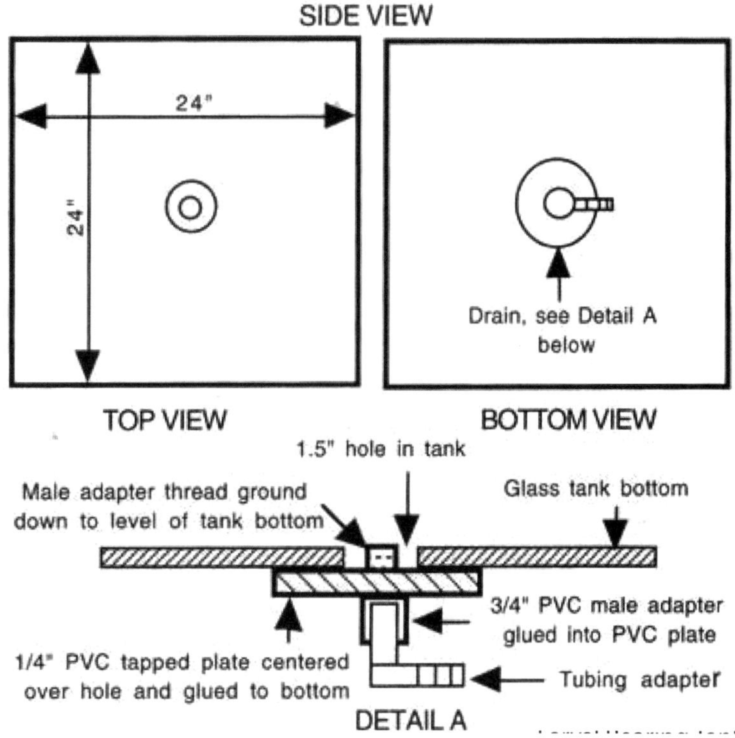

of two small nylon airline valves with a 1/8" NPT (National Pipe Thread) mounted on either end of the tank. Between the small valves we mounted a 1/2" NPT, plastic, angled spigot valve equipped with a tubing adapter that accommodated a 1/2" ID delivery tube. The smaller valves were for continuous water exchanges while the larger valves were used to quickly fill the tank or when making the daily 30% water exchanges. The air manifold was mounted above the water manifold and had two 1/8" NPT airline valves mounted above each back corner of the tank and one in the center. One was used for an airstone to each tank and the others were auxiliary outlets for driving the airlift pump hatchers or additional airstones for providing circulation to new nests. In addition, we provided ground fault electrical outlets to each tank which were mounted about 2' above the tank. These were used for aquarium heaters when needed.

Photo 20 - Larval rearing tanks at IOH

A total of 80 tanks were used in the larval rearing systems (Figure 49 & Photo 20). The system consisted of 4 rows of tanks located above four 1400 gallon subgravel biological filters. Three of these filters (24' x 4' x 2' deep) were the standard subgravel biological filter plate with a central row of 1 1/4" (3 cm) diameter airlifts along the entire length with 4" deep carbonate gravel. The fourth or primary filter was significantly different in that the gravel was 12" deep. The primary filter received all the effluent from the larval rearing tanks. Effluent first went into a prefilter basket lined with polyester cloth for removing all larger particles. Water then flowed into the deep bed filters. The concept of the deep filters was to actually create an anaerobic filter in an attempt to limit transfer of protozoans, ciliates, or other plankton blooms and reduce nitrate buildup. From the primary filter the effluent flowed by gravity into the three standard, subgravel, down-flow, aerobic biological filters. All three of these filters were connected together with self-leveling interconnecting valves. Water going to the larval rearing tanks was drawn from underneath the filter plates, passed through a large commercial DE (diatomaceous earth) filter, then through a 120 watt UV sterilizer. Carbon was used intermittently when needed. A specially adapted 5 gallon bucket (Figure 29) was filled with about 7 quarts of carbon. A small drain valve on the bottom allowed for complete drainage of the bucket when not in use.Water exchanges were limited to about 10% per month or less. This system was not connected to the adult breeding systems. Total capacity of this system was about 9,000 gallons (34 m^3).

Other Commercial Tank Considerations

A common practice in Southeast Asia is the use of fine mesh net troughs framed and supported by poles inserted into the bottom of a pond (Figure 50). These nursery containers are used to help concentrate live food organisms, control larval rearing and maintain water quality. Usually they do not use any aeration simply because it is not available yet, it would be advantageous. Courser netting can be inserted as the fish grow larger or when the mesh clogs with detritus.

Another option when using large rectangular tanks is to make a fine mesh rectangular gate that is lodged in place with foam strips mounted on each edge. Again you have the advantage of the large tank which helps provide more uniform water quality and the larva are concentrated for optimal feeding conditions. We have used this method on common coffin vault tanks used in the freshwater tropical fish industry. As the fish get larger the gate is simply moved further back and eventually removed.

The shellfish industry uses a larval rearing combination tank with common five gallon buckets utilized for food storage. These are modified by removing the bottom and mounting a specific

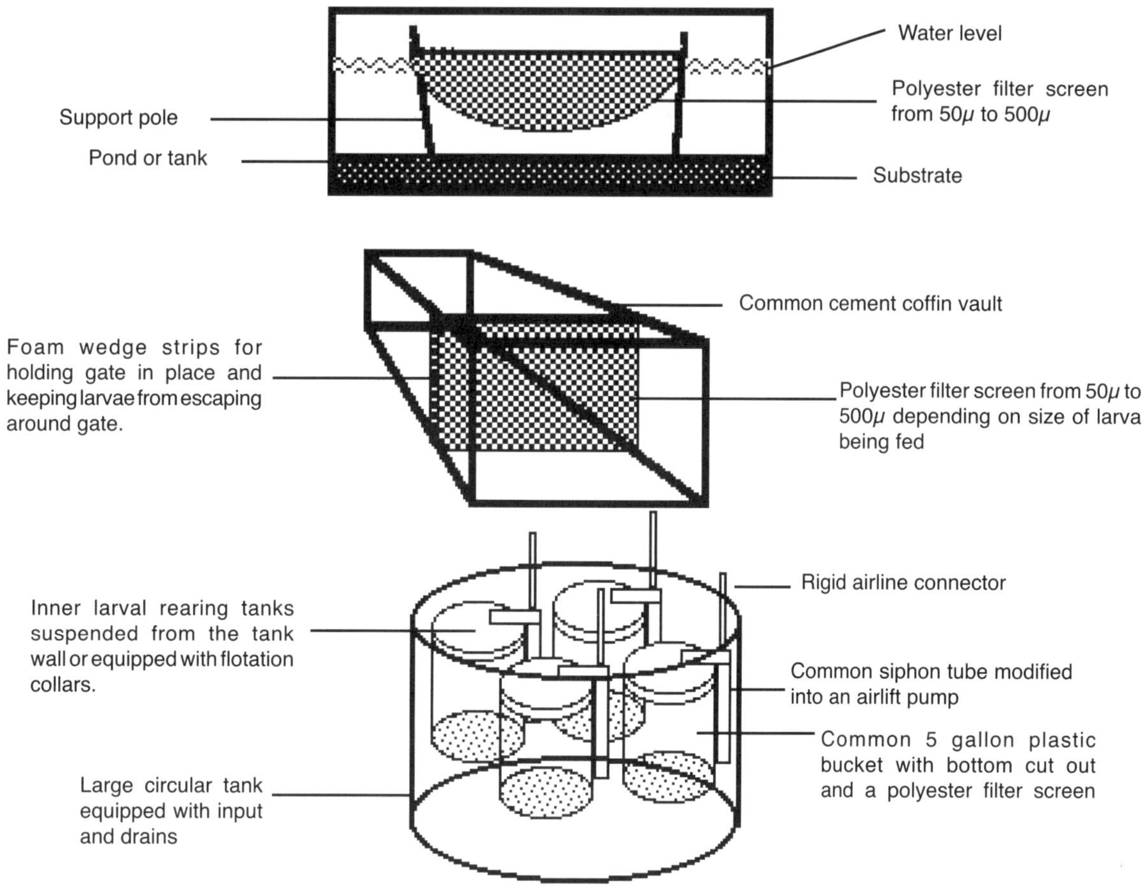

Figure 50 - Optional larval rearing containers used in commercial facilities.

micron filter screen on the bottom. Commonly they use both "up-wellers" and "down-wellers" (flow generated from either the bottom or from the top). For fish, "down-wellers" are more suitable. Common aquarium siphons are modified with a rigid air tube and hung over the side to create an airlift pump. Water is removed from the large tank, passes through the bucket and exits out the bottom. Food is retained in the bucket. As the fish grow and feeding increases buckets are exchanged with others having larger mesh bottoms.

Hobby Level Tank Configuration

Although the commercial tank configuration is elaborate and not suitable for the hobbyist, you must keep in mind that valves, drains, etc. are utilized for ease of operation for day to day production. Any modifications or adaptations aquarists make to help in their particular situation should be considered.

For a small scale hobby level production I feel a standard 10 or 15 gallon (37 to 56 L) rectangular tank about 10" (25 cm) high will do but a deeper model might be better. I prefer a tank with a hole drilled in the bottom and equipped with a stand pipe but, it is not necessary. Again, the tank should be wrapped with black plastic or normal cardboard. The outside bottom can be painted flat white or light tan paint or just mount the tank on a piece of white styrofoam. Lighting should be timed, not intense, and located well above the tank. Yet, lighting should be sufficient to easily see the larvae and the food particles. One airstone should be provided in the corner of the tank. No other habitat or filtration is needed.

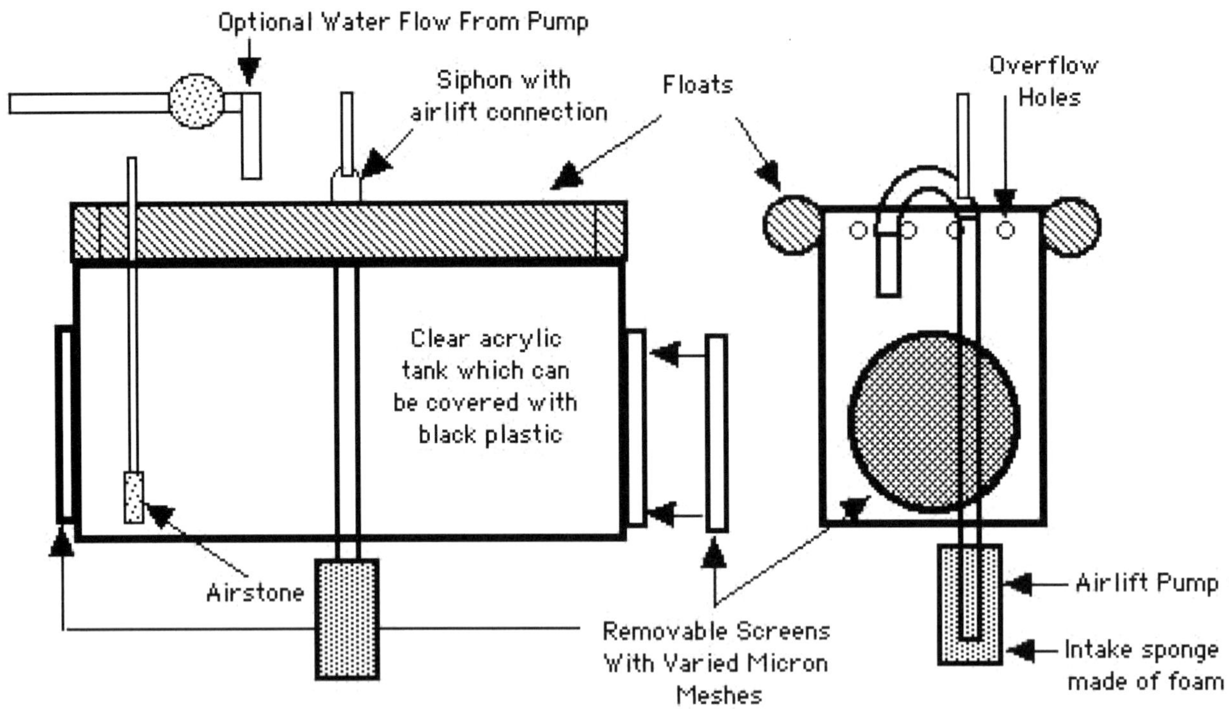

Figure 51 - Floating larval rearing tank designed by Florida Aqua Farms Inc.

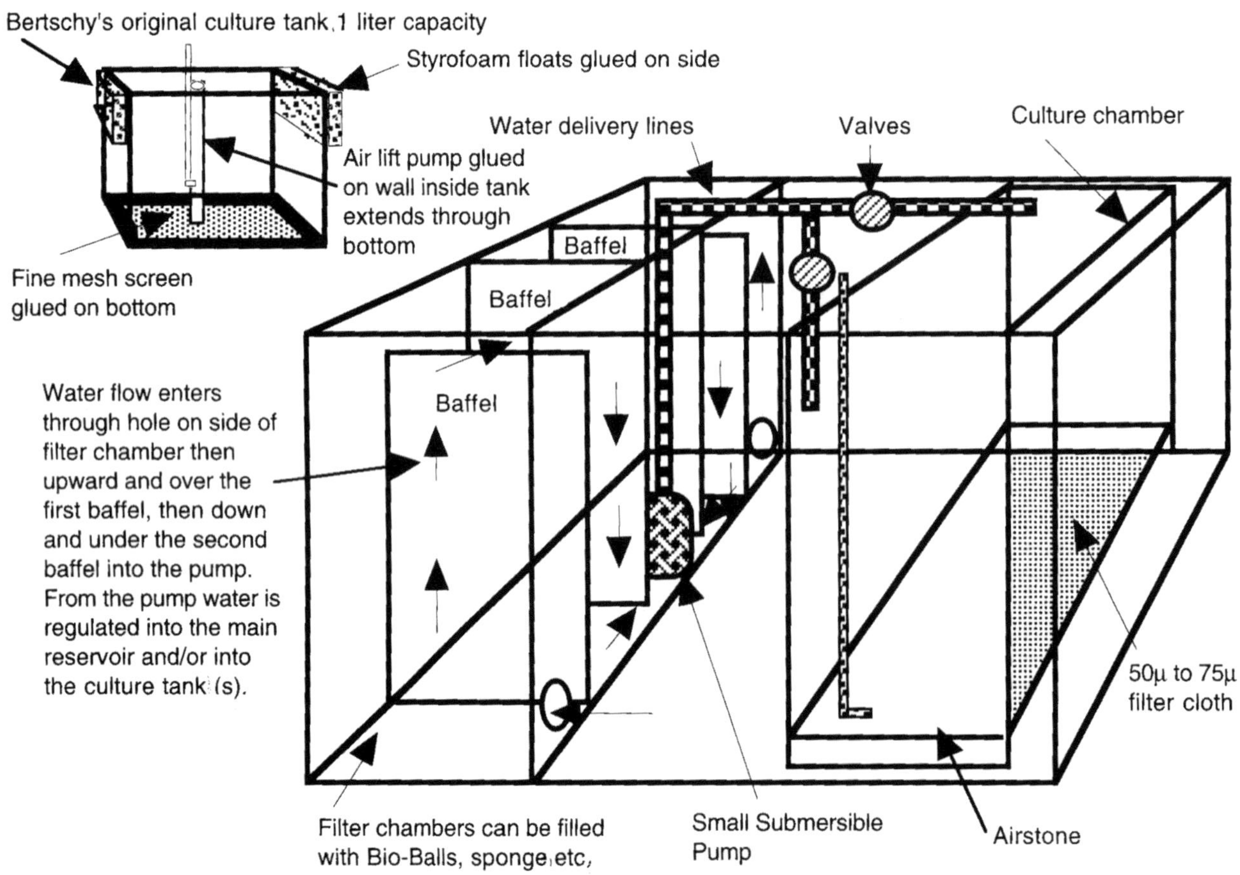

Figure 52 - Small self-contained larval rearing system developed by Allen Bertschy, 1979.

A second alternative is a floating tank or side mounted tank that fits into the breeding tank. Figure 52 provides a successful model that worked for a hobbyist (Allen Bertschy) in France. Florida Aqua Farms offers a similarly designed tank (Figure 51) which come in 3 and 6 gallon sizes with flotation, airlift pump, and removable multi mesh screens.

Allen further expanded his concept into a larval rearing unit or system which is shown in Figure 52. In this design the larvae are contained in a 6.7 gallon (25 L) compartment within a modified 53 gallon (200 L) aquarium specifically setup with contained filtration. Since the larvae are retained within the compartment they are not sucked into or come into contact with the filtration apparatus. In this design you could use an airlift pump or a small powerhead to drive the return flow. The tee mounted on the input pipe is to regulate the amount of water going into the larval rearing compartment. He simply twisted the tee at the angle which provided the desired flow into the compartment. This input line could be modified with valves. The chambers he shows were for filtration.

This concept could be further simplified by using an airlift or powered sponge filter (Figure 51). Since live food is contained within the larval rearing chamber, other mechanical filters and/or UV could be easily installed without killing the rotifers or *Artemia*. The mesh on the bottom of the tank should be around 50 to 100 microns. More than one rearing compartment could be used in this system. Also having spare compartments is advised since the screens often clog and, if necessary, larvae could be poured or scooped into another clean compartment.

Again, the sides or rearing compartment would be best painted a dull matt black or cardboard tan color. The outside bottom of the rearing system could be painted white. The preconstructed tanks from Florida Aqua Farms could also be utilized in this arrangement.

Larval rearing tanks do not have to be made of glass or acrylic although I prefer the ease of cleaning and the ability to view fish horizontally. Rearing tanks can be made of dark colored plastic buckets, waste containers, or similar containers. The advantage of using these is that they are readily available and easily outfitted with drains. When using any plastic containers make sure they are nontoxic and thoroughly cleaned and soaked before using. If you can smell plastic it indicates a gas is being released. Ideally these items should be cleaned and soaked until there is no odor. In essence many containers and sizes can be used to rear larvae as long as you can provide quality larvae, suitable amounts of food, good circulation, water quality, and the ability to service the unit. However, the number of larvae you rear may be limited per unit volume.

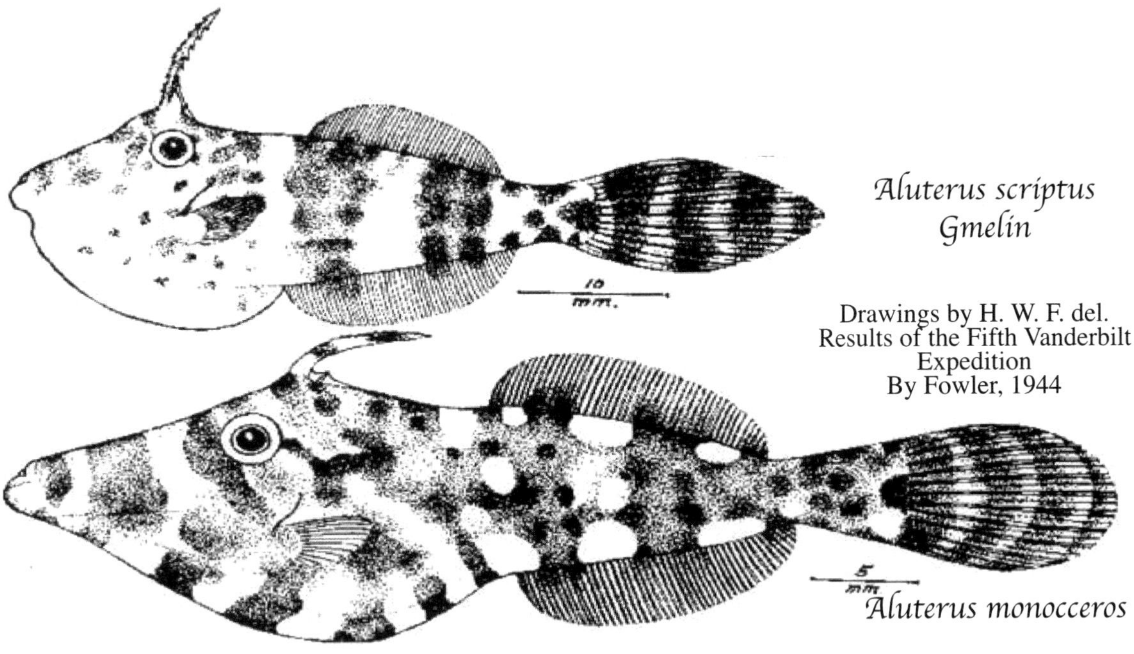

Aluterus scriptus Gmelin

Drawings by H. W. F. del.
Results of the Fifth Vanderbilt Expedition
By Fowler, 1944

Aluterus monocceros

LARVAL FOODS AND FEEDING

Larvae Prior to Metamorphosis

Under hatching procedures in the broodstock section, I outlined several ways to obtain new hatched larvae. Now that we have larvae and sufficient live and dry foods we are ready to initiate the rearing process.

Recent research (Brown & Colgan 1984, 1985; Browman & O'Brien, 1992a, b; Coughlin et al., 1992) on the foraging of zooplanktivorous (zooplankton predators) fish larva is aimed at assaying search behavior, and analyzing both predator movement, and prey detection. Three basic foraging behaviors are utilized by predators to encounter prey: cruise, pause-travel, and ambush (O'Brien et al., 1990). Examples include: planktonic fish larvae (pause-travel), anglerfish (ambush) as well as tuna and later stage clownfish.

Newly hatched clownfish range from 3.8 mm to 4.2 mm long TL depending on the species, egg quality, and time from spawn to hatch. At hatch, clownfish, like most benthic larva, have fully developed eyes equipped with the necessary reflective eye pigments for focusing and dentition suitable for capturing prey. Their fleshy fins are developed and ready for swimming and their gut is ready for digestion. They have little pigmentation and are basically clear in color except for pigmented regions around the eyes and gut. Clownfish larva have very little energy reserves (yolk and oil deposits). If not fed properly, mortalities will be significant within three days whereas many pelagic larva survive as long as six to seven days without food.

Coughlin et. al. (1992) reported on the swimming and search behavior in *A. perideraion* (pink skunk) larvae. They stated that clownfish, similar to other foraging animals, display at least two search modes after the onset of feeding: a highly linear ranging mode used to locate patches of food and a highly complex, convoluted searching mode used to exploit those patches once they are located. Unlike other foraging animals, clownfish larvae increase swimming speed when they encounter patches of high food abundance. This response is proposed to be the result of hydrodynamic constraints imposed upon the fish because of their relatively small size.

Coughlin (1993) studied prey location of the clownfish *A. perideraion* (pink skunk) feeding on rotifers (*Brachionus plicatilis*). Compared to other zooplanktivorous fish larvae, clownfish have excellent visual acuity at the first feeding. As opposed to many fish, clownfish have binocular vision which permits overlap of visual fields, thus allowing them to locate prey from many positions including those located directly ahead. Prey are not located exclusively at the boundary of their search space but, throughout a three dimensional search space of 0° to 120° away from the horizontal axis and 0° to 150° from the transverse plane. During this study larvae had a total body length of 4.20 ± 0.18 mm (n = 7). Most prey (25 out of 39) were not located until the larvae were <0.5 body lengths from the prey, although some prey were as far away as 1.3 body lengths. Prey density apparently did not affect location distances. Mean values of prey location were: at high density, 3.48 ± 1.04 mm (n = 26), and low food density, 3.13 ± 1.27 mm (n = 13). These were not significantly different (t-test, P > .05).

Healthy larvae are a clear yellowish color with dark pigmentation around the stomach and eye region. Stressed larvae stay dark along the entire body regardless of feeding. However, dark pigmentation of the entire body is common in the morning prior to the initial feeding. If the larvae are healthy this condition should disappear within an hour after feeding rotifers. Body shape of well fed larvae is oval or round depending on how it is observed. Underfed or starving larva are cone-shaped with a large head and tapering body down to the tail.

Clownfish mortalities go through typical patterns which should be understood. Mortalities are more common at hatch, day 2 after hatch when the yolk sac is almost gone, day 7-9 at metamorphosis, and around day 12 (see Table 38, page 141). The most significant losses are at day 2 and

8. Clownfish larvae can quickly starve or loose strength when food is improperly administered or of low nutritional value. For the first 9 to 15 days they are basically pelagic (swim within the upper water column). They scatter throughout the column and normally orientate themselves head first toward the current created by the rising bubbles from the airstone.

Early stage clownfish larvae (day 0 to 3) are typically stationary followed by intermittent bursts of swimming followed by another pause. This attitude is more common in pelagic larvae. While within the water column they swim in a sort of jerking motion. When food is visually located, they form an "S" shape, and propel themselves forward to capture the prey. Those that are day 0 to 2 slowly sink towards the bottom in a motionless manner after each burst and then swim back towards the top. By day 2 to 3 they are usually horizontal in the water column with the heads slightly elevated higher than the tail. Larvae that continue to swim in an almost vertical position are inferior and will no doubt die. Those that remain constantly against the sides of the tank are either inferior or are distracted by improper lighting and/or improper side cover.

Metamorphosis

Later stage larvae, days 3 to 6, actively swim or "cruise" and cease the jerking motion. Prior and during metamorphosis, days 7 to 11, they drop to the bottom and cluster in groups just off the bottom. Swimming changes from a jerky, darting motion to a smooth paddle-like motion. Clustering is more evident towards evening. During daylight the cluster formation breaks and they swim up into the water column or swim at midstream in counter clockwise circles. After metamorphosis, post larvae swim back into the water column in typical clownfish manner by swishing their caudal fins back and forth and paddling with their pectoral fins.

Metamorphosis is the most traumatic event in the life of a clownfish or any fish. It is the "Jeckel and Hyde" transformation (metamorphosis) from a simple larval form to a complex juvenile. This is essentially the transition from a true larva into a post-larvae (early stage juvenile). Larvae have large mouths, eyes, and teeth, virtually straight intestines, poorly developed fins, air bladder, and no scales. During metamorphosis the body completely changes. Scales appear, full pigmentation develops, fin rays are evident, air bladders become fully functional, and intestines become convoluted increasing the absorption area and efficiency of digestion.

Healthy larvae in good quality water conditions usually transform over a three day period. Early juvenile coloration is first detected by the development of the pale translucent white head bar. This occurs around day 7 to as late as day 15. Bertschy (1979b) did a magnificent job in recording development of pigmentation (Figure 53) on early metamorphosing post larvae to three month old individuals. Shortly after undergoing metamorphosis they begin to drop to the bottom and cluster in groups. At metamorphosis larvae are about 3/8" (9 mm). Stressed post larvae do not fully develop complete solid coloration. Most often the white vertical bars do not fully develop or become pigmented but appear clear. At times, both the body and bars are not fully pigmented. It is during poor metamorphosis that "mis-baring" occurs and is not a genetic trait as some speculate. *A. clarkii* (sebae) have the unusual ability of remaining in a pre-metamorphose state for several months and still grow.

Mortalities at metamorphosis are directly related to water quality and larval quality. High mortalities are usually evident just before, during, and just after metamorphosis (see Table 38). Simply put, if you fail to properly feed and care for early stage larvae they will not have the energy to withstand this extreme transformation. Healthy larvae in healthy water conditions usually transform within a three day period. Delayed metamorphosis is a serious problem often resulting in weak and defective individuals with semi-opaque pigmentation. Ideally you want the larvae to transform as soon as possible and develop solid opaque pigment.

Post-larvae (after metamorphosis) tend to stay together or huddle together when startled or in the early morning. It is not unusual to see large numbers of them swim in counter clockwise circles in the center of the tank. Healthy individuals remain basically motionless for hours in strong currents created by rising air bubbles. Others huddle behind the central stand-pipe as if relaxing by removing themselves from the current.

Pigmentation development in A. ocellaris

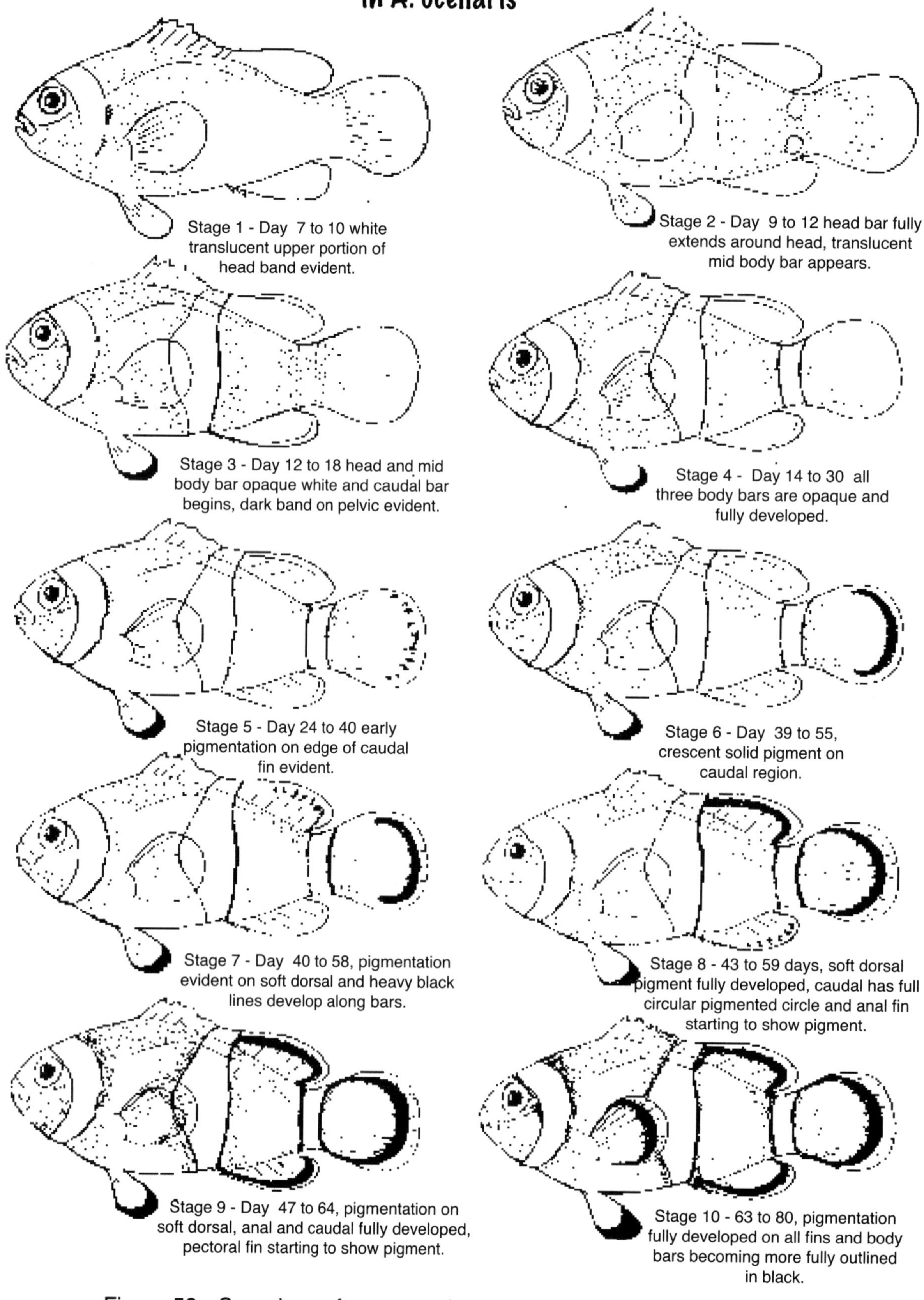

Figure 53 - Cronology of metamorphic pigmentation stages for A. ocellaris (Pomacentridae) by Bertschy 1979b.

In later stages of development, around day 15 to 20, the post-larvae become aggressive and dominance of fast growing individuals becomes evident. Around day 25 to 35 post larvae should average about 1/2+" (14+ mm) long. At this age we considered them true juveniles and they can normally be easily handled for transfer into juvenile growout tanks.

Feeding Strategy and Regime

Larvae hatch with an enclosed energy power-battery in the form of yolk and/or oil globules. Pelagic larvae are less well developed at hatch and this power-pack is utilized to a large degree to complete eye, mouth and fin development. Their power-pack usually includes both yolk and oil globules. Normally the yolk is utilized first followed by the remaining oil globule. Typical pelagic larvae survive up to six days without food. Benthic larvae usually have limited power-packs consisting mainly of yolk. Benthic larvae typically survive three days without food. However, regardless of their power reserves, larvae that are deprived of adequate initial food, become weaker quickly and may not recover.

If not properly fed and maintained prior to metamorphosis, high mortalities at metamorphosis and/or delayed metamorphosis will result. Many authors have stated that you should maintain 3 to 10 rotifers per milliliter. Yet, there is a distinct lack of solid data to support their recommendations. After two years of trying to maintain constant levels, we can say that this is not exactly the right approach. Natural zooplankton levels are normally considerably lower than 3 to 10 per cubic centimeter in tropical reef conditions. It is unrealistic to think that wild clownfish larvae are subject to constant levels and specific types of live food. Between wave action, tidal flows, temperature fluctuations, and diurnal light and dark periods, food types, particle size and densities will change hourly. There is no doubt they are subject to periods of feast and famine. It is estimated hatched larvae from pelagic eggs have only 1% survival whereas larvae hatched from protected demersal eggs have significantly higher survivals. Survival in the wild is usually not due to inadequate food supplies but from avoidance of predation in the egg and early larval stages.

Under culture conditions, larvae that do not feed well for one day often completely stop feeding and die. Whether this is due to overall weakness or just a physiological response is not known, but it can be become highly detrimental. There are several things which might turn off larval feeding including: poor quality larvae, bacterial infection of the gut, inferior water quality, poor quality food, too limited food supply, or poor transition from one food to another. Lasker *et al* (1970) found that irreversible starvation occurred when *Engraulis mordax* (anchovy) were denied food for more that 1.5 days after yolk sac absorption.

Gas bubbles often appear when feeding a exclusive diet of live feed (Photo 19a). This can result in death if not corrected quickly. Constant high concentrations per/ml of live rotifers or *Artemia* actually seem to turn off feeding. Larvae become lethargic and lazy and just don't really grow. Yet, if lower levels of live feed is periodically administered throughout the day in conjunction with micro dry feeds the fish are become alert, aggressive, active, and grow faster and the gas problem is reduced. Eriksen (1991) stated that the balance between underfeeding and overfeeding is extremely delicate. Underfeeding reduces the growth rate whereas overfeeding leads to a number of problems. The basic wild instinct is to fill their stomachs as quickly as possible when food is available. Yet, the stomach and intestine of most larval fish species is short and limited. Overfeeding, or too frequent feeding, does not allow for complete digestion and undigested food is just pushed through the stomach and gut see Photo 19b, page 130. Studies conducted on trout using open demand feeders verses sequential timed demand feeders yielded 24% better weight gain with a 20% lower feed coefficient for the sequential feeders.

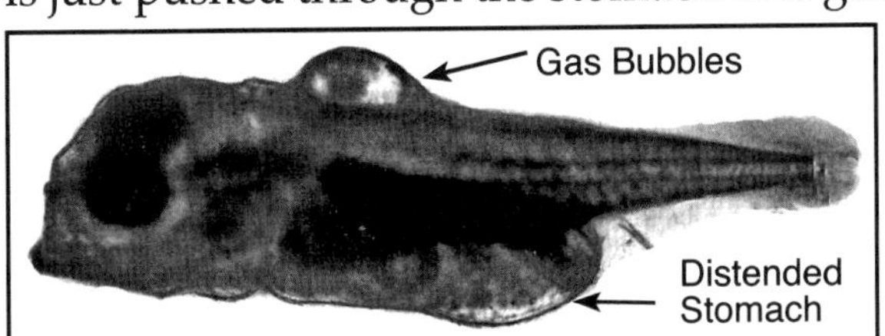

Photo 19a - Gas bubble deformity due to poor nutrition.

Theoretically, you can easily speculate what is going on. Growth and development in humans is a factor of good diet and exercise. If you are overly deprived of food or overly consume food you become lethargic, lazy, and unresponsive. Movement, either voluntary or involuntary, provides muscle tone and development. If a fish simply does not have to do anything but maintain itself in the water column and grab bountiful food it will not develop into a well rounded and "complete" individual. In humans, this can later develop into mental and physical inadequacies, so why not in fish? Native Indians were known for their ability to handle feast and famine. When food was plentiful they ate, when not they fasted, yet physically they were complete.

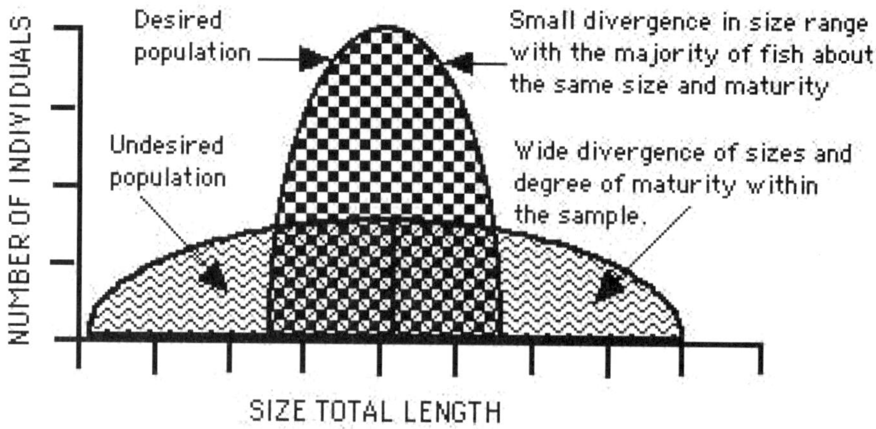

Figure 54 - Population curves based on good feeding and poor feeding procedures.

Children go through several obvious growth spurts which have a typical pattern of feast and famine. One minute they are eating all they can obtain. After several months, their feeding slows and they begin another growth period. Why not fish?

Larval Development In Relation To Food

Density of larvae in a rearing tank, how much food is fed, and how it is presented have direct relationships to uniform growth. Within a given population there are always individuals that do not grow very fast and those that do. Some differences can be attributed to genetics or illness. But, all to often the difference is due to poor feeding practices. A desirable feature within a population is to see at least 80% + of the larvae about the same size. Yet, if there is a wide separation of sizes within a population (Figure 54) this indicates several problems:

1. Tank is too large for the number of larvae contained.
2. Food is not nutritionally balanced.
3. Food is poorly distributed.
4. Not enough food is fed.
5. Dominance by larger individuals.

Table 27 - Daily need of microalgae and rotifers to raise 10,000 to 250,000 larvae.

PROBLEM:
1. Need: 300-500 rotifers/larvae/day.
2. Fact: 20 gal (75 liters) can yield 25 million rotifers (333/ml rotifers) in a 5 day cycle starting with 50/ml inoculant
3. Need: utilize rotifers for 7 days.
4. Need: Culture unit to support 1 million fry (200,000/tank) per day
5. Fact: Based on feeding 300 rotifers/day/fish you would need 10 million/tank/day
6. Fact: Total need is 60 million rotifers per day for one tank of 200,000 larvae
7. Fact: Takes about 100L maximum of 5.3 million cells/ml micro algae to grow 25 million rotifers.
8. Fact: one liter of microalgae water at 5.3 million cells/ml yields 250,000 rotifers/day

LARVAE							
No. of Larvae	**10K**	**25K**	**50K**	**75K**	**100K**	**150K**	**200K**
ROTIFERS							
No. of Rotifers	**3-5M**	**8-13M**	**15-26M**	**24-39M**	**30-52M**	**45-78M**	**60-104M**
Water Volume	**9-15L**	**24-39L**	**45-75L**	**72-117L**	**90-150L**	**135-234L**	**180-300L**
	2-4G	**6-10G**	**12-20G**	**19-31G**	**24-40G**	**36-62G**	**48-80G**
MICROALGAE							
Algae volume	**12-20L**	**32-52L**	**60-104L**	**96-156L**	**120-208L**	**180-312L**	**240-416L**
	3-5G	**9-14G**	**16-28G**	**26-42G**	**32-56G**	**48-83G**	**64-111G**

INDEX: K=thousand, M=million, L=liter, G=gallon

All of these factors are interrelated and we are assuming that environmental factors (water quality, light, etc.) and fish health are optimal. Trying to rear 3,000 larvae in 300 gallons requires significantly more food and manipulation of food particles than rearing them in 50 gallons. The larvae must be able to contact a food particle within a specific time span and after a certain expenditure of energy. If forced to seek food for long periods, the energy expense becomes out of balance and the larvae does not grow quickly. The larger the tank in relationship to the number of larvae being reared, the harder it is to administer and distribute food equally.

Live food, rotifers especially, lose their essential nutritional value in approximately six hours after feeding. Based on this fact it is best to feed newly "packed" rotifers several times a day and not more than can be consumed between feedings. Suppose you wanted to stock rotifers at 5 per ml. In a 300 gallon tank (1125 L) you would need 5.6 million rotifers per feeding for just 3,000 larvae. Yet, in a 50 gallon tank (188 L) you would only need 0.94 million rotifers. Clownfish larvae consume about 300 to 500 rotifers per day. Based on 500 rotifers/day, and assuming rotifers were only fed just once during the day, you would virtually waste 4.1 million rotifers per day in a 300 gallon tank. Suppose you intend to maintain 5 rotifers/ml over a 12 hour period and assuming that half the rotifers were consumed in 6 hours. You would need 8.4 million rotifers for the 300 gallon tank but only 1.41 million rotifers for the 50 gallon tank. Normally cultured rotifer densities are around 200 to 300/ml at harvest. At this level providing sufficient rotifers sounds like an insignificant problem. But, culturing rotifers takes time, costs money, which adds to your overhead. For example, a 20 gallon (75 L) rotifer tank can yield 18.75 million rotifers in 5 days (based on 250/ml). Just one 300 gallon larval tank with only 3,000 larvae will utilize 45% of your daily harvest whereas, the 50 gallon tank would take only 7.5%. This same problem exists when feeding live *Artemia* except the expense due to the loss of unused food may be even higher. In addition, eventual death of uneaten live foods adds to the deterioration of water quality. Table 27 is a guideline for determining the daily plankton culture volume (rotifers and microalgae) needed to raise a specific number of larvae based on consumption rates of 300 to 500 rotifers per day.

Dry foods present another problem with regard to tank size and shape. Ideally you want the food particle to float for several minutes but this is usually not the case. Again you want to present "X" amount of particles per larvae each time dry foods are administered. In a large tank with a limited number of larvae and large surface area you are forced to feed more dry food than needed to achieve good larvae-food contact. This additional unused dry feed directly reflects on water quality as it breaks down. Dry foods, like live foods, lose their potency over time. Leaching of nutrients begins as soon as the food absorbs water, so stirring the uneaten food on the bottom to force it back in circulation probably does nothing with regard to nutrition and only adds to degradation of water quality. If dry prepared foods are utilized significantly in large tanks, then a tall cylinder tank with a smaller surface area is best. This allows you to administer the food more evenly and allows more contact time in the water column before it settles to the bottom.

Another problem that occurs in large tanks with few larvae is the development of dominance. Within any given population of larvae there are those that are stronger and more aggressive. In a large tank, dominant individuals can successfully seek food and soon outstrip smaller individuals of food and space. Later, they begin to set up territories and become aggressive against others within their area. Obviously this results in poorer feeding habits within the smaller population as well as physical damage due to fighting. Yet, if larvae and juveniles are crowded, size becomes more uniform, territories are harder to establish, and aggression becomes significantly less.

Presentation is a major factor in initial stages of larval rearing. Prey movement is the key stimuli for initial feeding responses. Utilization of only dry diets is difficult. Food must be in motion and move slowly for the larvae to capture it within the water column. Dry foods quickly absorb water and sink to the bottom which is out of the normal feeding area for clownfish larvae. Floating dry foods are not suitable for clownfish larvae.

Larvae spend considerable energy in cruising, searching, and attempting to capture food items. Once an item is sighted, the larvae forms an "S" shape and then propels itself forward towards the prey. This is repeated over and over with limited success. Initially, fish larvae are polyphagous, sighting and gulping prey of the appropriate size. Initial feeding success is low, only 6% of the feeding strikes by herring larva are successful and 32%-62% by plaice larva (Hunter 1980). Larvae basically learn from trial and error. After many attempts to capture food they soon learn and understand their prey, and feeding success increases. Coughlin et al. (1992) studied *A. perideraion* larvae and calculated the daily feeding rate in strikes per hour by counting the total number of feeding strikes and dividing by the duration of all swimming paths on a particular day. On day two the feeding rate was 1-6 strikes/hour while on day three this rate increased to 6-65 strikes/ hour. The increase in feeding behavior at day three was reflected in growth patterns. At hatch the larva was 4.08±0.13mm, by day three 4.22±0.12 mm and by day seven the length had increased to 4.63±0.21mm.

Food density is much more important in the initial two or three days of development than in later stages. Larvae are generally visual feeders that eat during the light period. Reactive distance, the maximum distance that a larval predator can locate it's prey, is small. Beyond 0.7 to 1.0 body lengths, prey could not be detected by herring larvae whereas in plaice, pilchard and northern anchovy, the reactive distance is 0.5, 0.2, and 0.4 body lengths respectively (Hunter 1980). Coughlin (1993) reported the boundaries of the search space were 1.3 body lengths away along the axis of the fish and one body length away from the sides and above for *A. perideraion* (pink skunk) clownfish. This further illustrates the problem of using too large a larval tank coupled with your ability to maintain the proper high quality prey densities and correct presentations.

Table 28 - Clownfish larval feeding regime modified slightly from original IOH data.

Type	**Size**	**Age of Larvae In Days** (1 2 3 4 5 6 7 8 9 10 11 12 13 14 15 16 17 18 19 20-30)
Rotifers	50-150μ	Days 1–9: ++++++++++++>>>>>>
Dry Food	50-100μ	Days 3–12: ++++++++++++++++>>>>>
Artemia	300-400μ	Days 5–15: +++++++++++++++++>>>>>>
Dry Food	100-200μ	Days 8–16: +++++++++++++++>>>>>
Krill Meal	100-200μ	Days 8–18: ++++++++++++++++++++>>>>>
Dry Pellets	100-600μ	Days 10–19: ++++++++++++++++++++++++++++++++
Gelatin Mix	100-600μ	Days 17–20-30: +++++++++++++

++ = normal feeding period during normal growth and development.
>> = extension beyond normal feeding range due to size variation, slow growth rates, etc.

Table 29 - Daily larval feeding schedule

AGE	**0800**	**0930**	**1100**	**1300**	**1400**	**1645**
0-3 days		R		R		R
4-7 days	D#1	R & A	D#1	R & A	D#1	R & A
8-14 days	D#2	A	D#3	A	D#2	A
15+ days	D#2		D#3	D#2	D#4	D#2

CODE:

R = Rotifers
D#1 = Dry fine flake type food
D#2 = Semi-moist salmon mash
D#3 = Dry krill meal, shrimp meal, or Spirulina micro algae
D#4 = Salmon growout food or gelatin based mixed foods

Integrated Diets

Typically larval diets should constitute high protein levels, 35 to 60%, and high fat 15 to 25%. Lipids (fats) serve as an important source of dietary energy especially for cold-water or marine fish, which have limited ability to utilize carbohydrates for energy. As fish grow to true juveniles, the diet switches to higher protein and lower fat levels. As opposed to juveniles, pre-metamorphosis larvae have limited digestion ability and require a relatively constant, varied food supply, to gather all the necessary vitamins ,minerals, fatty acids, amino acids, etc.

A mixed larval diet is another key factor to success. Wild larvae are basically opportunistic feeders, in other words, they attempt to eat what they see and receive food from what floats by or might be attached to a surface. Tides, waves, currents, temperature fluctuation, and light intensities bring a constant array of new food types. Since larvae feed on a wide variety of foods,

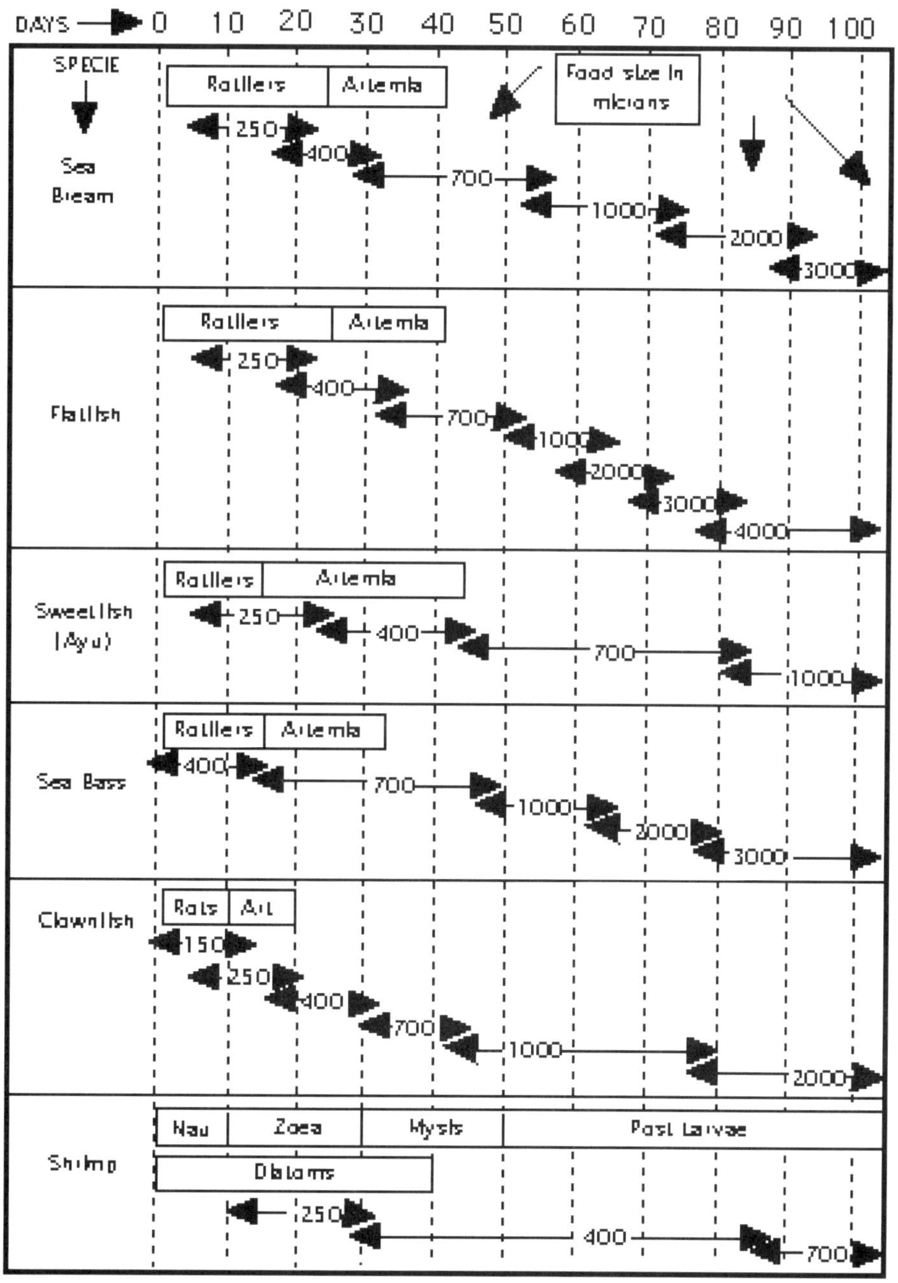

Table 30 - Feeding regime of several marine fish and shrimp, utilizing live phytoplankton, zooplankton and dry feeds. Note average particle size.

there is never an extended period of time between introduction of new types of food for them to try and consume. Their ability to utilize a variety of food items helps survival. Occasionally they strike something new, try it, and maybe like it. If suitable, this item may become the newest addition to their diet and other items will be dropped because they are either too small or too large or not as abundant.

Most people propose very rigid live diets consisting of rotifers followed by new hatched *Artemia* (brine shrimp). Often they say feed rotifers up to day seven and then switch to *Artemia*. They will tell you to maintain 10 rotifers per ml constantly then switch to 3 per ml of *Artemia*. Both the diet and the feeding procedures they recommend are much too rigid. Fish in general have a tendency to try or test new foods and usually do not quickly accept anything that has not been part of their previous diet. They will strike food, take it in, taste it, spit it out, and look for something else. Juvenile fish can tolerate a day or so of testing and fasting for a new food, but larvae cannot. A strict rotifer/*Artemia* diet is too rigid and nutritionally lopsided for consistent good results.

All diets used should have sufficient overlap so there is no lag in larval feeding. You cannot abruptly stop one type of food for another. For example, feeding success in northern anchovy larvae dropped from 80% to 40% at day 17 when their food was changed from rotifers to *Artemia* (Hunter 1972). After two days the feeding rate returned to its previous level. Clownfish demonstrate the same characteristic, but this type of feeding procedure can become detrimental. One day without sufficient food may cause early stage larvae to stop feeding. If your previous diet is marginal, limited, or nutritionally inadequate, larvae may not have enough energy or time to adjust to a new food.

Overlapping live diets in conjunction with a continuous dry, "binding diet", are keys to feeding success. As a dry "binding diet", I prefer a specific diet which changes only in particle size from hatch and into juvenile growout (approximately 40 days). Overlapping period for live foods (rotifers and *Artemia*) is usually 3 to 4 days but may vary with regard to the attitude and condition of the larval batch. For example, if there were a lot of small larvae within a batch, use of rotifers would be extended yet, integration of *Artemia* would continue but at lower levels. Table 28 gives a modified feeding regime used at Instant Ocean Hatcheries. Table 29 provides the daily larval feeding schedule at various ages. Note the sequential, continuous, use of dry followed by live feeds at ages 4 to 14 days.

Table 30 compares feeding models for various fish, including clownfish and shrimp. Shown are duration of feeding live foods (rotifers, *Artemia,* and diatoms) and particle foods. The arrows bracket the average particle size in microns during that particular stage of feeding. For example clownfish start off with food particles around 150μ and progressively get larger up to about 2,000μ (2 mm). Notice the overlap to dry feeds. These regimes are not designed to be absolute but serve as a guideline. What is striking is the short period benthic clownfish are maintained on live foods compared to other fish which are pelagic larvae.

Problems With Feeding Live Foods

Larvae, like small children, will easily eat "goodies" and pass up important foods. Live rotifers and *Artemia* are "goodies" while dry feeds constitute the primary food. Dry, ground and sized, quality flake foods, specially prepared dry diets and gelatin-mix diets constitute important intergradation factors in a feeding regime. They are carriers of specific fats, vitamins, minerals and trace metals that we often cannot incorporate into live foods. If you offer live and dry foods at the same time larvae will automatically opt for live foods and the dry diets will be wasted. Therefore, in the initial stages of rearing larvae, you have to force them to accept dry foods. To accomplish this, administered nonliving feeds when the fish are hungriest. Their first feeding of the day should be a dry food, followed by another before the noon and evening feedings. These should be fed at least an hour before introducing live feeds. Initially these constitute a light feeding of fine particles which increases in intensity and size as the larvae are weaned off live foods. Table 29 provides a look at daily feeding schedules versus age of the larval clownfish. However, as you can see, dry and mix diets played a more important roll in rearing clownfish at IOH than regimes used by others. As mentioned, sizing and the amount administered are important.

A very important warning, feeding *Artemia* can be detrimental to clownfish larvae and many others species I have worked with. If fed too long and/or not fed in conjunction with a dry diet or moist mix diet, larvae can die. *Artemia* should never be an exclusive diet. Symptoms of over feeding *Artemia* include nervousness and reaction to any disturbance. Simply cleaning a tank can cause them to go into shock and die. We often used a crude test to see if post larvae were suitable for transfer into juvenile tanks. We would capture several with a net, place them in our hand, and throw them back into the water. The effect was immediate, the shocked animals, spun around in circles, sank to the bottom, and laid on their sides. Many died but some would revive. If this condition was evident the amount of *Artemia* should be restricted and more dry diet administered. After one week of dry feed this condition usually ceases.

Food Quality Problems

Food quality can become a real problem and difficult to detect. When you first start rearing larvae there are so many pitfalls which can kill them that it is easy to blame one or another factor as the causative agent. Later, patterns develop that help you determine what is really causing mortalities. Poor quality food can cause sudden death, mortalities within a few days, or even drawn-out mortality after accumulative amounts are consumed.

Early stage larvae and juveniles require higher levels of essential fatty acids, yet fat can easily become rancid and detrimental if not properly preserved. Flake foods, if not properly stored, can be one of the poorest, most rancid foods on the market. Rancidity tests conducted at IOH revealed that of all the food types and ingredients utilized, a well respected flake food was the most rancid. Make sure flake foods you purchase have been stored properly and are high in vitamin C, E and A which act as anti-oxidizers and natural preservatives. Do not purchase more than you can use in a six to eight month period and if possible store in the freezer or cool dark area.

Dry meal feeds can also become a problem if prepared with a grain base contaminated with aflatoxins. Aflatoxins are caused by a fungus that attacks stored grain. In high heat and humid conditions this is quickly manifested. Storage and obtaining only what you can use within a specific time is essential. Freezer storage is the best in most cases.

Live food can also become detrimental. Several years ago vast quantities of brine shrimp eggs obtained from the San Francisco Bay region were contaminated with a hydrocarbon. When distributed, these became a world wide problem when fed to larval fish but not necessarily juvenile fish. High larval mortalities occurred several days after the new hatched *Artemia* were fed. In this case the toxin was accumulative before becoming detrimental.

Another toxic problem may be caused by dinoflagellate blooms. At IOH we experienced uncontrollable blooms of *Oxarius marinas* a very small motile dinoflagellate that looks like sparkling flashes under a microscope at 100X or less. When rotifers were fed to the larvae there was not a significant problem unless the *Oxarius* levels became high, started to die, and released toxins into the culture water. Another problem with their presence was that *Artemia* consumed the dinoflagellate, became toxic and thus became toxic to the larvae. Mortalities were evident 36-72 hours after the introduction of *Artemia* to a tank containing dinoflagellates. Therefore, live food cultures and culture waters should be visually monitored under a microscope if unusual mortalities occur. *Oxarius* can be killed using UV at 2,530 angstroms.

Detrimental effects can also come from sources other than toxins within a diet. Deficiencies of essential vitamins, minerals, fats, or amino acids can kill larvae. A classic example occurred while rearing clownfish larvae at IOH. Vitamin B_{12} is an essential vitamin in the fertilizer medium used to rear microalgae. A technician noticed that the particular microalgae we were culturing would grow without the vitamin and thus eliminated it from the formula. In the larval rearing area we started to experience significant early stage mortalities while the rotifers were being fed. Larvae surviving to the introduction of live *Artemia* showed a significant drop in mortalities thus, indicating that there may be a problem with the rotifers being fed. Within days after the missing vitamin was replaced, initial larval survival leveled off.

There are two important factors to learn from the missing vitamin problem. Changes of an ongoing procedure should not be made without the consent of all involved and without at least small scale experimentation. A chain of command, rules, and historical logs are essential. Secondly, this is a classic example of the statement "you are what you eat." Rotifers are excellent carriers of what they eat. Although the microalgae would grow without vitamin B_{12} it apparently was incorporated within the cell when present in the fertilizer medium. Rotifers then consumed microalgae and passed this essential vitamin or essential compound made from this vitamin to the larvae.

Food Size

Food size is an essential factor throughout the larval rearing process. Under a microscope, larvae often display a voracious set of teeth. Yet, these are basically for capture rather than for ripping or biting food from a large particle. I have often heard people say just feed them new hatched *Artemia,* they will eat the legs if they can't eat the whole body. Even if they managed to tear off the legs, what nutritional value do legs have over the entire body? Nutritious food of the right size particle, presented in the right conditions, is essential.

Optimal particle size increases as the fish grows but should not exceed 20% of the mouth opening (Dabrowski & Bardega, 1984). This statement is misleading and may have more bearing on the species being reared. I feel foods as large as 50% or less are acceptable (see Table 29). Usually proper food size is relative to the mouth gap of the larvae. Larvae usually have tremendously large mouths in relation to their body size. Initially most larvae prefer food particles that are 50 to 100 microns wide. Hunter (1980) showed that the upper limit of prey size eaten increases faster than the lower limit. Most larval fish expand the upper size range of prey selected as their mouth size increases, but continue to consume small prey as well.

Particle size of dry foods (Table 28) should correspond in size or be smaller than the live food being fed. Micronized dry foods are available but can be created by using a mortar and pestle and suitable sized plankton screening. Simply grind high quality flake foods into small particles and size utilizing plankton screening or sieves. Particle foods can be easily administered by using a common spice jar equipped with a tight fitting lid and porous screen. Store all dry prepared foods in tight fitting jars in the freezer except what you are currently using. Quality micro-particulate diets formulated for marine fish and/or shrimp usually contain fish meal, fish oil, crab, yeast, chicken egg yolk, clam or scallop meal, krill meal, and spirulina. Specially formulated dry larval diets may also be obtained which can drastically reduce the time involved in preparing these foods.

Adult female rotifers commonly used in marine aquaculture are about 250 microns long (1,000 microns = one millimeter) and 100 microns wide, whereas the smallest strain of new hatched *Artemia* nauplii are 400 microns long and 250 microns wide. New hatched rotifers are about 1/3 the size of new hatched *Artemia.* Smaller strains ("S" strain) of rotifers are available and adults are about 2/3 the size of regular strains. For rearing clownfish the normal rotifer strain is ideal. Within any population of rotifers there are smaller individuals. If they were constantly sieved out of a normal population and cultured, you could conceivably develop a small strain culture. It may have to be sieved several times to accomplish this.

In general, smaller food particles are always better. If in doubt, continue using smaller particles. Within a population of fish there are those that grow quickly and those that do not. If survival is to be optimal you must be capable of feeding all individuals within a population.

Frequent feeding is important to all larval fish, food can be offered 10 to 24 times a day or almost continuously and in excess (Charlton and Bergot, 1984; Charlon *et al* 1986). This is a misleading statement and applies when only dry feeds are administered since this food will not stay within the water column more than a few minutes. In addition, this statement was based on rearing carp larvae which have no stomachs which may require massive feeding. This practice is not recommend since more feed does not always make it better especially to clownfish.

Feeding Live Foods

Rotifers constitute the most essential initial diet for most marine larvae. Rotifers, as stated before, "are what they eat." These are living packages of initial nutrition. They must be packed with "goodies" before being fed to the larvae. I can't emphasize this enough. Water quality and proper feeding prior to metamorphosis (transition to an adult looking juvenile) are going to reflect directly on future juvenile development. The larval food chain actually starts with adult broodstock. Adults that are fed properly produce strong, quality larvae. If quality larva are properly fed they will metamorphoses into strong vigorous juveniles. Like humans, nutritional deficiencies during early development directly result in physical and mental problems in the future.

The following batch culture procedure for preparing rotifers for feeding larvae was used at IOH. Early in the morning all rotifer culture vessels were fed a yeast-based mixture. Culture vessels were then assessed for rotifer density. Harvested cultures were split and a determined amount were set aside to inoculate new batch cultures. Remaining rotifers were split into three equal amounts for the three daily feedings. A gallon of microalgae was given to all three containers. After an hour the first bucket of rotifers was fed to the larvae. At noon a second bucket was given more microalgae and fed to the larvae and in the evening the remaining bucket was treated the same way. In each case the rotifers were "packed" before being fed to the larvae.

I often hear that you can raise all the rotifers needed on baker's yeast without having to fool with microalgae. This is correct, yet the nutritional value of yeast fed rotifers is lacking significantly in essential omega three fatty acids (HUFA) such as eicosapentaenoic acid. Comparison of microalgae fed and yeast fed rotifers showed up to 25% higher eicosapentaenoic acid. Common practice is to feed microalgae 12-24 hours before harvest. Utilization of products like Roti-Rich™ or Culture Selco™ are special yeast-based foods that can be used to culture rotifers with only supplementation of relatively small amounts microalgae daily to enhance reproduction. See our companion manual on Plankton Culture for more details on rotifer culture.

Color of rotifers is an important factor in larval feeding. Starving or yeast fed rotifers are basically clear or almost totally transparent. Larval fish seek food by sight not smell. Therefore, a clear transparent body will be harder to locate than a rotifer that is packed with dark microalgae.

Enrichment products (Selco™, squid oil, fish oil, etc.) contain essential fatty acids such as 20:5 omega 3 and 22:6 omega 3 as well as specific vitamins such as B_{12}, C, and A can also be utilized and easily incorporated into this regime. Maximum efficiency for incorporating fatty acid enrichment takes 1 to 6 hours of feeding. Therefore, these products could be suitably fed to rotifers utilized in the noon and evening feeding. Enrichment diets are limited and are fed only when needed. Constant feeding of fatty acid enrichment diets to culture rotifers is not recommended and does nothing with regard to nutritional value except for what was fed on the day of harvest.

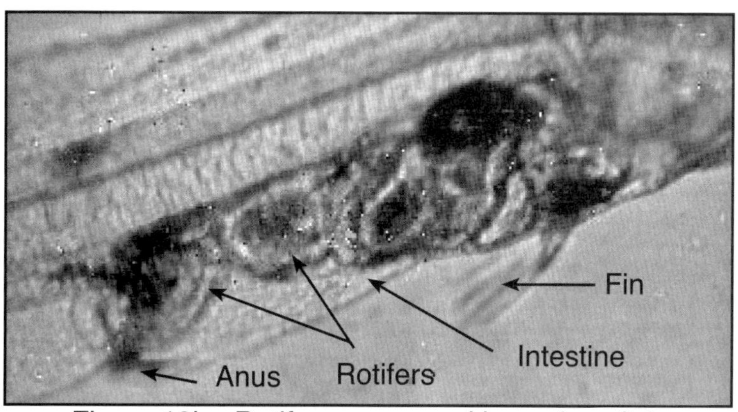

Figure 19b - Rotifers consumed by a clownfish.

Once enriched with fatty acids, rotifers can quickly lose 1/3 to 2/3 of their nutritional value within 6 hours. In less than 24 hours they revert to their original level.

A rotifer's life span is a short seven days and geared towards procreation. Nutritional qualities are quickly transposed into the development of enriched eggs for continued growth of the population. During harvest, many eggs are knocked off and do not contribute to the nutritional value of the rotifer prey. Emphasis to keep the rotifers "packed" before feeding them is important. In addition, light, routine, additions of small amounts of microalgae to the larval tank can also be used to keep rotifers well fed and help maintain water quality. However, use of microalgae can promote growth of unwanted protozoans and invertebrates.

Harvest rotifers to be fed that day through a 50 or 60 micron plankton collector. Dispose of old culture water. Do not add old culture water to your larval rearing tank or reuse. Immediately after collecting rotifers, reconstitute them into a liquid culture of microalgae. Analyzed culture water from which rotifers were harvested have high nitrates, high organics, low pH, and perhaps high ammonia. This is not what you want to be adding to your larval rearing tank.

How many rotifers do you feed to a specific size larvae? If you have several larval rearing tanks, determining how many rotifers to be fed to each tank becomes basically a judgement call. Ideally you want about 3 to 5 rotifers per ml but this may be altered depending on the number of larvae, their age, and size. In the morning about an hour after feeding rotifers, observe the larvae by looking down into the tank. Larvae that have not fed during the night look dark and thin, with a large head and small body. Basically this is the sign of a stressed larvae, sort of the way we all look in the morning before our jolt of coffee. If the larvae are healthy and feeding well, they will become clear yellow with mixed dark pigments within an hour or so after the rotifers are added. Their stomachs will be swollen and bulge out evenly on both sides. The best way to feed rotifers or *Artemia* is to do the following:

1. Add an equal liquid measure, 2 oz., 3 oz. etc., to each tank.
2. Allow the rotifers or *Artemia* to spread evenly in the water column.
3. Visually evaluate the density with regard to the larval density.
4. Add more rotifers or *Artemia* as needed.

Kitajima (1993) proposed a formula in a effort to determine how many rotifers were needed to raise the pelagic larvae sea bream to 10 mm TL. Rotifers consumed (F) and the total length of the fish (L) relate to: $F = 0.3927 \times L^{3.676}$. He calculated it takes 57 billion rotifers to raise 2 million sea bream with a maximum initial need per day of 8.4 billion rotifers. Culture period of sea bream on rotifers is about 20 days (see Table 30) or about 1,000 to 2,000 rotifers per day per fish. I feel the rotifer needs could be reduced if more overlapping of dry feeds were utilized. At IOH we utilized only 20 million per day to raise about 30,000 larvae or about 666 rotifers (range 500 to 1,000) per larva per day over a 5 to 9 day period.

What you are trying to avoid is shortchanging live food supplies to one tank for another. Discrepancies in obtaining proper densities during the morning feeding can be easily rectified at the noon feeding. As you develop an understanding of what a well fed larvae looks like compared to an underfed one, you can adjust levels during later feedings.

It is important to understand that new hatched *Artemia* are 2/3 larger then new hatched rotifers and equivalent to approximately 10 adult rotifers in volume and 30 rotifers in weight. Therefore, you do not need as many new hatched *Artemia* to achieve proper feeding levels. Enrichment of *Artemia* is not as easily accomplished since newly hatched *Artemia* do not feed for at least the first 10 to 12 hours. If enrichment is to be used it must be used on day old brine shrimp which are considerably larger (see Plankton Culture Manual for more details).

Artemia are considered a panacea with regard to obtaining an immediate live food source directly out of a can. But, *Artemia* are not considered a live food source of most wild fish environments. They live in high saline bodies of water where virtually nothing but algae survives. Nutritionally they seem adequate but there appear to be certain trace nutrients missing. Utilization of mixed and overlapping dry diets seems to meet this need. Fatty acid and/or microalgae enrichment prior utilizing them as prey is helpful. Whether nutritional change is due to enrichment or a result of changes they under go during their initial 12 hours of development is not clearly defined.

LARVAL CARE AND MAINTENANCE

Larval quality, food quality, and feeding procedures are the most important factors in rearing any larvae. Routine maintenance is closely coupled to the rearing procedure. A rigid program must be employed to maintain acceptable water quality and to understand what is happening to your larvae. Again, as stressed in the broodstock section, I feel that daily records are important and should be kept regardless of how small or big the hatchery.

Water Exchanges And Flow Rates

Water exchanges and/or continuous flows not only control water quality but also provide stability. However, early stage larvae, prior to metamorphosis, do not respond well to significant, partial or constant water exchanges. It is as if a certain balance of dissolved metabolites, organics, or whatever are required for stability. In addition, continuous flow rates drastically reduce live food levels and perhaps live food quality. In general water quality in the early stages of larval rearing (up to 6-8 days) should not be problem since the primary food items are alive. Obviously overfeeding of even live feeds can result in a water quality problem. On day one, it is important that all the dead eggs and larva are removed immediately. Routine vacuuming of tank bottoms on a daily basis is essential for water quality stability. Water flows at IOH were administered through standard, threaded, white nylon, aquarium air valves (Photo 19 & Figure 48). Two were mounted above each tank. Table 31 shows the water exchange procedure used at IOH that provided continuously good results.

Table 31 - Water Exchanges In Larval Rearing Tanks

FLOW TYPES			DAY - 1-2-3-4-5-6-7-8-9-10-11-12-13-14-15-16-17-18-19-20>30
NF	=	No Exchange	++++
X	=	30% Exchange	++++++++++++
D	=	Drip Flow	+++++++
S	=	Slow Flow	+++++++++
M	=	Medium Flow	++++++++
F	=	Full Flow	+++++++++++
2F	=	Double Full Flow	+++++++

Flow rates were based on the following criteria:

1) Tolerance of early stage larvae towards water exchanges.
2) Maintain larvae and live food in tank and preventing them from being caught on outflow screens.
3) Maintain microalgae, antibiotic or trace element levels utilized in early stage larval rearing.
4) As a rule of thumb, we used 30% exchanges until 70% of the larvae have head bars (starting to metamorphoses). After this point progressively faster flows were administered.
5) Additional considerations include general development characteristic of specific species such as A. clarkii which may need a longer 30% water exchange since they often metamorphoses more slowly.

Daily Larval Maintenance

In commercial facilities, routines occur daily regardless of holidays, weekends, or illness. Daily maintenance can be broken down into three parts: mandatory early morning care, weekly care and evening care. Early morning care is the most important and consists of the following:

1. Make initial dry food feeding to tanks that are 3 days or older.
2. Harvest rotifers and place in microalgae for enrichment.
3. Count, record, and remove dead larvae by siphoning them out.

4. While siphoning remove large accumulations of food and detritus.
5. If nest was hatched inside the larval tank, remove hatchers and evaluate percentage hatch, dead eggs, and dead larvae. If nest is not completely hatched reposition hatcher for another night. (See Broodstock Section for Hatching Procedures)
6. Evaluate older tanks for head bar formation, preconditions for metamorphosis. Estimate percentage. Pre-metamorphosis conditions are usually evident from days 6 through 10.
7. Clean bottoms completely with "Vacuum Cleaning Siphon" (Figure 56) on Mondays and Fridays or more often if needed. Slowly drag the vacuum flatly across the bottom to avoid stirring debris and sucking in larvae. Control flow rate by pinching siphon tube with the other hand.
8. Make water exchanges or adjust flows as needed. Code data with the following: X for 30% exchange, "D" for drip, "S" for slow, "M" for medium, and "F" for fast. Check and adjust air flows. (See Table 31)
9. Adjust air flows to a point where a 1/4" rise occurs on the surface.
10. Once the dead are counted, tanks cleaned, and water exchanges have been made, you are ready to feed the first live foods for the day. Usually this occurs between 9:00 to 9:30 am. Keep in mind the rotifers were harvested earlier and "packed" with microalgae and other nutrients.
11. Second dry feeding around 11:00 am.
12. Second rotifer and/or *Artemia* feeding around 1:00 PM.
13. Third dry feeding around 2:00 PM.
14. Set up nests to be hatched overnight around 4:30 PM.
15. Third rotifer and/or *Artemia* feeding around 5:30 PM.
16. Check air and water flows.
17. Go home and get ready for another 364 days of the same.

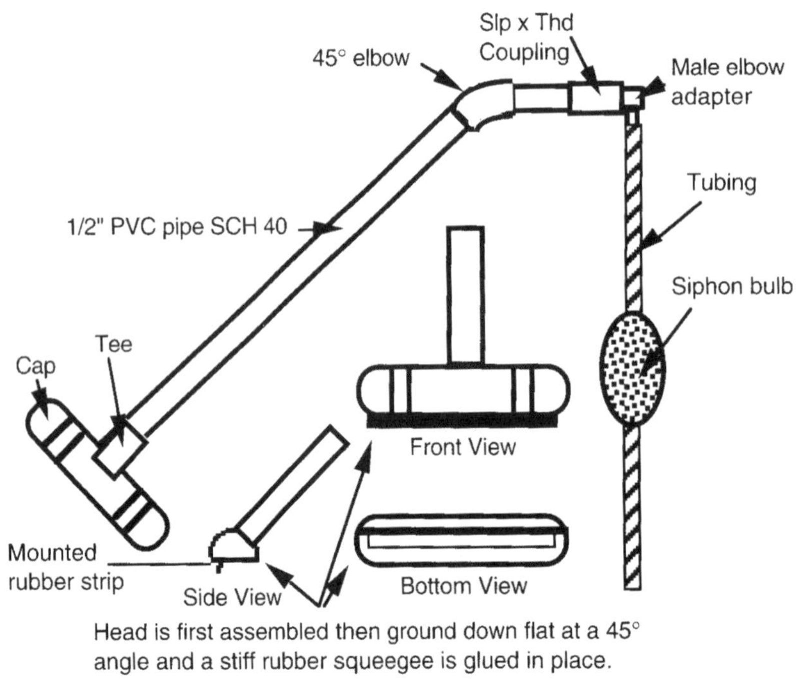

Figure 55 - vacuum assembly for cleaning bottoms of larval rearing tanks

Disease, Deformities, And Treatments

Larvae, for the first 20 days, are basically resistant to common diseases such as *Oodinium* and *Cryptocaryon*. However, early stage larvae (day 1 to 3) are highly susceptible to bacterial infections of certain types (not clearly identified). Newly hatched larvae will often not initiate feeding. This seems to be caused by either poor quality larvae and/or a bacterial infection of normally good larvae. As a prophylactic measure, we commonly hatched nests in tanks treated with a single dose of "Furan II", a nitrofurizone compound. Dose was 1/2 teaspoon (2.5 cc) per 40 gallons (150 L). This practice was not routinely administered, butonly when there was a prevalence of non-feeding larvae hatching. The first morning a tank is treated is considered "Day 0." The normal 30% water exchange was not made until the morning of "Day 2." Water removed during exchanges on "Day 2 and Day 3" was flushed through activated carbon to reduce any effect "Furan II" might have on the biological filter. Treated tanks were marked with a red tag which hung on the water inputs to prevent accidental water exchange.

Use of "Furan II" was administered intermittently and only when larvae or post-larvae stopped feeding or when mortalities increased especially at pre-metamorphosis (Day 5 to 9). Normally a tank was not retreated within one week. Utilization of other bactericides such as neomycin and streptomycin were not suitable for closed system use and had an adverse effect on biological filters.

Occasionally copper sulfate solution (0.15 to 0.20 ppm) was administer to late stage post-larvae for treatment of *Cryptocaryon* or *Oodinium* but this was not common practice. Note, 1 ppm = one milligram per liter. Low dosage copper treatments were administered in the morning and in the evening to tanks which were on flow.

"Bent-back" or lordosis is often a common problem in pelagic larvae and at times in benthic larvae. Often this is thought to be a mechanical injury whereas others believe it to be genetic. While culturing pompano we experienced this quite often. Initially some felt that this was due to mechanical injury from air bubbles. To prove this was not the case, we fit a 6' diameter tank with an aeration collar around the entire circumference of the tank with air outlets every 6" thus creating a bubble screen. Air was provided at a vigorous rate and the tank stocked with fertilized eggs. Results showed no more lordosis than our control. Kunihiko (1985) conducted a similar experiment on sea bream in 10 ton tanks (10,000L) and found that lordosis (bent-back) was scarce in fish groups reared at aeration rates of 50-100 ml/minute while deformities appeared at higher frequencies with non-aeration or at more that 500 ml/minute. In addition, he found moderate and uniform currents in rearing water led to a lower percentage of lordosis and a higher survival rate even without aeration. However, survival, and percentage of fish with inflated swim bladders, were low in tanks with non-aeration and no current. Attempts were also made to see if lordorsis was related to nutrition, environment, or genetics. In another study eggs from a hatchery with prevalent cases were transferred to another with limited cases. Deformed larvae were found in the parent hatchery at a rate of 14.1 to 35.4%. Whereas, larvae reared in the second hatchery had no deformities. Therefore, lordosis does not appear to be genetic and is probably due to environmental conditions that occur at hatch. Low levels of vitamin C are known to cause lordosis in catfish. In clownfish, we found this was more prevalent with slow or late hatching larvae. We attributed lordosis to low temperatures and/or poor adult nutrition.

Uninflated swim bladders may occur in pelagic and benthic fish. Fujita and Kitajima (1978) found this is affected by the quality of initial feeds. Larvae fed yeast fed rotifers tended to have a higher percentage of juveniles with uninflated swim bladders whereas, microalgae fed rotifers yielded a significantly lower deformities. However, occurrence of lordosis was not as defined. Higher percentages of uninflated swim bladders were higher in non-aerated tanks. It was thought that abnormally-developed swim bladders were caused by a failure of air gulping at the surface (Iseda, 1982). They tested this theory by sealing off the surface of one tank with paraffin and the other had an open surface (Kitajima el al., 1981). More than 90% of the control group of ,milkfish larva had normal bladders whereas, in the sealed tank none were inflated.

Improved environmental conditions have greatly reduced the occurrence of lordosis and swim bladder problems. These conditions include: 1) supplying moderate current and aeration, 2) removal of surface scum to allow air penetration and gas exchange, 3) supplying exchange water in a showering or splashing fashion, 4) introducing newly-hatched larvae at a fairly low density, 5) selecting broodstock with well-developed swim bladders; and 6) feeding rotifers with high nutritional feeds.

"Edema" is characterized as large, clear, fluid-filled bubbles that develop on the dorso-lateral abdominal section of the body. Concomitantly, larvae begin whirling about the surface and are unable to feed. With pompano and black seabass this seems to be correlated with increases of nitrite and ammonia (Hoff et. al. 1977). Shelbourne (1964) described two edema type conditions in plaice larvae: yolk sac edema and "water belly" abdominal edema. He suggested high metabolite levels as a possible cause. It is possible that the creation of this problem may also correlate with low temperature or extreme temperature fluctuations.

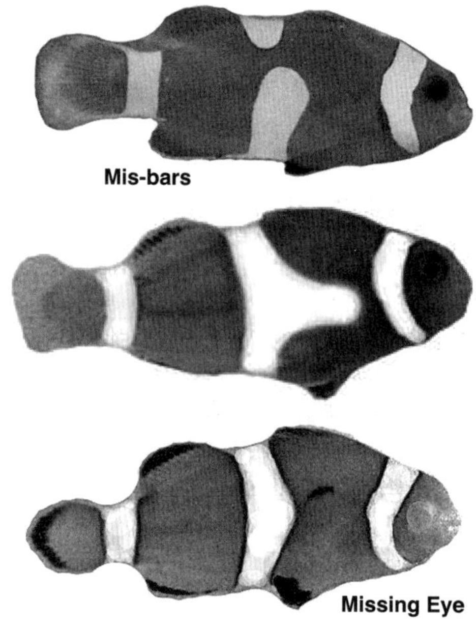

Photo 20 - Deformities due to either genetic or environmental conditions during metamorphosis.

Hydrogen sulfide is normally not a problem in shallow culture tanks, but those who use deep cylinder tanks and low aeration should take this into consideration. It is imperative to remove fecal matter and uneaten food from the bottom on a daily basis. Pockets of waste material can create low levels of hydrogen sulfide. Coupling this with low circulation creates a stagnant toxic bottom area. Larvae that enter into this area can be killed quickly.

Mis-bar (Photo 20) is a deformity caused by poor water quality and/or delayed metamorphosis. Missing vertebrae and eyes may be genetic but attempts to produce progeny from mature individuals usually resulted in significant numbers of natural progeny. However, these deformities may be induced by mechanical injury during the egg stage or early larval rearing. All aspects of development of mis-bared individuals were basically natural. However badly mis-bared individuals grew slower and had an unusually long metamorphosis. Culture of *A. clarkii* often resulted in long metamorphosis and pigment deformities. Pigment deformities also occur in nature.

Use of Iodine

Iodine has been reported essential for development in many aquatic species. Hsiao (1969) demonstrated the necessity of iodine in metamorphosis and reproduction of a marine brown algae. OKelly (1974) cited several studies reporting requirements of iodine in marine algae. Iodine is necessary for strobilation in scyphozoans, and its relation to thyroid hormones is well documented. Freiden (1962) reviewed the effects of iodine-based thyroid hormones and found them essential for amphibian metamorphosis. Belsare (1976) described development of thyroid tissues in larvae of two fish species. First stages were apparent by 40 hours and in five days active cell division and follicle development were noted. He concluded that these cells began trapping and synthesizing iodine early in the larval period. Lam (1980) found that immersion in 0.1 ppm thyroxine enhanced development and survival of larval talapia.

Varying extended periods of metamorphosis were noted at IOH. Tests conducted by (Frakes, F. Hoff, and W. Hoff 1982) clearly demonstrated that metamorphosis is dependent on trace quantities of iodine. The primary source of iodine appears to be direct assimilation from the water, but an alternate route via the food chain exists, allowing metamorphosis to occur in low-iodine waters. When iodine concentrations are below natural sea water (NSW) levels, temporary delay can occur, presumably because of a lack of thyroid hormones. However, factors such as high

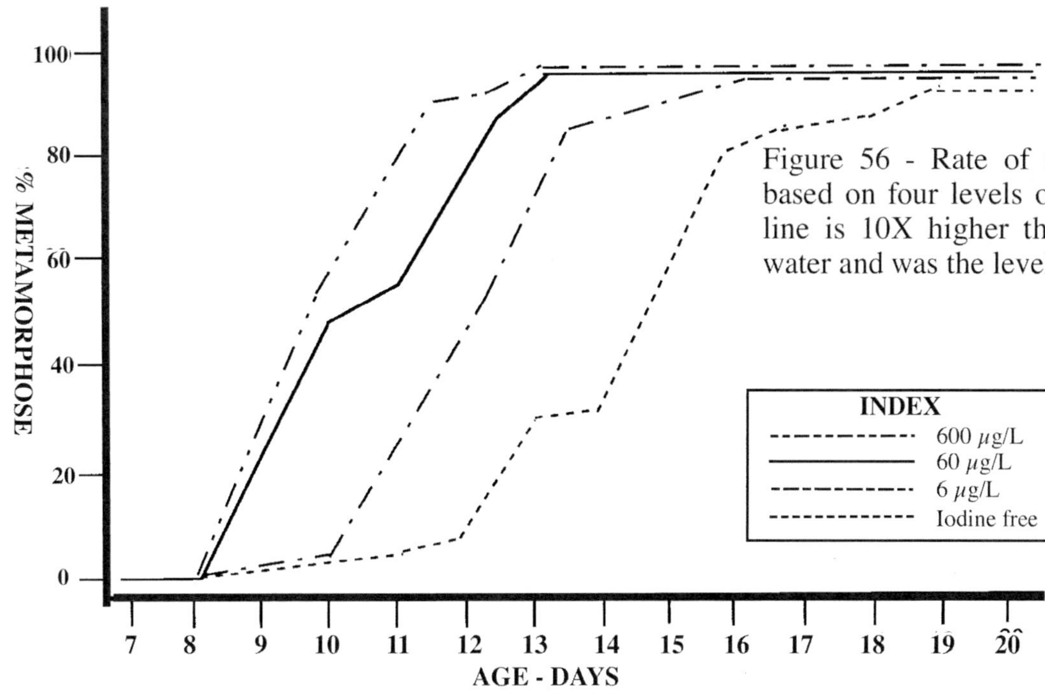

Figure 56 - Rate of metamorphosis based on four levels of iodine. Solid line is 10X higher than natural sea water and was the level used at IOH.

nitrate-N concentrations (100 mg NO_3-N / L) can also have an effect (T. Frakes and F. Hoff 1982). Young and Langille (1958) list iodine at 8μg / L in Nova Scotian waters, a level that could be sufficiently low to affect metamorphosis.

In closed recirculating systems, iodine can become limited since so many organisms including algae and invertebrates have demonstrated a need for iodine. Bioaccumulation by algae or in algal denitrification utilizes tremendous amounts of iodine. For example, 1,000 L of culture water with 50 mg NO_3-N / L and 60 ug iodide / L can be reduced to near zero by the growth of just 14.2 Kg (31 lbs) of macroalgae. Mechanical and chemical losses of iodine have been reported including photochemical oxidation of iodide to iodine gas by exposure to ultraviolet light (Miyake and Tsunogai 1963). Seto and Duce (1972) studied the loss of iodine on sea salt particles caused by bursting bubbles. Organically-bound iodine was shown to be actively lost to the atmosphere. Common aquaculture practices of trickle towers, heavy aeration, and foam fractionation using protein skimmers in conjunction with UV sterilizers can drastically reduce iodine levels. Ozone may also have an effect but has not been studied.

Experiments conducted on clownfish (T. Frakes, F. Hoff and W. Hoff 1982) resulted in desired levels of 60 ug / L. Figure 56 clearly shows the effect of iodine-free water versus the effect of various levels up to 600 ug I / L (10 times NSW) . Juvenile clownfish were tested for several weeks at levels 10,000 times natural seawater level concentrations with no observable adverse effects. Utilization of a full trace solution including enhanced iodine levels is recommended regardless of the artificial salt mixture being used.

Iodine was routinely administered to all batches of larvae on days 2, 4, and 6. Dosage was calibrated to be 60 ug / L (60 ppb = parts per billion). Equations for determining iodine loss are not available and would be complicated but, overdoses do not appear to be a serious problem.

Water quality is a very important factor with regard to larval development. Although larvae tolerate a fairly wide range of certain factors, they can undergo abnormal physi-

Table 32 - Effect of high and low nitrate on A. ocellaris larvae.

Group	Survival		Mean Length	Metamorphosis	
	Day 21	Day 72	Day 72	Day 21	Day 33
Low nitrate	37%	37%	14.4 mm TL	97%	100%
High nitrate	34%	12%	13.2 mm TL	34%	88%

ological and physical changes which will be permanent and perhaps detrimental in later stages of their lives. Maintaining quality water conditions is extremely important prior to metamorphosis. "Mis-barring" or irregular formation of the bands that occurs in clownfish is not genetic but caused by irregular water conditions during metamorphosis. "Flared gills" is another external characteristic caused by irregular water conditions. "Notch-gill cover" may also be caused by improper water conditions.

Frakes and Hoff (1982) researched the effect of high concentrations of nitrate nitrogen on larval and juvenile *A. ocellaris.* Kincheloe et al (1979) reported nitrate levels of 5-10 mg NO_3/L as mildly toxic to developing eggs and fry of freshwater trout. Chinook salmon were affected at 20 mg NO_3/L. Spotte (1970) suggested that nitrate levels for saltwater aquarium fishes be kept below 20 ppm (20 mg/L). Spotte (1979) later provided clarification that 20 mg NO_3-N/L and levels as high as 400 to 1000 mg/L NO_3-N (400 to 1,000 ppm) were noted for various aquatic animals. Otte and Rosenthal (1978) reported levels up to 1,100 mg NO_3-N/L (1100 ppm) in brackish-water tilapia culture, with normal concentrations between 200 and 400 mg/L (200 to 400 ppm). This wide variance made us question what levels are considered safe and what is toxic for clownfish.

Two identical, 213 gallon (800 L) systems, were set up and maintained for three months. Nitrate was elevated to around 100 mg/L (range 85.4 to 119.0 ppm) in one system by the addition of sodium nitrate and the other was maintained at around 16 mg/L (range 11.5 to 23.4 ppm). In each system, two, 38 liter tanks, were each stocked with twenty one, 72 day old fish, a third tank was stocked with six, 40 day old fish, and two tanks with forty newly-hatched larvae each. The results are shown in Table 32. The effects of elevated nitrate were most pronounced in larvae.

Table 33 - Range and average water quality parameters in the larval rearing room at IOH.

Parameter	Range	Average
Temperature	23° to 28°C(73.4 to 82.5°F)	27°C (80.6°F)
Salinity	26 to 30 ppt(1.0191 to 1.0222)	28 ppt
pH	7.9 to 8.2	8.0
Ammonia	0.0 to 0.1 ppm NO_3-N	<0.025 ppm
Nitrite	0.0 to 0.025 ppm NO_2-N	<0.025 ppm
Nitrate	15 to 30 ppm NO_3-N	20 ppm
Phosphate	0.025 to 3.0 ppm PO_4-P	1.5 ppm

Although survivals to 21 days were comparable for the low and high nitrate systems (37% and 34% respectively), metamorphosis was retarded in 97% of the fish in the high level, compared to only 40% for the low level. Normally metamorphosis in clownfish at 27°to 29°C (80.6° to 84.3°F), begins on day 7 and extends to day 10.

An outbreak of *Amyloodinium sp.* at day 21 necessitated treatment of both systems with quinine sulfate (1 g/100L). The treatment appeared to further retard metamorphosis until the drug was removed with activated carbon a week later. Within 5 days after removal of the treatment, metamorphosis in the high nitrate tanks rose to 88%.

After 72 days, survival in the high-nitrate system was only 12%, whereas survival in the low-nitrate systems remained at 37%. Mean total lengths for the high group was 13.2 mm and 14.4 mm for the low group. Juvenile tanks had 100% survival under both conditions. Growth was restricted, but not significantly, in the high levels. Color in all groups was better in the low-nitrate tanks.

Basically this is the end result of physiological and physical changes caused by poor water quality which delayed metamorphosis. Delayed metamorphosis results in inferior, weak, larvae which in turn succumb to irregularities in growth and survival at a later date.

Histological examination of post-larvae from the high-level group showed evidence of tubular necrosis and destruction of the kidneys. This destruction may have been caused singularly or combined by: high nitrate, use of quinine, and/or a result of extreme delay in metamorphosis. The main factor is that damage does occur under abnormal water quality conditions at metamorphosis. Examination of juveniles used in the experiment indicated they were normal.

Effects of other water quality parameters on larvae are not as clearly pictured as nitrate-nitrogen and iodine levels, but no doubt have an effect. Ammonia, nitrite, and pH are important factors and probably have a more pronounced effect on metamorphosis than nitrate. There are indications that pH levels above 8.2 may have a detrimental effect on larva. Maintaining calcium levels of 300 to 400 ppm (normal seawater) may be advantageous.

There are ways to determine water quality without the aid of a test kit. You can smell the water for fouling or breakdown of food. Usually smell just over the air bubbles. You can feel the water and see if it is oily between your fingers or "thick looking" at the surface. You can see if it is clear or colored. Milkish white cloudy water indicates high bacteria and/or ammonia. Yellow water indicates high amounts of dissolved organics. Determining water quality this way takes time to learn, but it can be done.

Nitrites are probably more deadly than ammonia or nitrates in saltwater. When present, it indicates a breakdown of your biological filters to convert ammonia to nitrate. Something is killing or limiting the nitrobacter populations in your filters. Usually this occurs when drugs, copper sulfate, etc. are used in excess and not filtered through carbon when draining water from treated tanks.

I do not recommend the use of carbon on a continuous basis, especially in larval rearing systems. As pointed out, larvae respond to minute levels of trace metals and perhaps dissolved organics. This is especially true of early stage larvae prior to metamorphosis. Carbon strips vast quantities of specific trace metals to a point where they become limiting. This also goes for ion exchange resins, and other ion clarifiers. You are working in a closed system where everything is competing for the trace nutrients, there is no natural input, so eventually there is depletion. Routine utilization of trace nutrients in addition to enhanced iodine levels is recommend.

Water quality in the larval area is really not hard to maintain, since the biomass you are working with is minimal and usually live foods are involved. Fouling is caused more by over feeding of dry foods or in some cases live foods. People try to incorporate a filter into their larval rearing tank or system. Filters in conjunction with small larvae and live planktonic foods just do not work out well. They will remove the live foods from the tank and entrap larva. Prior to metamorphosis, daily 30% water exchanges and routine cleaning of bottoms 2 to 3 times a week will usually maintain proper water quality. If you experience a heavy load of detritus increase cleaning. If water quality starts to deteriorate increase the exchanges to twice a day, once in the morning and in the evening prior to feeding.

Records at IOH during the last four most productive years were averaged and are presented in Table 33. Water exchanges in this area were about 20% or less per month. Carbon was utilized about once a month for a 7 day period. In addition to water exchanges, trace metal solutions were added once or twice a month. This same trace solution is currently available under the name Macro Algae Grow™ from Florida Aqua Farms Inc. In addition to the trace metals, iodine was added either directly to the rearing tanks or into the system. About 200 ml of one molar solution of potassium iodide was added monthly to the 10,000 gallon (37.5 cu meters) system.

Transferring Post Larvae

Post larva are classified as those that have completed metamorphosis and have the overall appearance of a small juvenile fish. Consider fish that are 25+ days old as "true" transferable juveniles. Before transfer from the larval rearing area, there are several factors that must be met. Criteria, testing, and transfer methods are discussed in the juvenile rearing section.

Petroscirtes azalea
(Jordan)(Albemarle I., 1937)

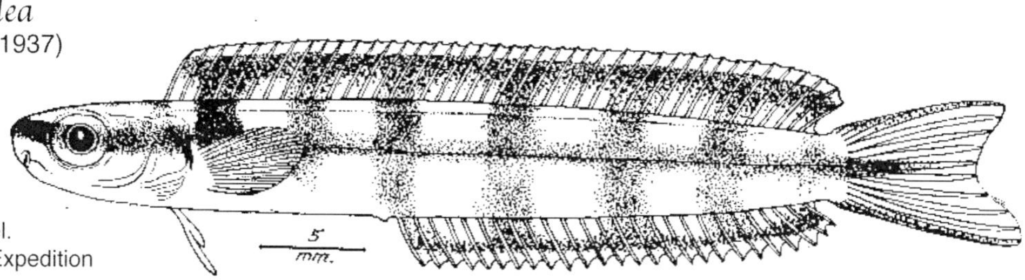

Drawing By H. W. F. del.
Results of the Fifth Vanderbilt Expedition
By Fowler, 1944

LARVAL RECORDS AND DATA

Again data and record keeping becomes important as with adult broodstock. Two forms were used in the larval rearing process, Daily Larval Log sheet and the Monthly Larval Production form. Daily sheets were made each evening for the next morning. One sheet served each row of tanks (4 rows, 12 tanks/row). Below (Table 34) is an example of the daily larval data recorded. The headings on the data sheet are defined below.

Tank = Assigned larval rearing tank where nest is hatched. Based on data on the top right of the data they are in row 4.
Sp = Species - O = ocellaris, F = fernatus, P = polymnus, M = melanopus, A = akallopsis, FM = frenatus x melanopsus cross, FE = frenatus x ephippium cross, B = biaculeatus.
Age = Age of larva in tank. Day 0 is the morning after hatch and updated each day a new daily log sheet is filled out.
Dead = Number of mortalities which were counted, removed and recorded in the AM and PM (2 columns).
Wtr = Types of water exchanges or flows. Two columns for AM and PM. X = standard 30% exchange, D = rapid drip flow from one valve only (larval tanks had two small air/water valves). S = slow flow from one valve, M = medium flow from one valve, F = fast flow from one or two valves depending on the number of fish, their age and their condition.
Treat = Indicated if tank was being treated and with what drug and how long (treated water was never put back in the system.
Temp = measured usually during the morning.
Nest = Letter is the broodstock row (A) and the tank number (29). These white pair tags hung on the broodstock tank, and removed when a spawn was found in the morning, recorded and place on a running calender board on the estimated date of hatch. When ready the tag and nest were assigned to a larval tank and info recorded. Hatches occur during the evening, and the tag (see page 78) was returned to the proper broodstock tank the following morning.
Hatch = estimated percent of hatch based on the number of dead and/or living eggs still on the nest site plate or pot. Dead larvae are also considered part of the hatch since they hatch and died.
DE = dead eggs on bottom tanks or on the spawning substrate used
Met = percent of metamorphosis of fish in tank. Normally flows were started and 30% exchanges stopped when 70% of the larvae metamorphosed.
KI = Addition of potassium iodide to enhance metmorphosis.
Comments = General comments of interest and transfer of post-larvae and which tank in the juvenile area.
PLs = Number of post-larvae (20-30 day old fish) transferred into the juvenile grow out tanks

Points to consider on this sheet.
Tank 8 = These akallopsis hatched (80% with 77 dead eggs), during the evening and are assigned Day 0 the next day.
Tank 8 & 12 = Fish are being treated with nitrofuazone to aid a non-feeding problem. One dose only.
Tank 2 & 6 = Nests of frenatus from tank A29 and ocellaria from tank C6 were set up for hatching that evening.
Tank 5 & 11 = Fish 25 & 30 days old are transferred to two different juvenile growout tanks, 1 & 2 in row G. and row F, tank 6 in the juvenile growout section.

Table 34

DAILY LARVAL LOG

System LC
Date 10/31/80

Tank	Sp	Age	Dead am	Dead pm	Wtr	Treat	Temp am	Temp pm	Nest	Hatch	Dead Eggs	Met	KI	Comments	PLs
1	O	8	1	3	D/		27.9						5		
2	F				/		27.6		A29					New nest	
3	FM	12			M/		27.5							Good size	
4	FE	11	3		/X		27.9					100			
5	P	25	1		F/		28.0							160 to G2 350 to G1	510
6	O				/		27.8		C6					New nest	
7	O	6	9		X/X		27.9					5%	5		
8	A	0	66		/	FUR II	27.6			80%	77				
9	B	14		1	M/		27.6							Vacuumed 2x	
10	O	1	14	27	X/		27.6								
11	B	30			F/		27.6							407 to F6	407
12	O	0	25	53	/	FUR II	27.8			50%	121			Poor nest	

Sub Total	817
Balance	3050
Total	3867

Table 35 - Example of the montly post larval report used at IOH

March , 1982 POST LARVAL REPORT

	AK	BI	CL	EP	FR	FE	FM	EM	ML	OC	PR	PO	RU	NE	Total
No. of Pairs	3	-	10	3	13	5	2	-	3	24	1	1	-	1	66
No. of Post Larva	725	-	5242	1127	6642	2004	352	-	260	4604	0	942	-	165	22063
% of Total	3%	-	24%	5%	30%	9%	2%	-	1%	21%	-	4%	-	0.7%	99.7%
Post Lava/Pair	242	-	524	376	511	401	176	-	87	192	-	942	-	165	334
Nests Used	7	-	22	6	20	7	2	-	4	39	1	1	-	1	110
Nests Lost	-	-	-	-	2	-	-	-	-	-	-	-	-	-	2
Avg. Nest Size	333	-	763	683	551	536	302	-	205	170	23+	1217	-	217+	432
Total # Eggs	2330	-	16780	4096	11023	3750	604	-	818	6626	23+	1217	-	217+	47484
Dead Eggs	153	-	1266	875	2216	640	144	-	436	505	NR	56	-	NR	6291
Hatched Larva	2177	-	15514	3221	8807	3110	460	-	382	6121	23+	1161	-	217+	41193
Percent Hatch	93%	-	92%	79%	80%	83%	76%	-	47%	92%	-	95%	-	-	87%
Percent Survival	33%	-	34%	35%	75%	64%	77%	-	68%	75%	-	81%	-	-	54%

NOTES: 3/1 - Phased out non-feeding of Artemia in LB 1-12
3/3 - Basket particle filters cleaned.
3/17- Basket particle filters cleaned.
3/17- Installed timer on #2 filter Floodlight on at 0800 - off at 2200 attempt to enhance Caulerpa growth
3/20- Started new bag of Salmon mash- seems to have a red tint + finer grind than before.
3/30- Basket filters cleaned.

INDEX: AK = akallopisos, BI = biaculeatus, CL = clarkii, EP = ephippium, FR = frenatus, FE = frenatus x ephippium, FM = frenatus x melanopus, EM = ephippium x melanopus, ML = melanopus, OC = ocellaris, PR = percula, PO = polymnus, RU = rubocinctus

Table 36 - Post-larval production at IOH over four years which clearly shows season differences regardless of controlled conditions maintained in the larval rearing area.

	Jan	Feb	Mar	Apr	May	Jun	Jul	Aug	Sep	Oct	Nov	Dec
1979	18070	16487	1391	16180	11713	18142	24111	18097	10309	10362	10150	17528
1980	22193	23823	20198	28165	36503	32254	22392	11954	23566	21941	17150	4532
1981	8439	15540	23320	19796	20569	21229	24129	18373	17701	17738	14680	14880
1982	11215	22458	22063	27160	34059	27103	22161	19787	8093	22069	-----	-----
Average	14979	19577	19893	22825	25711	24682	23198	17052	14917	18027	13993	12313

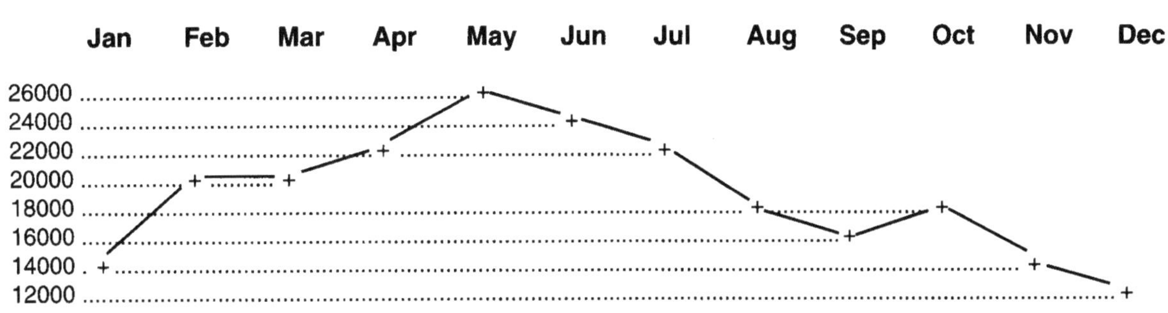

Table 37 - Average post larval survival on a seasonal basis. Data based on 2 years of records at IOH.

	Jan	Feb	Mar	Apr	May	Jun	Jul	Aug	Sep	Oct	Nov	Dec
80/81	22.6%	49.8%	50.5%	52,3%	59.1%	56.7%	50.4%	45.5%	35.6%	36.4%	35.6%	21.7%

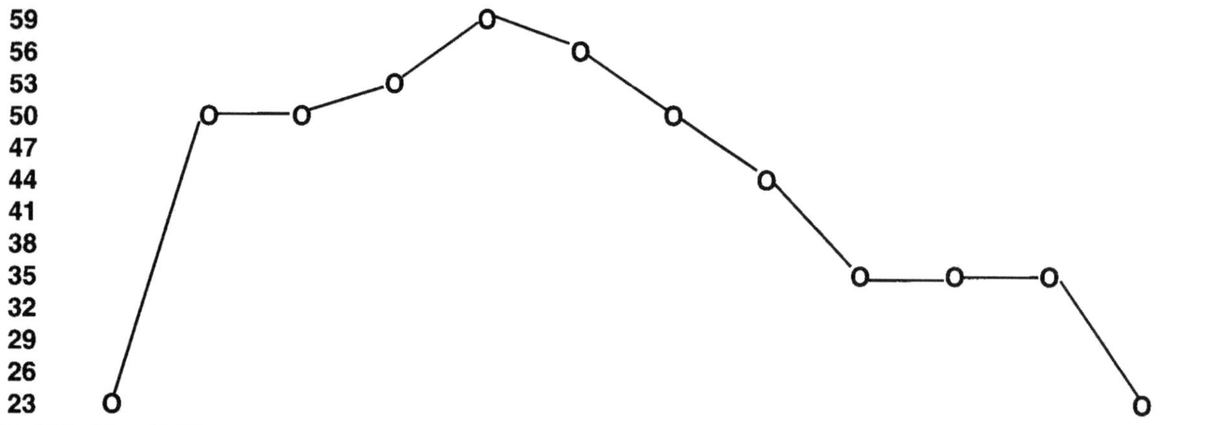

Table 38- Typical daily larval mortality pattern based on the average of 20 radomly selected hatches from *A. ocellaris* from October 1981 to January 1982.

Day	0	1	2	3	4	5	6	7	8	9	10	11	12	13	14	15	16	17	18	19	20
Avg	8	21	28	16	5	9	12	18	23	20	16	16	16	11	4	2	2	1	0	0	0

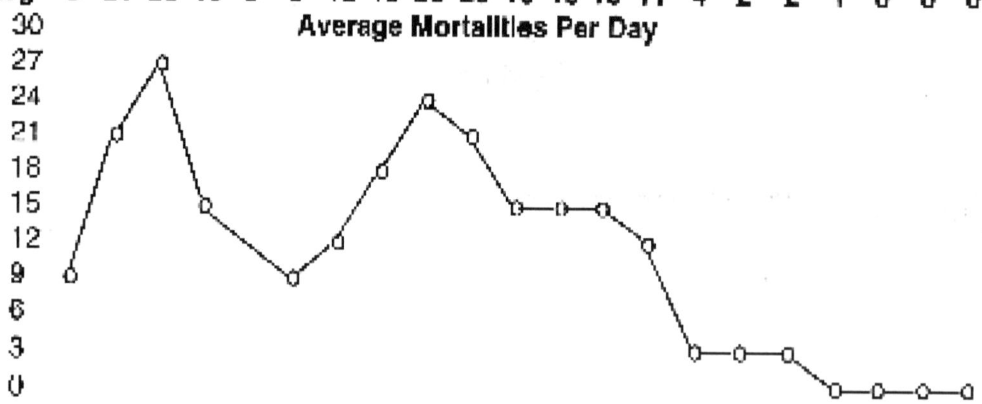

Table 39 - Percentage hatch and survival of post larvae (20-30 day old) of various species of clownfish reared at IOH. Data based on 20 months of records and on actual counts of unhatched eggs, dead and surviving post larvae.

	HATCH			SURVIVAL		
SPECIES	**LOW**	**AVG.**	**HIGH**	**LOW**	**AVG.**	**HIGH**
A. akallopisos	49%	86%	99%	0%	38%	64%
A. clarkii (sebae)	62%	83%	94%	4%	21%	42%
A. ephippium (red saddle)	34%	73%	92%	4%	29%	60%
A. frenatus (tomato)	44%	65%	87%	36%	60%	75%
A. melanopus (cinnamon)	21%	48%	90%	17%	54%	85%
A. ocellaris (ocellaris)	82%	88%	94%	37%	65%	77%
A. percula (percula)	88%	95%	100%	0%	57%	84%
A. polymnus	54%	85%	99%	0%	48%	85%
A. rubrocinctus	28%	68%	92%	0%	58%	88%
P. biaculeatus (maroon)	30%	70%	99%	0%	36%	89%

This daily raw data was condensed into monthly reports. Although time consuming and without the analysis capabilities of computers, these reports represented the very heart of our operation. Table 35 is an example of our monthly Post-Larval Report.

This data was further analyzed into graphs and yearly production assessments. Table 36 provides data on the post-larval production at IOH over the last four years. As shown, there are clearly peak and low seasonal periods of production even though light period, temperature, water quality, and feed were fairly consistent. This suggests natural biorhythms may strongly influence year-round production yield but not stop reproduction when proper environmental conditions are maintained. Average percent survival of larvae hatched against those that survived to post-larvae (20-30 days) before transfer to juvenile growout tanks is shown in Table 37. This graph closely follows the curvature of post larval production in Table 36 without regards to egg production. Again it appears that the season greatly influences production even when maintaining good conditions. Comparison of egg productions (Table 19) shows a similar conditions with regard that lowest production in the winter months and higher during April through September. Peak nest production and yield occurred in May through September in all respects.

Egg production (Table 19) was fairly consistent at 70,000/month with declines in the cooler months from mid to late October to March however, average post-larval survival of 50%+ occurred from February through July only. This probably indicates general degradation of egg quality during the warmer months (July - September) when water temperatures can approach 28° to 31°C. When comparing post-larval production from 1979 to 1982 we found that higher production was achieved when temperatures were maintained close to 27°C.

Some may look at this data and speculate that the natural light period and intensities may cause the annual rise and decline in production. This might be true since June 20th is the longest day and December 20th is the shortest day in this latitude. In Florida our shortest day is about 10 hours and longest is 14 hours whereas at the Equator day/night period is 12 hours year round. Our production and survival curves increase as natural light period increases before the longest day. In addition to natural light through the ceiling, larvae were subject to 14 to 16 hours/day of artificial light located about three feet above the tanks. However, daily average temperatures increases quickly in Florida from June through late September and then drops. Our data seems to follow this trend.

Table 38 is the average mortalities of 20 random hatches of various *A. ocellaris* form October 1981 to January 1982. As shown, there are defined mortality peaks day 2 after hatch and another at early metamorphosis around day 7-8-9. By day 13 mortalities are very low. Initial high mortalities at day 2 is usually due to poor egg quality resulting in weak larvae or delayed hatch due to poor water quality or temperature. The second high mortality is usually due to the inability of weaker larvae to initiate proper levels of feeding. Metamorphosis is a traumatic experience requiring a large expenditure if energy. If conditions are minimal, mortalities can be significant at this point.

Table 39 shows the percentage hatch and survival of post larvae (20-30 days after hatch) of various species of clownfish reared at IOH. Note there are distinct differences between species. With regards to hatch you can see that *A. melanopus* (cinnamon) consistently had a lower hatch, 48% average those that hatched and survival to post-larvae averaged 54%. Looking at survival is clearly shows that *A. clarki* (sebae) were the hardest to rear (21%) but not hard to hatch (83%). Whether these differences were do to species problems or related to our culture conditions is not well understood.

JUVENILE REARING

Introduction

The hardest point to realize in fish culture, is that all your efforts to get to this stage are controlled by what you did in the beginning. Problems that incur in juvenile rearing are directly related to broodstock treatment, nest and egg quality, and the ability to properly rear larvae prior to metamorphosis. As emphasized before, quality is carried right down the line and later problems are either genetically or physically induced at a very early stage of development. Older juveniles, those that survive, can alter their life style to compensate for physical or genetically induced deformities. Yet for pet market purposes, these individuals are not usually accepted. External physical factors such as mis-baring, flared-gills, notched operculum, and pug-nose are physical deformities related to poor larval rearing practices prior to metamorphosis. One eye, shortened body, and color variances appear to be genetic problems that occur during fertilization and egg development.

Post-larval (10 to 30 days) development reflects directly on juvenile development as the larval period (0 to 10 days) reflected on early post-larval development. Both are critical stages that seem to program future development and success. During this period growth and food adjustments are very important, and seem to dictate how fast true juveniles (30 days to six months) will grow.

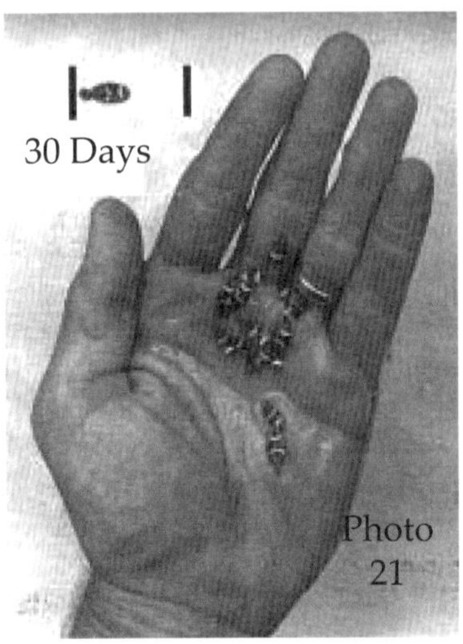

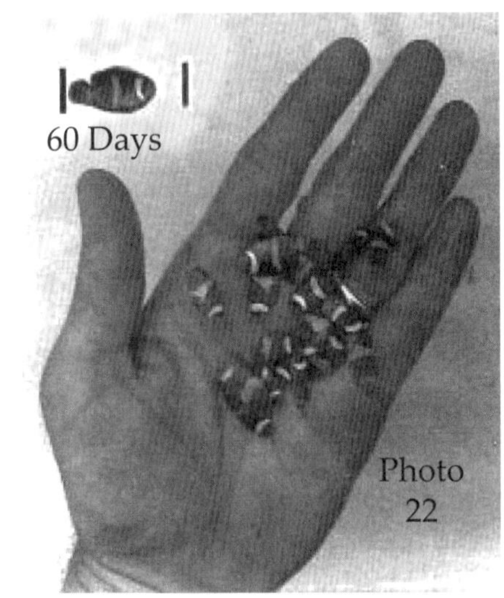

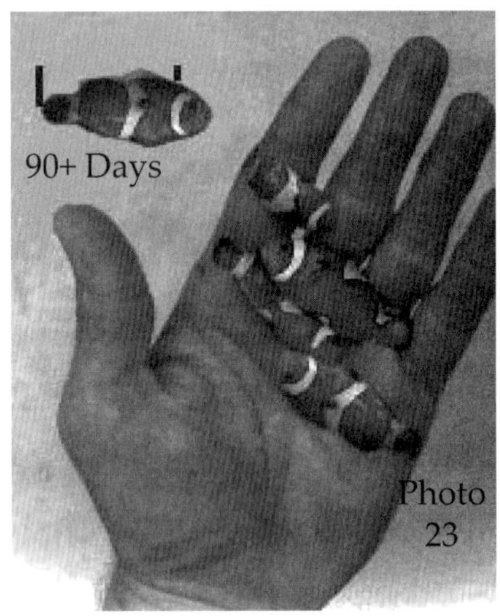

Photos 21, 22, and 23 - Juvenile A. ocellaris growth stages. Photo 21 is an example of a 30 day old post-larvae which is the size and age when they were transferred into juvenile growout tanks. Photo 23 is the size when A. ocellaris were considered safe

In commercial operations, it is considerably cheaper to correct problems in the broodstock, early larval, and post-larval rearing stages than to attempt to rectify resulting symptoms in the juvenile stage. It is not unusual to spend three months rearing fish to a marketable size and then see them slowly die from unknown reasons after subjecting them to relatively low levels of stress. Internal organ defects are usually suspect and no doubt occurred during early development. Prime causes of internal physical defects are: genetic, feeding procedure, food types and water quality. The latter factor is probably the most important especially around metamorphosis. I cannot emphasize the importance of a rapid and full metamorphosis starting at the proper age. Delayed metamorphosis is not natural, causes future internal and external physical problems that may or may not become detrimental. Following metamorphosis, growth rates of post-larvae and juveniles should be virtually linear.

Juvenile rearing is the most rewarding aspect of your efforts. It constitutes the polishing of a salable product. Growth accelerates at a phenomenal pace and color intensifies. Juvenile (post larvae) clownfish at transfer should be around 1/2" (13 mm) at about 30 days old (Photo 22). Generally it takes a total of four months to rear a salable 1 1/2" (38 mm) TL (total length) fish. Yet, it is not impossible to obtain a 1 1/2" to 2" TL (38 to 50 mm) individual in just three months from birth (Photo 24). Juvenile growth and development are strongly influenced by water quality; secondly, by food quality; and lastly, by the amount of food fed. Water quality is basically the essence of obtaining good results. During juvenile rearing the biomass and BOD (biological oxygen demand) increase significantly and your efforts should be directed to make sure your growout systems can meet the demand. System designs which allow 1.5 to 2.0 turn-overs per hour and good particle and biological filtration become extremely important in juvenile rearing.

Pigmentation is an important factor in marine tropical fish. Yet, keep in mind that if larvae are properly reared, resulting juveniles have the inherent potential to develop full vibrant pigmentation. Again this is highly dependent on early stage larval rearing techniques and practices prior to metamorphosis.

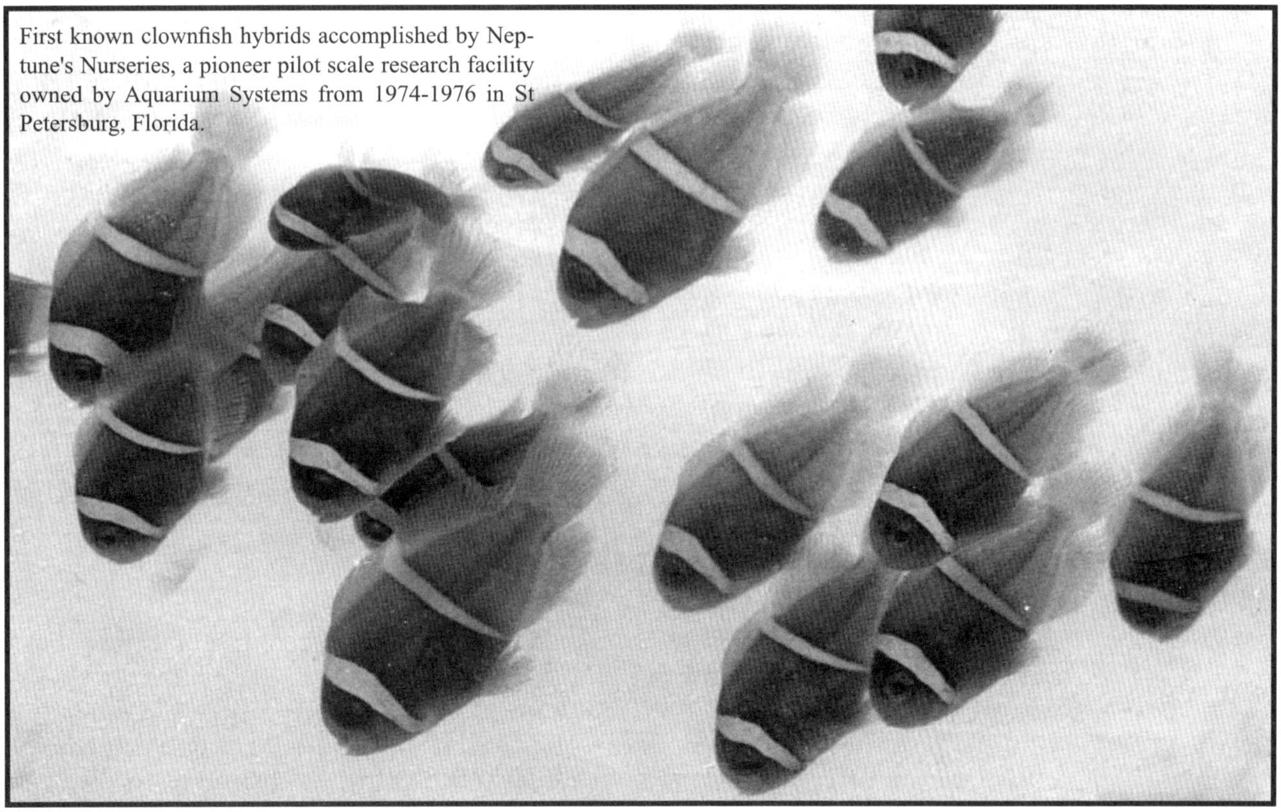

First known clownfish hybrids accomplished by Neptune's Nurseries, a pioneer pilot scale research facility owned by Aquarium Systems from 1974-1976 in St Petersburg, Florida.

Photo above is one of three species of hybrids cultured at Instant Ocean Hatchery in November 1975.. These shown are a cross between *A. frenatus* (tomato clown) from the South China Sea and *A. ephippium* (red saddle clown) indigenous to the adjacent Northeast Indian Ocean. They were listed and sold as FE. Further crosses were made between *A. frenatus* (tomato clown) and A melaopsus (cinnamon clown) listed a FM and between *A. ephippium* (red saddle clown) and A melaopsus (cinnamon clown) listed as EM.

JUVENILE REARING FACILITIES

Rather than completely rehash the rearing facilities and filtration systems described in the adult broodstock section, I will only emphasis important aspects to consider in juvenile rearing. A significant portion of the juvenile culture facilities should be dedicated toward filtration. It is during juvenile grow-out that filtration capabilities become critical. Increasing bioload, consisting of growing fish constantly demanding more food, increases BOD (biological oxygen demand) required for respiration and oxidation of waste. If not properly handled, the buildup of waste particles becomes detrimental. Expansion of my thoughts in the juvenile area in particular does not mean these designs discussed cannot be utilized in broodstock and larval rearing facilities. In fact, I encourage all that is possible to create clean "wholesome" water but, not to the point where water becomes nutrient free and virtually sterile. Presently, there are sufficient filtration equipment and chemicals consisting of: carbons, ion exchangers, ammonia and nitrate removers, UV and ozone sterilizers, reverse osmosis, protein skimmers, macro and micro-filters, and de-nitrifiers to virtually produce water so pure that it could not support life. This is simply not the objective you want to achieve. Fish, plants, and invertebrates live and grow in a complex mixture of dissolved salts, organics, and inorganic trace nutrients. This "biological soup" is the essence of life but in closed systems it can quickly become excessive or out of balance. Your objective is to balance this "soup" without stripping out essential nutrients. I have listened to so many aquarists brag about all the gadgets, fancy lights, and chemical filtration they have, then turn around and ask me why they can't grow plants or corals.

The best approach with filtration is to be simple, excessive, but not to the point where the water is virtually sterile of organics and inorgainics. Design systems that are flexible and easy to modify. Don't glue pumps and filters permanently in place but rather, clamp them or use unions so they can be easily and quickly cleaned, changed, or exchanged. Always allow additional space in your filtration design to expand.

Your main objective is to provide adequate mechanical filtration to remove all particulate matter floating in the water or settled on a surface. Biological filtration does not depend on particulate detritus, but only on dissolved nutrients. The biggest mistake in filter designs has been to trap particulate matter and biologically break it down into dissolved nutrients within the same filter. Ideally, you want to remove and dispose of all undissolved particulate matter before it breaks down any further thus reducing excessive levels of dissolved nutrients. Simply stated, biological filters work best when all particulate matter is eliminated first.

Growout Tanks

Juveniles can be reared in about any size or shape container. I have seen juvenile clownfish maintained in containers roughly in size from five gallon aquaria to 500 gallon circular tanks. In selecting tanks there are points to consider.

1. The tank should be easy to clean, manage, service, and fish should be easy to capture.
2. Access to the tank should not be too high or too low, but comfortable.
3. Tank size should really be based on the ease of maintaining water quality, circulation, and even dispersal of food.
4. Valves, lighting and aeration should be conveniently located and easy to adjust.
5. How many juvenile fish you are trying to rear per tank?

At Instant Ocean Hatcheries we used the same size tank as our standard broodstock tank (Figure 17 & 30). It was made of 3/4" "plyform" plywood, epoxy coated, and had a single 12" x 12" (30 x 30 cm) window, The inside dimensions were 23" x 23" by 20" deep with a 48 gallon capacity (180 L). Normal filled capacity was 42 gallons (158 L). The top of the tank was located so that it came to chest level. This allowed for ease of cleaning, feeding, and handling fish. On the bottom a plastic spigot valve was installed for draining and cleaning. To keep the tank uncluttered, the water level was maintained by an outside stand-pipe.

Access to the tanks was via a boardwalks which allowed for all plumbing to be located out of the way and allowed for dryer conditions for employees. The entire floor of the hatchery was tilted so that water drained on the floor went to a central drain then out to outside holding tanks for reconditioning (see Figure 64).

For small home level culture, virtually any size tank or vessel can be used provided you can maintain water quality. Hanging baskets (Figure 35) maintained within the adult tank have been used (Wilkerson 1993 personal communication). These come in fine and course mesh. You can start with the fine mesh and then advance to the larger mesh as the fish grow. Use of a mesh basket allows good water circulation and quality. Caution should be considered with regard to how many fish you try to rear in a single basket. If they constantly fight then there are probably too few, and if they are lethargic and not growing, there are probably too many.

Filtration At IOH

At IOH, all juvenile tanks were connected to a central recirculating system consisting of 60 tanks, 30 in each of 2 rows, 42 gallons each (158L). Each was equipped with a subgravel biological filter with a single 1 1/4" (32 cm) diameter air lift in the center. Below the rows of tanks was a second large subgravel biological filter 60'L x 4'W x 24"H (18.28 x 1.22 x 0.6 meters) with an internal capacity of 2,630 gallons (9.8 M^3).These were equipped with 8 airlifts per filter. Normal total growout capacity per system was maintained at around 2,300 gallons (7.5 M^3). Total water capacity of one juvenile growout system was 7,720 gallons (30 M^3). The biological substrate was a mixture of calcite, aragonite, shell, and coral mined from the ocean floor in the Bahamas.

Besides the subgravel, we used a 160 watt UV and at times a DE (Diatomaceous Earth) filter. Water exchanges were usually not made during the entire growout period of 4 months.

Current Filtration Suggestions

Currently, I would recommend the same tank and reservoir configuration, but not the filtration arrangement. The concept of utilizing a large volume of recirculating water provides a margin of safety with regard to how quickly both water quality and temperature change. In small, confined, low volume situations, water can quickly change in quality and become toxic. Whereas, in large volume situations, changes are slower and can often be rectified since you have a measure of time to work with. A larger volume acts as a heat sink by modifying hatchery temperature and preventing quick temperature changes. Stacking culture tanks above reservoirs and accessing them by a boardwalk conserves space and allows for clean, easy, plumbing situations. It is basically like having two floors in the building.

With regard to biological filtration, I do not advocate using conventional submerged, subgravel filters either in the grow-out tanks or in the reservoir below. Keep in mind that within the juvenile grow-out system, more food will be fed than in any other aspect of the hatchery. Gravel, crushed coral, oyster shell, etc. is just too hard to clean and maintain and can easily clog which can result in "toxic tank syndrome." Moe (1982) was one of the first to discuss this problem and noted that it was a common occurrence and could not be attributed to unknown pathogens and normal diseases. At IOH we also recognized this problem. Basically this is caused by the sudden "death" or reversal of the biological filter from an aerobic to an anaerobic situation. There are no known measurable indicators like ammonia or nitrite but there are some physical clues with regard to the visual buildup of detritus in the filter. The final stage usually happens at night and death of the fish can occur within six hours. Anaerobic bacteria can release toxins such as hydrogen sulfide, cresol, phenol, indole, and skatole (see diseases for further discussion).

Since sub-gravel biological filter collapses can be a prevalent problem in salt or freshwater aquaria, I now recommend simply using a bare tank with a single large air-lift sponge filter. In a 50 gallon (188 L) tank I would used a sponge filter with an external surface area of around 125 square inches (806 cm^2). The air lift should be around 3/4" (1.9 cm). These are available from Florida Aqua Farms. These portable biological filters are very unique and easy to use within a hatchery. They can be pre-conditioned using special nitrosomus/nitrobacter bacteria solutions and then placed in the tank ready to go. They can be easily removed and cleaned by simply squeezing them in freshwater and while damp, replaced back into the tank. New reconditioned filters could also be added as needed. In addition, they can be soaked in chlorine solution to remove all pathogens or washed in a washing machine (without soap). A word of caution is needed. New sponge material should be soaked for at least a week in several baths of freshwater before using. Once they are conditioned there is no problem.

I have been working with semi-closed and closed saltwater systems for over 27 years, and feel that multi tank systems are superior to static tanks. Juvenile grow-out is no exception. This is the only way to keep good, equal, water quality, and reduce maintenance. Figure 18 outlines what I feel is an ideal setup for growing juvenile clownfish.

All return flows from the growout tanks would be combined into a common manifold. This manifold would enter at the first chamber of the reservoir. The effluent would first pass into 50 to 100 micron (1000 microns is equal to a mm) polyester filter bags for removal of heavy detritus. These are convenient, cheap filters, and if installed properly, can remove 80%+ of the particulate matter coming from growout tanks. The water then goes into the reservoir below. To further filter out particulate matter a set of baffles can be installed through which the water flows up and down these barrier walls (Figure 18). Inside these chambers, can be suspended mesh bags of carbonate material for restoring natural buffer capacity. In the final chamber a pump withdraws water and half the flow goes into a fluidized bed filter which flows into a wet/dry trickle biological filter and back into the reservoir. The remainder of the flow goes into a set of Rainbow™ cartridge filters which reduce particulate matter down to 35 and/or 5 microns. At this point an optional set of cartridges can be installed for the use of carbon on an intermittent basis. Water would flow into carbon chambers only when needed. Continuous use of carbon or other specific ionresins are very selective for certain trace metals and other necessary minor nutrients, hence replacement maybe necessary. After passing through the Rainbow™ cartridges, the water goes into a UV (ultra violet sterilizer) unit. The UV will help reduce bacteria and protozoan contamination. In order for a UV to be effective, all detritus material should be removed from the water. Selection of the right size UV is based on return water flow rates, but I feel a unit twice the desired flow rate is best. This is the final stage before returning to the growout aquaria. In lieu of, or in combination with, cartridge filters, other mechanical filters such as sand and DE filters can be utilized in larger systems.

Ideal tanks for growout are plastic wash sinks available at most builder supply houses at very reasonable prices. Windows can be easily installed and they are already equipped with sloped floors and drains. They can be sanded and painted with epoxy to achieve a suitable color for growth. See the section on System Design and Filtration.

Opisthognathus scops panamensis

Drawing By H. W. F. del.
Results of the Fifth Vanderbilt Expedition
By Fowler, 1944

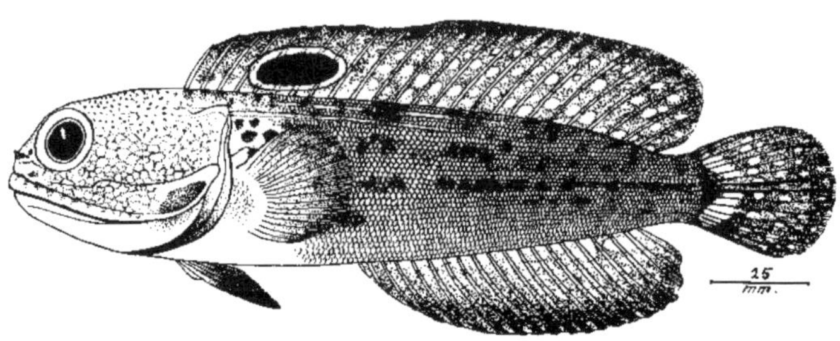

JUVENILE REARING PROCEDURE

The key factors in rearing juveniles are providing enough food, keeping the tank clean, maintaining water quality, and early detection of diseases. At this point the goal is to grow the fish as fast as possible to a salable state. Water quality, nutrition, and disease control become significantly more important in the juvenile grow-out area than anywhere else.

Criteria For Post-Larval Transfer

Transition from their first home, the larval rearing tank, to juvenile growout tanks is a traumatic experience and can become a disaster if not properly executed. Post-larvae can be transferred from larval rearing tanks beginning at about day 14 to 16. However, best results are obtained at an age of 20 to 30 days. Age alone should not be considered the only criteria for transfer. Size, health, and behavior are also strong considerations. Post-larval clown fish should average 1/2"+ (13+ mm) total length or better. If this is not the case, retain them longer in the larval rearing area. Proper growth size seems important especially with regard to *A. ocellaris* and other torpedo-shaped fish rather then deep-bodied fish like *A. frenatus*. Deep-bodied fish seem to have more reserve power and can handle stress more easily. Before transfer, they should be fully adapted to an exclusive diet of dry and/or mixed foods being administered three or more times per day. Their behavoir should be vigorous which constitutes charging towards the surface when you approach the tank. If they dart back into a corner and huddle most of the time, they are probably under stress. Swimming should be strong, and vigorous.

Before transferring post-larvae, test to see if they can tolerate dramatic changes in environments. After their first feeding, stick a net in the tank and chase them around a little, hand test them for shock and reaction (see section on feeding). Come back after an hour and see how they look and act. At this age they should tolerate netting with little difficulty. A good time to transfer fish is about 1 to 2 hours after the first feeding.

Transfer Method

If they pass all the criteria for transfer, fill a gallon (3.75 L) bucket with water from the larval rearing tank or system and provide moderate aeration. Use of slime and stress protectors such as Stress Guard™, Poly-Aqua™, or Probiotics™ may help. Carefully net out fish with an 8" to 10" fine mesh net, don't chase them but approach slowly and corner them . Using two nets will help if they are especially active. Gently remove only 20 to 30 at a time and count them. Place a wet hand under the net bag to provide a flat surface which helps facilitate quick counting. Lower the net into the bucket to release the fish, don't dump them. Capture only the number of fish you need to stock one juvenile grow-out tank.

Once the desired number have been harvested, place the bucket on the floor of a juvenile tank that has been previously filled with one or two inches of water in it (Figure 57). Continue to provide aeration to the bucket and tank. Very slowly flow water from the juvenile system into the bucket so that the waters mix. Eventually the water level in the tank will be high enough so that the bucket will float and tip. Usually the fish will continue to huddle inside the bucket and can be poured into the tank the following morning. The longer the acclimation period, the better.

Besides slime protectors, wide spectrum antibiotics are at times added to the transfer bucket to reduce bacterial infection created from netting and stress. Care should be taken when using these drugs since they can affect biological filters. Normally, if post-larvae are active, full bodied, and large enough they can withstand transfer without use of drugs.

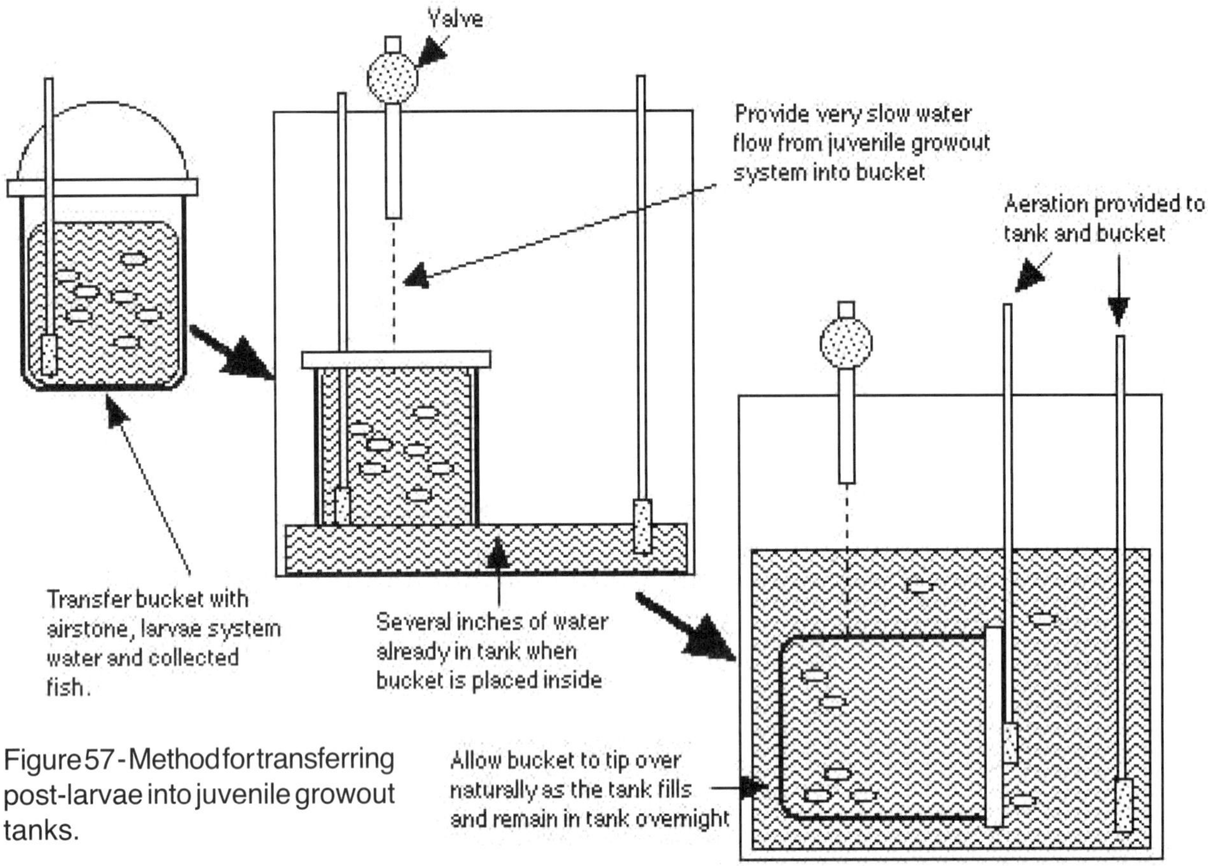

Figure 57 - Method for transferring post-larvae into juvenile growout tanks.

An information tag (Figure 58) is then set on the tank indicating where they came from with regard to the specific larval rearing tank and how many are stocked. This tag remains with these fish until the tank is finally harvested. If fish are removed for sale, the number and date are placed on the tag. At final harvest small, unsalable, and large salable fish are all noted on this ID tag for calculating total survival.

Stocking Density

Clownfish are very territorial and defensive, a characteristic that develops at a very early age. Prolonged fighting and territorial displays can become a real problem and can cause differential growth, deformed mouths and heads, short tails and pevic fins, and eventual death. There are four ways to stop or slow territorial problems: (1) provide each fish with it's own tank, (2) provide a highly diverse habitat, (3) increase the tank volume so each fish has several gallons of water (extensive culture), or (4) crowd them so there is no territory to defend (intensive culture). Of these suggestions, the latter two are utilized in commercial situations. For obvious practical reasons, the intensive method is normally preferred when comparing costs, space requirements, and maintenance problems.

We tried stocking densities from 2 to 20 fish per gallon (3.75 L) and consistently got better survival, more uniform growth, and less fight-induced deformities at 7 to 9 fish per gallon (3.75 L). Regardless of the clownfish species and age up to four months, this ratio seems to work best. Li and Brocksen (1977) noted that behavioral characteristics of certain ages and species of fish are related to dominance establishment. When fish densities are too low, dominant fish chase subordinate fish and cause injuries that occur during fighting. When densities are in an optimal range for the species, the incidence of dominance chasing and fighting is reduced. Stocking densities at IOH were 350 to 400 fish per tank, 21,000 to 24,000 fish per system, or 8.33 to 9.52 fish per total tank volume (42 gallon filled) or 2.72 to 3.50 per total water volume of an entire growout system.

Growth rates are directly related to your ability to administer food evenly and at the same time maintain water quality. Selecting larger tanks because you think the fish will grow faster is simply not the case. You can grow a five pound fish just as fast in less then two gallons providing you can properly maintain water quality. What you are trying to achieve is maximum efficiency and uniformity. When fish density is low per unit volume, food is wasted which accumulates and reduces water quality. In addition, fish have a tendency to spend more time fighting, seeking dominance, and establishing territories which in turn results in wide growth variance and injury. At too high densities, growth is basically even, but slower due to water quality management and the necessity to administer more food at a given time. Selecting a suitable density that provides relatively equal, fast growth for all the fish is what you want to achieve.

Tank Identification

Each tank should be marked with an identification tag (Figure 58) which states the specie, number of fish transferred from specific larval rearing tanks and date of transfer, and total originally stocked. Tags should also provide data with regards to how many fish are removed for sale and the date. A daily log was used to recorded mortalities, problems, feeding, and other trends in behavior. Mortality trends were evaluated weekly to evaluate pending problems, apply medication and/or change the environmental conditions.

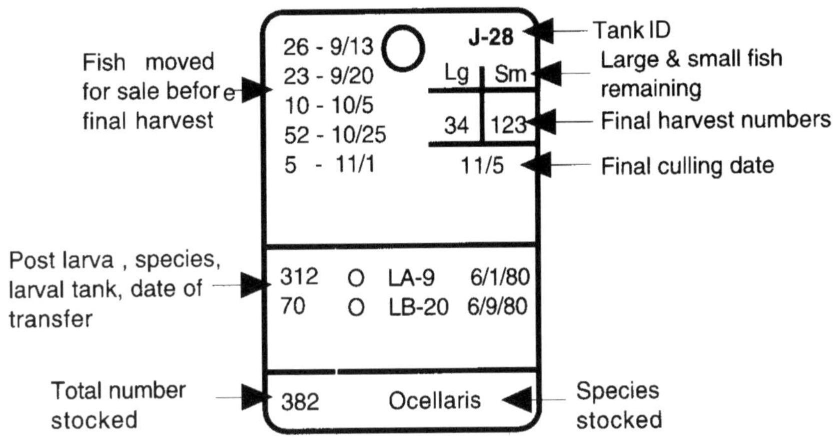

Figure 58 - Tank information tag used on juvenile growout tanks at IOH.

Feeding

Juveniles should be fed a minimum of three times a day to obtain maximum growth. Gelatin mix was fed at IOH two to three times a day and supplemented with one to two feedings of

Table 40 - Feeding schedule for juvenile clownfish used at IOH

	Food	8:30am	11:00am	1:30 pm	3:30pm	4:30pm
Young Juveniles	Mix	-	X	-	X	-
(up to 3 weeks)	Dry M	X	-	X	-	-
	FDP	-	-	-	-	X
Older Juveniles	Mix	-	X	-	X	-
	Dry P	X	-	X	-	-
	FDP	-	-	-	-	-

Mix = Gelatin mix (see pages 69, 70, & 71)
FDP = Freeze dried plankton, Krill
Dry = Salmon mash (M) or pellets (P) #1, #3, or #4

salmon mash size #1 (850μ) or #2 (1.18-2.00 mm) pellets and/or freeze dried plankton. Salmon mash or pellets were fed initially in the morning prior to counting mortalities or cleaning tanks. Gelatin foods were fed the remainder of the day at 11:00 am, (1:30 pm sometimes), and 3:30 pm. Usually juvenile clownfish were fed four times on week days and twice daily on the weekends. Salmon mash is a semi-moist dry diet with moderately high fat levels (15%) and high protein levels (50-55%) while our gelatin-based mix was low in fats (4.44%), moderate protein (26%), but high in macro and micro algae, vitamins, minerals, and pigment enhancers. Table 40 outlines a typical juvenile feeding regime used at IOH.

Initially, it is good practice to make sure food particle size is conducive to mouth size. This should be nothing more than a straight transition of foods, food sizes, and feeding practices that were being administered in larval rearing for a week or longer prior to transfer. Initially, the gel diet was often sized by passing it through a sieve. Size can be regulated by how hard the mix is pushed through or rubbed on the sieve screen. This practice is obviously detrimental to water quality and should only be used to establish good, solid, feeding and feed recognition by newly transferred juveniles. Two to three weeks is usually enough.

Older juveniles were fed a 1/2" cube of mix suspended on a de-barbed hook (Photo 24). Although strange, this practice worked very well. More aggressive individuals attack the food and tore it apart while less aggressive individuals gathered below and captured falling uneaten food particles. Dry foods were administered using large spice shakers equipped with grid size corresponding to juvenile mouth size. All foods were stored in a refrigerator or freezer. Sieves and containers were sterilized each day. Feeding was based on about 3% total body weight per tank, per day. Feeding rate is basically a judgement call based on the number of fish in the tank, their daily attitude, and water quality degradation. Sticking to a rigid amount of food to each tank, each day, may be detrimental if fish are not feeding well. However, excessive feeding does not result in faster growing fish even when they are eager eaters. It is really a matter of consumption and efficiency of digestion. As mentioned before, fish have a habit of consuming what is offered but do not adequately utilize excess food. See Nutrition and Disease section for more feeding information.

Maintenance

Uneaten food and fecal matter should be removed each morning using a siphon or turkey baster with a short piece of semi-rigid tubing mounted on the tip. In addition, a count and removal of dead individuals should be conducted. This was accomplished by turning off the master air supply valve for an entire row of tanks then using a viewer to look over the entire bottom of the tank. The viewer was made of an 8" long section of 4" diameter PVC pipe with a clear plastic plate glued on the end. Two 1" diameter holes were drilled close to the top for gripping viewer. In addition, water flow (1.5 to 2 turnovers per hour) and air supply (heavy) should be adjusted daily.

Windows and sponge filters should be cleaned every two weeks. Sponges should be submerged in freshwater and squeezed several times to remove heavy particles, squeezed again to remove excess water and placed back into the rearing tank. Or they can be hosed, squeezed, and placed back into the tank. Do not put new unconditioned filters into rearing tanks since they have to recycle and can be toxic. Aerobic bacteria in the sponge filter will not be killed quickly by placing it in freshwater.

Figure 59 shows results from an experiment comparing the effect of cleaning procedures with regard to nitrifying bacteria destruction and water quality after restarting a system. Our normal procedure for cleaning a juvenile system was to drain the entire system, wash-down the filter gravel and tanks with freshwater, fill the entire system with freshwater and circulate for three to seven days. It was then drained again, refilled with saltwater and stocked with fish. This procedure

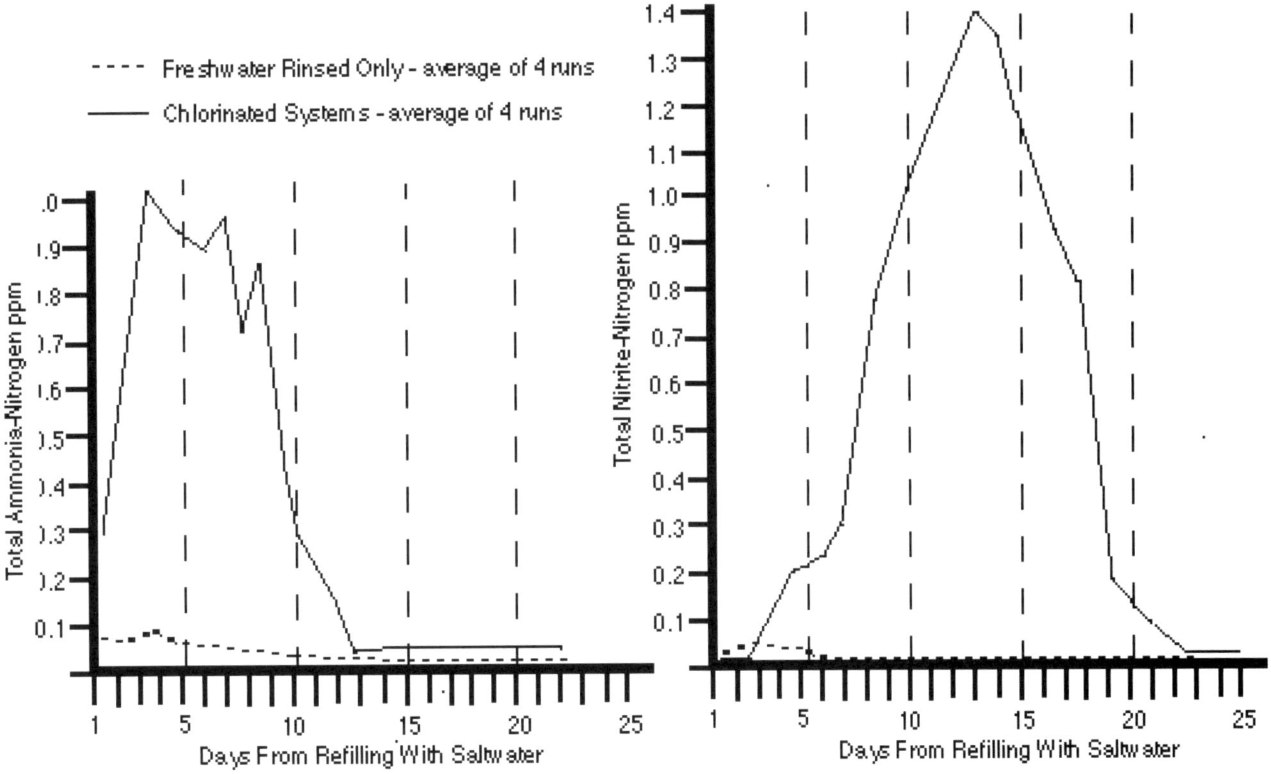

Figure 59 - Comparison of ammonia and nitrite levels in eight juvenile systems that were drained, and cleaned. Four were filled with freshwater no chlorine and four included chlorine. They were drained again after 3 days, rinsed and refilled with saltwater.

was later modified to include chlorine during the rinse cycle, circulated, neutralized with sodium thiosulfate, tank drained and refilled with saltwater and fish. Monitoring water quality showed very little rise in ammonia when compared to systems cleaned with chlorine and freshwater, indicating that nitrifying bacteria are not overly effected by switching from salt to freshwater and vise versa. Keep in mind, grow-out systems at IOH had sub-gravel airlift filters in each tank and a reservoir located below the tanks which also had a sub-gravel filter.

Chlorinated systems showed the typical rise and fall of nitrites and ammonia over a two to three week period. Fish were stocked immediately after these systems were cleaned regardless of whether they were chlorinated or not. Ironically, over a 37 day period we had lower mortalities, 2.63%, in chlorinated systems compared to 13.20% for straight freshwater even though the chlorinated systems went through a complete nitrogen cycle. Ammonia and nitrite concentrations reached levels considered toxic by some. Histological examination of these fish did not show any adverse lesions.

Table 41 - Typical culling data sheet use at IOH.

JUVENILE CULLING DATA SHEET

ROW ___M___ YEAR _1982_

Tank	Specie	Original#	Date	Cull Date	Lg	Sold	Sm	Tot	%
1	Cl	400	4/26	9/2	43	0	111	154	39
2	Cl	401	4/26	9/2	38	0	105	143	36
3	Fr	400	4/28	9/2	71	20	107	198	50
4	Fr	350	4/29	9/2	31	79	80	190	54
5	Oc	358	4/25	9/2	39	75	54	168	47
6	Ep	360	4/25	9/2	8	0	34	42	12
7	FE	300	5/1	9/2	54	52	104	210	70
Totals	-	2569	-	-	284	226	595	1105	43

Criteria For Sale And Culling

Juvenile fish do not like to be handled often. Excessive handling in order to remove a few salable fish is not advised. Undue stress just aids possible disease. Setting a criteria of 50%+ salable fish per tank before harvest can be used as a guideline. Harvesting large individuals from a single tank can go on for several weeks and then it is advisable to cull remaining fish and condense them for a final growout and rest. As fish are removed for sale the amount were marked on the tank tag (Figure 59). Table 41 is an example of a culling data sheet. Culling sheets were further condensed into monthly reports for each juvenile system. Table 42 is an example of a Juvenile Survival Data Sheet.

Table 42 - Monthly juvenile grow-out data sheet which was completed after a row of tanks had been culled.

JUVENILE SURVIVAL DATA SHEET

ROW P DATE 9/2/82

DATA

Missing tags	**0**	0 out of 28
Culling date	9/2/82	
Stocking date	5/1/82	
Number of grow-out days	124	
Average stocking age	28	Range 25 to 31
Total fish stocked	11,139	
Average fish stocked per tank	399	Range 356 to 432
Number sold before culling	793	
Number salable at culling	1923	
Total number of salable fish	2686	
Number of small or sick fish at culling	1939	
Total number of survivors at culling	3862	
Total number of survivors plus those sold	4625	

ANALYSIS

Survivors sold before culling	17%
Sold before culling of those stocked	7%
Salable from total survivors at culling	50%
Non-salable from total survivors at culling	50%
Total salable from those stocked	24%
Survivors from those stocked	42%

CONCLUSIONS

Pale silver sides noted at culling, signs of TB? This system obviously suffered high mortalities early or late? If early, it might be attributed to the unusual high stocking density. Although, row "O", which is on the same system, had excellent survival and was stocked at the same density.

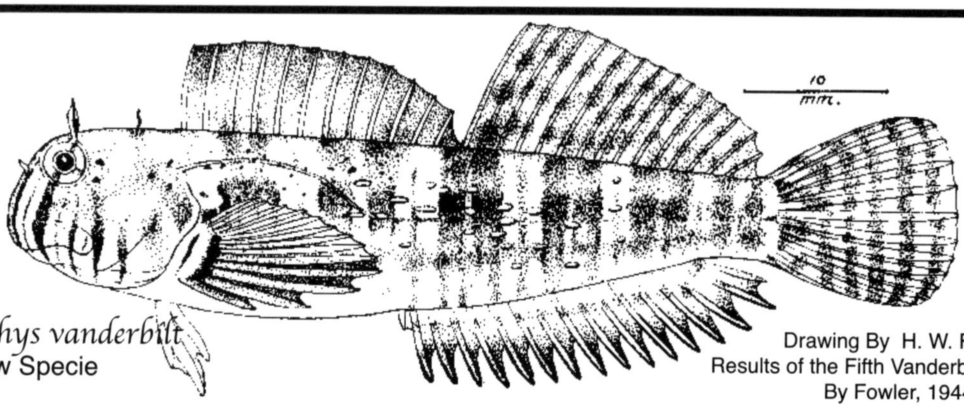

Salarichthys vanderbilt
New Specie

Drawing By H. W. F. del.
Results of the Fifth Vanderbilt Expedition
By Fowler, 1944

NUTRITION AND DISEASES

Ideally we want to achieve a feeding efficiency as illustrated in Figure 61. Nutrition and diseases are factors which when coupled with water quality will make or break you. I have lumped these two very important factors together since "diseases" are usually closely tied to nutrition and water quality. Water quality is basically a controllable factor providing we can monitor the factors that create problems. To date, I am still not totally convinced that measurements of ammonia, nitrite, nitrate, pH, alkalinity, phosphate, and temperature are everything we need to know about water quality. Recent use of minor trace elements such as selenium and iodine help support my apprehensions. I vividly remember discussions with my friend Stephen Spotte many years ago regarding trace elements. It was stated that oysters in natural unpolluted waters concentrate selenium but, was of no use and cannot be passed. I felt then as now that this is simply not the case. Concentrations of natural trace nutrients within animal tissues are usually beneficial and often seasonally necessary. I am truly an advocate of vitamins, trace elements and minerals when and whenever they might be needed, especially within closed systems.

Pathway Detection

Stress is a condition in which an animal is unable to maintain a normal physiologic state because of various factors adversely affecting its well-being. Stress is caused by placing a fish in a situation which is beyond its normal level of tolerance. An animal is able to adapt to stress for a finite period of time. During this period, the animal may look and act normal, but is depleting energy reserves because of the extra requirements placed upon it (Francis-Floyd 1991).

Most diseases are secondary results of poor nutrition compounded with a stressful environment. Fish are no different than humans, when we are tired, eat poorly, and are under extreme external stress, we become sick. Juvenile fish, like kids, are the most susceptible to nutritional disorders and can easily be stressed if they become weak. Children are most susceptible during or just after a growth phase when the entire body is changing. Young fish are also always on the verge of self-destruction during the traumatic initial growth phases.

Detection of vectors or sources of problems must be organized into planned investigations and experiments. The process of step by step elimination is certainly the most logical way to proceed. Yet, most of us lack the patience or data to eliminate one possible vector from another. Sacrifice of time and animals might be necessary to save your operation. Statistically valid procedures and results are the best way to eliminate problems. Yet, it requires space, time and sacrifice.

When dealing with suspected "diseases", most of us resort to the shotgun approach with little knowledge of the cause, what dose of medication is required, how long to treat, and what to suspect. Our knowledge and reasoning for resorting to this technique is usually based on hearsay, frustration, and impatience. Unfortunately we usually do not get the same result others claimed they did.

A "disease/nutritional" problem should be looked at with a process of elimination. Figure 60 provides many vectors to consider when trying to determine where or why a problem exists. Careful examination of all these vectors should be considered before making a final decision on how to induce an effective treatment program. Within this figure I have provided various widths of connecting lines. Heavy lines represent what I feel are the most important vectors and considerations. You will notice that food plays an important factor in my feelings toward eventual disease problems. Notice I feel that new introductions of broodstock are basically irrelevant providing normal practices of isolation and slow acclimation are used. Diseases are always present, regardless of what you try to do. It is your actions and reactions that create keys for opening up future problems. All too often you are providing the catalyst that put problems in motion.

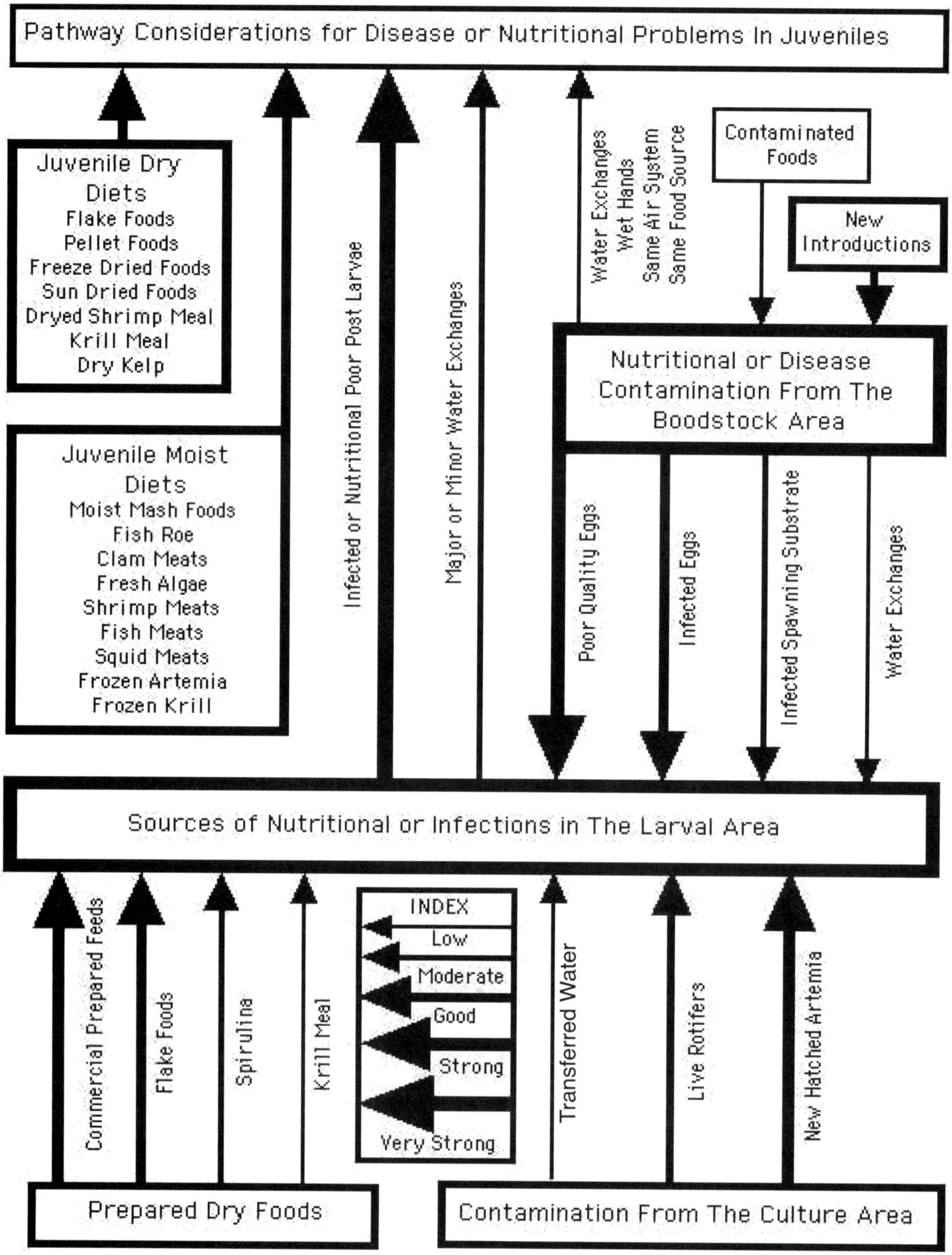

Figure 60 - Nutritional and disease pathway considerations in adults, larvae, and especially juvenile fish.

Notice I feel that commercially prepared food can be a significant source of problems in the larval area. Unfortunately, if these foods are not fresh or have not been properly stored they can present a minor to a very significant problem. Water exchanges are considered of less importance than possible nutritional deficiencies from contaminated foods.

Two items are most important to fish culture: water quality and food. I strongly feel, that in most cases, we are trying to overfeed fish under the assumption they will grow faster. Basically this sounds good but too many studies have also shown significant growth is achieved using less, but more expensive feeds. Seeing we have a problem just trying to determine what is good for our own bodies, how can we be so sure about fish?

Foods

Ideally we want to achieve a feeding efficiency shown in Figure 61. I have yet to find a food on the market that is truly suitable as a single diet for the proper growout of juvenile clownfish or many other ornamental fish. The main diet fed at IOH to broodstock and juveniles was a home made gelatin-based mix. I have personally utilized gelatin based foods since 1971 to raise and maintain many marine fish and invertebrates. Mixes are unique since you can control the quality and variety of ingredients (see pages 69-71). You can add specific vitamins, minerals, and trace metals as needed. You can add enhancers, attractants or even drugs for treatment. Gelatin based mixes take time to make but are more versatile and uniform with regard to quality.

Initially fish require higher fat and protein levels in their food, up to 15% and 50% respectively. As they get older (around 2 months) levels should be reduced to 3 to 5% fat and the protein levels should be around 30 to 40%. With regards to vitamins, I prefer a full spectrum, high concentration, with emphasis on vitamin C, E, B_{12}, A, and D. Likewise, minerals should also be full spectrum. Utilization of kelp meal is a good natural source for minerals. See Broodstock section for diet constituents and mixing instructions for gelatin diets.

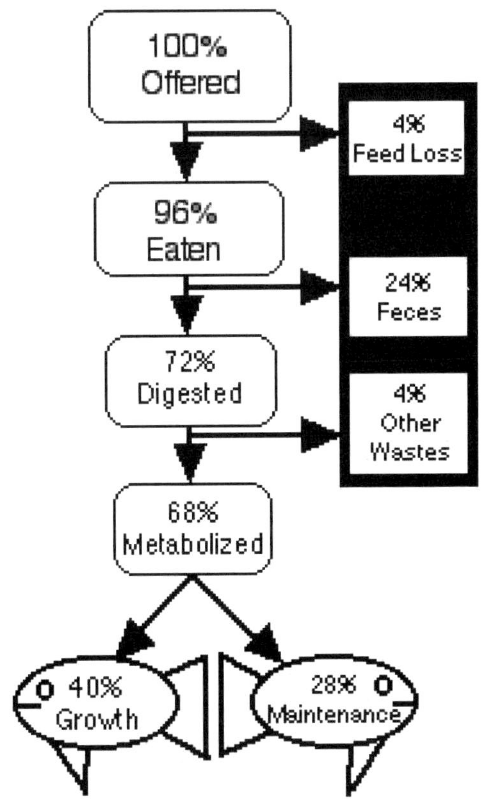

Figure 61 - Ideal feeding balance we want to achieve.

Feeding Strategy

Clownfish are basically surface and water column feeders and become lazy when forced to feed off the bottom. Therefore, to obtain effective, efficient feeding, efforts should be made to keep the food suspended on the surface or in the water column as long as possible. This can be accomplished by feeding foods that float, increasing uplifting currents in the tank or suspending the food in a feeding apparatus (Photo 24).

Over feeding is a serious problem in juvenile growout. Small fish consume a higher percent of their body weight per day than do larger fish. Juvenile clownfish can consume 3 to 4 percent of their body weight per day until they reach sub adult stage (7 to 9 months age). Feeding will drop to about 2 to 3 percent as they approach adulthood. As a guideline Masser et. al. (1992) provided estimated food consumption by size of typical warmwater fish. A nine gram fish consumes about 5%, 18 gram fish 4%, 27 gram fish 3.3%, 114 gram fish 3%, 340 gram fish 2.8% and 454 gram (1 lb) fish 2.2%. For culturing juvenile clownfish the largest salable fish at 3 months age will weigh less than 9 grams.

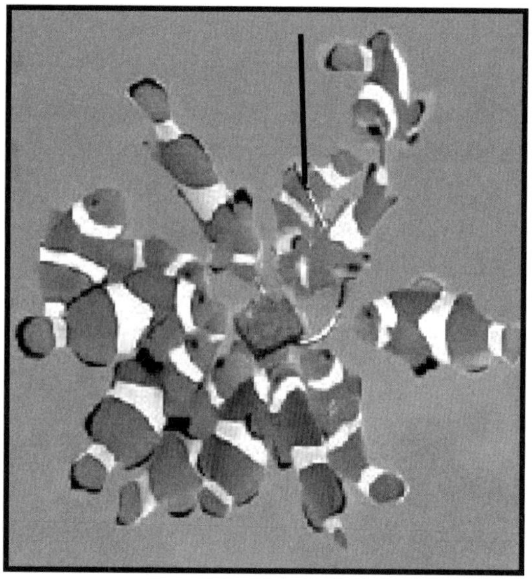

Photo 24 - Feeding method used at IOH. Gelatin based food suspended on a barbless hook.

So how many times and how long should you feed fish on a daily basis? Bryant and Matty (1980) studied the frequency on the growth of carp juveniles. Groups of fry collectively weighing 100 mg were fed trout fry feed at an optimal rate of 15% body weight per day via automatic feeders set at 2, 4, 6 and 8 hours and continuously over 24 hours. Fry fed continually and at 2 and 4 hour intervals showed almost identical growth rates whereas, 6 and 8 hour intervals showed poorer growth and food conversion. Best growth was achieved by fry being fed continually, but the most even population growth was among fry fed at two-hour intervals, 12 times a day. However, excessive feeding was detrimental even though no detectable deterioration in water quality could be determined. In addition, there was evidence of gill damage associated with bacteria. This experiment helps point out that too much feed may be detrimental and can become a negative. Masser (1992) states that research has shown small fish grow faster if fed three to four times a day. Feeding several times a day seems to reduce problems of feeding dominance in some species of fish. Multiple feedings spread out the waste load on the biofilter and help prevent sudden decreases in dissolved oxygen.

If using a gelatin diet, remove only enough food from the freezer for feeding for five to seven days. Allow it to defrost in the refrigerator compartment and cut into 1/2" to 5/8" (13 to 16 cm) cubes. At IOH, gel cubes were attached to stainless steel debarbed, dulled, hook suspended about 6" down into each tank which allowed the fish to attack the food as desired (Photo 24). Suspending food accomplished two things: 1) it prevented the food from quickly sinking to the bottom, and 2) it allowed fish to remove desired amounts of food. We found this method more effective than breaking food into particles. Clownfish are well equipped with frontal teeth for ripping off food particles so it is advantageous to utilize this attribute.

If properly prepared, gelatin diets will remain together for 12 hours. Usually it takes about 5% to 7% gelatin to accomplish this. There are distinct advantages in administering moist foods that can stay together yet, are soft enough to be eaten.

1. Water quality does not deteriorate as quickly with solid foods.
2. Uneaten solid foods can be easily removed.
3. Solid, moist foods are less prone to leaching of essential nutrients and vitamins than flake foods. Once any food is placed in water it starts to absorb water, break apart, dissolve and leach out essential nutrients and organics.
4. Populations of fish can be easily assessed for health status based on the amount of uneaten food. When using dry flake or granular foods, it is harder to distinguish the amount of uneaten food from fecal material.

Pigmentation

Factors you want to achieve in juvenile growout are: (1) obtain a good size fish in the smallest amount of time, (2) produce a morphologically perfect fish, (3) achieve intense pigmentation. Coloration is an important factor since it is related to a variety of things including: tank size, background coloration, growth rate, and nutrition. There are those who say that if the fish does not have good coloration at metamorphosis then it will not get any better. Others claim the sun or strong light is needed. Still others feel diet is the key factor. I have found that fish have sufficient pigmentation (chromatophores) to develop desired color intensities providing they are properly maintained prior to hatch and during metamorphosis. I have not found any evidence that young, muddy-red colored, tomato clownfish cannot be changed into a brilliant, vibrant red individual. Takeuchi (1960, 1968) noted that the color of the integument of *Oryzias latipes* depends to a large extent on the intensity of coloration of the original eggs. For more discussion see spawning section regarding nest coloration.

Table 43 - Astaxanthin content found in natural food items (in part from Torrissen, et al 1989)

Food item	Form	Content
Copepod (C. finmarchcicus)	oil	520 mg/kg
Crab, red (P. planipes)	oil	1,550 mg/kg
Crab, (Greyon quinquedens)	freeze dried	76 mg/kg
Crawfish, (Procambarus clarkii)	oil	750 mg/kg
Crawfish, (Procambarus clarkii)	meal	137 mg/kg
Krill, (Euphausia pacifica)	fresh frozen	80-120 mg/kg
Krill, (Euphausia pacifica)	dried meal+oil	200 mg/kg
Krill, (Euphausia pacifica)	oil	727 mg/kg
Krill, (Euphausia superba)	fresh frozen	672 mg/kg
Krill, (Euphausia superba)	dried meal	208 mg/kg
Shrimp (Pandalus borealis)	fresh shelled	20-128 mg/kg
Shrimp (Pandalus borealis)	silaged waste	74 mg/kg
Shrimp (Pandalus borealis)	vacuum dried	100 mg/kg
Shrimp (Pandalus borealis)	steam dried	192 mg/kg
Shrimp (Pandalus borealis)	oil	1,095 mg/kg
Yeast (Phaffia rhodozyma)	dry	30-100 mg/kg

Table 44 - Xanthophyll content in plants.

Plant	Form	Content
Algae, common	dried	2,000 mg/kg
Algae micro (Chlorella pyreniodosa)	fresh	4,000 mg/kg
Algae macro (Ascphyllum nodosum)	fresh	340 mg/kg
Algae macro (Fucus vesiculosus)	fresh	350 mg/kg
Algae macro (Fucus serratus)	fresh	920 mg/kg
Marigold petal	meal	7,000 mg/kg
Paprika, spanish	meal	275 mg/kg

Intense coloration is primarily developed through food. But, this alone will not do the trick. For example, fish grown in a very large tank (low density, extensive culture) often have better color than those in crowded conditions. Fish that grow slower than normal usually have more intense coloration even under intensive conditions. Fish grown in tanks with dark backgrounds develop dark, muddy, coloration while those in light colored or just glass tanks have light pale coloration. Fish with excellent coloration that are moved to clear glass aquaria like those found at a wholesaler or shop often become dull or light in color in a matter of weeks.

Stress plays a very important factor in coloration. Fish are like chameleons, quickly changing color based upon background and physical conditions. For example, when you go to feed a fish, it's color brightens, if sick or highly stressed, it's color darkens. When a fish is close to dying it's color may become intense but darker than usual. Observing clownfish coloration at night is significantly different. They become very pale and the white bars are almost transparent. Pigment cells expand and contract according to light intensity and background coloration.

Fish that grow slowly seem to have better coloration, but when dealing with economics, the objective is to grow fish quickly to market size. Yet, not to the point where they are basically weak and highly susceptible to stress. Good coloration becomes an acquired attribute over time.

Wild caught fish often have vibrant coloration but when retained in the average aquarium, the color becomes dull or "muddy." This has been attributed to diet changes but sometimes, a combination of diet, tank background, lighting, and/or water quality is responsible.

Freshwater and marine fish, like other animals, cannot synthesize carotenoids, thus their occurrence in such animals, and quantitative relations involved, depends on the amount present in the food ingested (Meyers 1983). Czecuga (1979b) and other researchers have shown that a variety of naturally occurring foods provide a better opportunity for accumulation of carotenoids in the body of salmonids (salmon) than when singular diets are presented. Tanaka (1978) has shown that ingested pigments may be deposited in fish tissues directly without modification, or after biotransformation, depending on the tissue and species specificity in the carotenoid assimilation. Of the latter, astaxanthin and canthaxanthin are most frequently end products of carotenoid metabolism in marine and freshwater fish.

Bright red, orange, yellow, and blue coloration of fishes originates from fat-soluble pigments (carotenoids) in the food. Two main carotenoid classes in the oceans are carrotenes and xanthophylls, which produce red and yellow coloration. Carotenoids, especially lutein, beta-carotene, and zeaxanthin, are synthesized initially by micro and macro plants, then assimilated directly or indirectly into the skin and flesh of fishes. Crustacea which feed on microalgae mats can metabolically convert beta-carotene to echinenone and then into cathaxanthin or astaxanthin. Cathaxanthin and astaxanthin, common pigments in fish, can be derived from a common precursor, beta-carotene which is biotransformed. Carotenoids of marine origin are reviewed by Liaaen-Jensen (1978). Table 43 lists natural sources of astaxanthin and Table 44 lists xanthophyll content of several plants (Rice & Overton, 1993).

Fish are thought to fall into three categories in their ability to transform carotenoid pigments

Table 45 - Pigmentation analysis of feeds being fed to clownfish compared to wild and cultured clownfish at IOH.

Carotenoids	GBM	FDP	MR	LE	WA	CJ
Unknown[1]	0.17	-----	-----	0.10	-----	-----
Beta-Carotene	-----	-----	-----	-----	-----	0.4
Echinenone	-----	-----	-----	0.24	-----	0.09
Omega-Cryptoxantin	-----	-----	-----	-----	0.12	-----
Cryptoxanthin	0.06	-----	-----	0.70	0.19	0.15
Isocryptoxanthin	-----	-----	-----	0.46	-----	-----
Canthaxanthin	-----	-----	-----	0.40	-----	-----
cis-Lutein	-----	-----	-----	0.06	-----	-----
trans-Lutein	0.07	-----	-----	0.18	trace	10.39
cis-Zeaxanthin	-----	-----	0.09	-----	-----	-----
trans-Zeaxanthin	0.59	-----	4.11	0.05	78.65	trace
Astaxanthin diester	-----	22.68	-----	6.61	trace	0.31
Astaxanthin monoester	-----	-----	-----	2.11	-----	-----
Astaxanthin	-----	-----	-----	19.02	-----	-----
Capsanthin	0.73	-----	-----	-----	-----	-----
Unknown[2]	-----	-----	0.35	-----	1.35	-----
Total Carotenoid	**1.62**	**22.68**	**4.55**	**29.93**	**80.31**	**11.38**

1 = Hydrocarbon
2 = Epoxide carotenoid
GBM = Gelatin-based mix, FDP = Freeze-dried plankton, MR = Mullet roe,
LE = Lobster eggs, WA = Wild adult, CJ = Culture juvenile

(Simpson and Kamata, 1979): (1) those that make changes including biotransformation of beta-carotene to astaxanthin, (2) those that can make no changes but only deposit the pigment ingested, and (3) those that can biosynthesize astaxanthin but not from beta-carotene. The first category includes goldfish and some fish that have distinct color differences from juvenile to adults. Hata and Hata (1972, 1976) showed that goldfish convert the yellow pigment, zeaxanthin to astaxanthin and they can metabolize some beta-carotene but no lutein. The California sheephead (*Semicossyphus pulcher*), change from yellow to red is controlled genetically and a dietary carotenoid is not a factor (Crozier, 1969). The second category contain many species that deposit whatever pigment is fed, and the intensity of coloration is a function of the carotenoid concentration in the food. Included are salmon which also have the unusual ability to deposit pigmentation in the flesh and skin. Yellow pigment deposition in the flesh of channel catfish is produced by zeaxanthin and lutein (Lee, 1987). Pigmented flesh in catfish fillets is considered undesirable and is caused by administering feeds containing 11 mg/kg+ xanthophyll. The third category includes a few fancy ornamental carp and apparently clownfish, only a different pathway may exist. Yellowtail and red sea bream are also included in this category since they convert dietary astaxanthin into tunaxanthin.

Common, off-the-shelf pigment enhancing products can play a significant role in maintaining good coloration. Yet, paprika, beta-carotene and roxanthin, often added to prepared foods to obtain coloration, are not very effective with regard to clownfish.

Over a two-year period, Dr. Simpson (University of Rhode Island) consulted with IOH regarding clownfish pigmentation. During the first series of tests, newly captured wild and tank-raised *A. ocellaris* were sacrificed for tissue analysis. Tissue from wild fish consisted of zeaxanthin as the primary pigment rather than the expected astaxanthin. Tank-raised fish mainly had lutein due to the general lack of either pigment in their diet. Tests on other species revealed more significant levels of astaxanthin (20%) but still high levels of zeaxanthin (55%). Zeaxanthin is not a rare pigment in animals and plants, but it usually occurs as a minor pigment. The fish analyzed contained all-trans-zeaxanthin indicating it was derived from another pigment source. Table 45 lists the initial analysis of carotenoids in wild adults versus cultured juveniles (*A. ocellaris*) and various foods being fed at IOH. This work set the stage for our efforts to enhance coloration of cultured clownfish. Carotenoid composition is expressed in ug/g wet weight, except the freeze dried plankton which is dry weight.

Further analysis revealed the following carotenoids in four species of wild *Amphiprion* (*ocellaris, biaculeatus, frenatus and clarkii*): beta-carotene, cryptoxanthin, lutein, tunaxanthin, all-trans-zeaxanthin, 9-cis-zeaxanthin, 13-cis-zeaxanthin, and astaxanthin.

An experiment was conducted using three diets consisting of: 1) low pigmented salmon diet (control), 2) enhanced salmon diet with 20 mg.% wet weight astaxanthin and 3) 20 mg.% zeaxanthin wet weight. Results showed: group 1 (control) deposited no astaxanthin, 5.8 mg.% zeaxanthin, 8.9% beta-carotene, group 2 deposited 17.8 mg.% astaxanthin, 11.1 mg.% zeaxanthin, 13.0 mg.% beta-carotene, and group 3 deposited just a trace of astaxanthin, 25.8 mg.% zeaxanthin, and 8.0 mg.% beta-carotene. Visual observations of color were as follows: group 1, poor to fair, pale yellow; group 2, very good, bright orange; group 3 good, bright yellow.

From this experiment it was determined that astaxanthin is the key pigment in clownfish and is converted to zeaxanthin which in tern is converted to beta-carotene (diet 2). When zeaxanthin alone was fed, it was directly deposited in the tissue, no astaxanthin occurred in the tissue but beta-carotene did. Previous experiments using beta-carotene pigment showed no improvement in coloration over control fish. Astaxanthin directly adds to the intensity and red-orange tone of pigmentation in clownfish.

Other tests with other species showed that pigment levels of 10 mg.% astaxanthin and 10 mg.% zeaxanthin resulted in coloration equal to or exceeding wild coloration. Products containing significant amounts of astaxanthin are the most effective in enhancing pigmentation in clownfish and many other marine fish. Frozen, freeze-dried, and meals of the planktonic krill (*Euphausia pacifica*) and (*Euphausia superba*) are some of the best food sources for astaxanthin. Lobster eggs, freshwater crayfish eggs or shell extract, and macrobrachium eggs are also good sources. Pigment analysis on lobster eggs showed primary amounts of astaxanthin. Synthetic and extracted forms of astaxanthin and zeaxanthin are available but currently hard to obtain in the USA in small quantities. Recently a special cultured yeast (*Phaffia rhodozyma*) containing 30-800 mg/kg astaxanthin is being cultured. However, the digestibility of yeast and shrimp (*Panaeus sp.*) by-product meals for extraction of astaxanthin is low, unless offered in a predigested form.

As you can see, coloration is influenced by several factors. Fish are born with the necessary pigment cells to obtain any intensity of coloration you desire. You only have to enhance or stimulate these cells to produce the desired color. Manipulation of diet, exterior environments, lights, water quality, and maintaining healthy and unstressed fish all contribute. All you have to do is figure out the combination that works best for the fish you are working with. Usually diet is the most important factor.

What role fish coloration may play with regard to health is not clearly documented. Foremost among this is a beneficial effect on the endocrine system in terms of gonadal development, egg maturation, egg viability, and larval survival and growth (Meyers 1983). Experimentation and discussion presented in the broodstock section regarding pigmented eggs clearly indicates the effect on hatch. Tacon (1981) postulated that these compounds may perform a biological role similar to that of alpha-tocopherol, a naturally occurring antioxidant, thus protecting sensitive tissues and reactive compounds from oxidative damage. Gross and Budowski (1966) noted that astaxanthin may add a provitamin A complex. Salmonid species exhibiting a higher carotenoid

Table 46 - Symptoms that indicate pending water quality, disease and/or nutritional problems.

EARLY INDICATORS

1. Lack of appetite.
2. Discoloration, including off-white (gray) tones on white bars.
3. Lethargy or listlessness, such as lying on or near the bottom.
4. Rapid breathing of gills.
5. Bunching together in groups and in corners.
6. Excessive scratching against walls or gravel.
7. Dorsal fin not erect.
8. Cloudy water.
9. Fecal matter hanging from anal region.
10. Gulping air at surface.
11. Gills flared out.

LATER INDICATORS

1. Swimming with a locked open mouths.
2. Swimming with head in upright position.
3. Clamped side fins (pectoral fins).
4. Using only tail fin (caudal fin) for swimming.
5. Sunken belly region.
6. Head size large in proportion to body.
7. Sunken dorsal crest region on top of head.
8. A few white spots on caudal and pectoral fins.
9. Excessive mucus or whitish film on dorsal portion of the body.
10. Dry looking skin as if mucus coating is gone.
11. Mucus pores, dark circles around head region and gills are open or infected.
12. Heavy breathing.
13. Sores do not heal quickly.
14. Frayed fins which do not regrow quickly.
15. Resting on bottom of tank.
16. Bloated abdomen or raised scales.

SYMPTOMS NORMALLY RELATED TO WATER QUALITY

1. Increased respiration.
2. Increased slime production.
3. Irritability.
4. Scratching.
5. Swollen gill filaments creating flared gill operculums.
6. Loss of color.

content of astaxanthin, showed greater resistance to bacterial and fungal diseases. This has been attributed to the function of vitamin A in strengthening the cell membrane (Czexzuga, 1979a). Muller, et al (1980) postulated that pigmentation may serve as a protective agent in absorption of potentially harmful light energy or by quenching singlet oxygen. Just knowing that pigmentation responds to diet, light and water quality may be all we need to know about the overall impact on fish health.

Disease And Health Status

Diseases are controlled by many aspects. Host susceptibility is controlled by species, age, size, physiological condition, temperature effects on antibodies and immune responses, and natural resistance or immunological status. Environmental conditions can induce stress via high or low temperatures, low oxygen, crowding, water enrichment, poor nutrition, and seasonal changes. In most tropical species, May through October is the peak season for diseases.

Table 47 - Points to check and measures to take when a disease is suspected.

1. Observe water surface for oil film or slime on tank sides and surface.
2. Cup some water in your hands and smell if it is sower or different then usual.
3. Check water quality, especially ammonia, nitrite, oxygen pH, chloramines, and salinity in order of importance.
4. More recently there has been emphasis on other parameters of water quality including, calcium, magnesium, and trace metal levels. Although not well understood, these may play a significant role in saltwater fish culture as hardness is in freshwater culture.
5. Clean tank bottom and filters.
6. Make a 30% water exchange especially if water quality cannot be checked. Make a second exchange in 4 to 6 hours.
7. Check oxygen and/or increase aeration.
8. Make sure UV sterilizer is working and bulb is up to date (usually changed once a year or less).
9. Increase flow rates, and reduce light intensity.
10. Clean tank bottom and sponge filter.
11. If conditions do not improve, physically remove fish from tank and put into a clean tank.
12. Increase internal particle filtration rate and/or add a DE power filter (refer to DE info in book).
13. Increase in-tank biological filtration by using a conditioned large air lift powered foam filter.

Juveniles are the most prone to disease and health problems. Early detection of pending disease or health problems is absolutely essential to success. Health status is best determined early in the morning. Fish are more stressed and weaker in the morning and easily display certain signals or clues which indicate pending problems.

Disease is usually always a secondary result of prolonged stress. For example, low oxygen often results in bacteria infections a week later. When environmental perimeters are held constant and adequate diets are routinely provided, fish can resist or overcome most diseases and infections. The concept of total sterilization of water by elimination of all harmful bacteria and protozoans is virtually impossible. Harmful bacteria, protozoans etc., are always present in a stable aquarium to some degree but are insignificant to healthy, well-adjusted fish.

Stress can be induced many ways. Deteriorating water quality is the most significant external environmental form of stress to fish, as air pollution is to humans. Alteration of the physical environment by the addition of fish, change in habitat, routine tank maintenance, water exchanges, poor nutrition, etc. all constitute forms of stress. If these alterations are coupled with deteriorating water quality, disease will soon result, especially if stress is prolonged.

Mortality rate can also help indicate what the problem might be. Quick significant mortalities are usually due to deteriorated water quality conditions. Parasitic diseases cause slow constant death patterns whereas bacterial infestations are quicker. Nutritional mortalities are usually very slow and constant.

Methods of control and prevention of fish disease consists of several factors: (1) prevention of contact between pathogens and hosts is most important, (2) increase either natural or genetic resistance, (3) immunization of fish, (4) Manipulation of environment to favor the host, and Prophylactic use of chemicals or dietary supplements. Therapy after the disease occurs is also important.

To successfully treat diseases in juveniles it is important to recognize early signs of stress. Treatment applied in later stages of disease usually does not produce satisfactory results. Often badly diseased fish that recover are inferior in strength and quality throughout the remainder of their lives. Table 46 is a list of symptoms that may indicated pending water quality, disease, or nutritional problems.

It is prior to or during the initial feeding that you can determine the health status of your fish. Young fish like to eat! Healthy fish quickly swim to the surface or towards the front glass every time you approach the tank. Fish that huddle in a corner, do not attack food aggressively, even after you move back from the tank, need attention. Immediately check water quality since this is usually the main cause. Surface gulping or alignment in currents may represent low oxygen either by interference of gill membranes or low oxygen levels in the water. Gulping may also indicate high nitrites and chloramines.

Observation of fecal deposits and fecal material hanging from anal pores are good indicators of health status and are usually tied into nutritional imbalances. Normally fecal pellets should be oblong brownish-green balls which stay together. If you notice fish excreting liquid fecal material or have white strings hanging from the anal pore, you have a pending health problem. The "hanging white strings" indicate constipation or a gut infection. In either case improper food may be the initial main problem.

When suspected disease is first noted there are several points to check and measures that can be taken to immediately reduce stress (Table 47).

Nutritional And Internal Disorders

Nutritional and internal disorders are often mistaken for diseases and when treated, resulting in little effect. Externally these disorders look similar to many disease problems but internally

Table 48 - Symptoms normally associated with nutritional disorders.

1. Stressed attitude, hiding in corners.
2. Nervous, forming tight balls which do not break up quickly when food is offered.
3. Continuous poor feeding, mainly in mid-water as apposed coming to surface.
4. Fish feed but slowly get thinner and "waste away".
5. Flared operculum (gill covers).
6. Abnormally short operculum.
7. Deformed spines.
8. Pale gills.
9. Skin looks "dry" especially around head and mucus pores look like holes.
9. Depigmentation in the head region.
10 Heads up orientation on bottom of sides rather then level.
11. Poor attitude yet no obvious tissue inflammation, body sores or fin deterioration.
12. Whitish-yellow fecal strains which hang from the anus and on the tank bottom.
13. Normal fecal deposits are dark brown or greenish and ball shaped.
13. Equal bilateral malformation or sunken forehead and body in general.
14. Fatty deposits throughout the interior body tissues including abdominal cavity.
15. Fatty globules deposits on lower portion of the fins and caudal fin.

there are many differences. Nutritional disorders usually occur in fish that are around two to three months old. A distinguishing characteristic of nutritional disorders is usually equal body changes on both sides, whereas in disease situations the physical body changes are usually unequal. Causes of nutritional disorders are usually vitamin or mineral deficiencies or prolonged feeding of rancid food and/or fats. Blazer et al (1989) conducted feeding experiments using two commercial diets on freshwater catfish (*Ictalurus punctatus*). Their results indicated that diet consistency may indeed play a significant role in influencing a fish's immune system and its ability to resist infection. Typical symptoms of nutritional disorders are listed in Table 48.

Nutritional disorders are basically two types: 1) lack of essential vitamins, minerals, etc. during growth, 2) accumulation of a toxic factor to a point where it becomes detrimental. In both cases, usually a vital organ is eventually destroyed or malfunctions, resulting in deformities and possible death. When working with specific populations of fish, there are patterns which might indicate nutritional disorders rather than diseases. If a disease or water quality problem exists, usually the entire tank is affected. Nutritional disorders are typified by a certain percentage of weaker, smaller individuals breaking down first. Under toxic conditions, larger, more aggressive, heavy-feeding individuals will be the first to show symptoms. In general, fish with nutritional disorders often do not stop feeding but continue to slowly waste away. Those that do survive are permanently disfigured since often muscle or bone tissue is totally destroyed. Internal essential organs are usually injured and as these fish grow larger, the injured organs cannot function properly with regard to food intake, excretion, and/or digestion, and the fish slowly suffers.

Food Storage

It is essential that all foods, including freeze dried feeds, flakes, and dry pellets be properly stored. To often I have seen food stocks being stored in hot, humid, buildings or in the hatchery itself. Fish food should be handled no differently than the food you eat. Fish food, like human food, have a shelf life which can be easily extended if properly stored. Once a product is sealed, nutrient destruction is reduced, however chemical changes in fats and vitamins continue at different rates. Vitamins such as B_{12} (riboflavin), folic acid, B_6 (riboflavin), C, and E are all susceptible to destruction. Diets with high fat contents are more likely to go rancid. Furthermore, the formation of rancid fats reduces the activity of fat-soluble vitamins such as A and E (Blasiola, 1991).

Beware of bargain priced frozen, freeze dried, or sun dried foods, since they may have already been stored improperly or longer than normal. Beware of cheap pellets and flake foods since many are nutritionally inadequate, poorly handled, and not suitable for growing juvenile fish. Food costs within a hatchery should not become a primary issue when rearing young fish. Good, well balanced foods, produce faster growing, quality fish and in the long run make up for the additional costs.

Fresh frozen seafood products, plankton, *Artemia*, shrimp, clams, etc., should be kept frozen until needed at -10° to -20°F. They should be packed in tightly sealed, good quality plastic bags. Additional precautions could include: (1) freezing products in 7 to 9 ppt sterile saltwater to make them osmotically balanced and reduce leaching, (2) addition of natural antioxidants such as vitamin E and C and synthetic antioxidants, ethoxyquin, BHT, or BHA to reduce oxidation and spoilage of food.

Dry food should be refrigerated or held in cool dry buildings away from humidity and all light. Preferred storage is in a freezer. Once a bag or container is open it is essential that it be tightly resealed. Do not buy more dry food than you can use within a 6 to 8 month period.

Vitamin Deficiencies

To fast-growing juvenile fish, inadequate levels of vitamins can become a limiting factor. Vitamins serve as catalysts to biochemical reactions that build or break down energy sources. Without proper levels of specific vitamins these reactions do not occur. Many vitamins are synthesized within the gut of fish provided they are in good health. Others are obtained through the foods they ingest. Vitamins are divided into two groups, water soluble and lipid soluble. Eleven water soluble vitamins are recognized as essential to fish. Required levels are not well known. Vitamins are easily destroyed or reduced by heat and oxidation therefore, proper storage is essential. Table 49 by Halver (1989) outlines the clinical signs and pathological effects caused by vitamin deficiencies.

Comparison of symptoms caused by vitamin deficiencies with "diseases symptoms" (see Broodstock Section), you quickly notice that nutritional disorders could indeed be the main problem.

Blasiola (1991) outlined several deficiencies and their pending results. Vitamin C (ascorbic acid) is a water soluble vitamin which can be unstable and destroyed rapidly in high temperatures. It is essential in the formation of collagen, the fibrous component of bone, cartilage and connective tissue. Other deficiencies of vitamin C include, abnormal skeletal development, eye lesions, weight loss, reduced healing and lowered resistance to disease. Under natural conditions fish can normally synthesize sufficient vitamin C to handle their own metabolic needs, but under crowded, stressful conditions found in a hatchery, they may not be able to meet the demand.

In addition, deficiencies of vitamin C have been shown to interfere with normal wound repair.

Table 49 - Clinical signs and pathological effects of some vitamin deficiencies. Including averaged suggested levels from various sources (Halver, 1985).

Symptoms of Vitamin Deficiencies

Thiamin...........Poor appetite, muscle atrophy, convulsions, instability and loss of equilibrium, edema, poor growth.

Riboflavin.......Corneal vascularization, cloudy lens, hemorrhagic eyes, photophobia, dim vision, incoordination, abnormal pigmentation of iris, striated constrictions of abdominal wall, dark coloration, poor appetite, anemia, poor growth

Pyridoxine......Nervous disorders, epileptiform fits, hyper-irritability, ataxia, anemia, anorexia, edema of peritoneal cavity, colorless serous fluid, rapid postmortem, rigor mortis, rapid and gasping breathing, flexing of opercula

Pantothenic....Clubbed gills, prostration, anorexia, necrosis and scarring, acid cellular atrophy, gill exudate, sluggishness, poor growth

Niacin..............Loss of appetite, lesions in colon jerky or difficult motion, weakness, edema of stomach and colon, muscle spasms while resting, poor growth.

Biotin..............Anorexia, lesions in colon, muscle atrophy, spastic convulsions, fragmentation of erythrocytes, skin lesions, poor growth.

Folic acid........Poor growth, lethargy, fragility of caudal fin, dark coloration, macrocytic anemia.

Vitamin C........Scoliosis, lordosis, impaired collagen formation, altered cartilage, eye lesions, hemorrhagic skin, liver, kidney, intestine muscle.

Myoinositol.....Poor growth, distended stomach, increased gastric emptying time, skin lesions.

Choline...........Poor growth, poor food conversion, hemorrhaging.

Suggested Levels

Vitamin A	**4000 mg/kg diet**
Vitamin D	**938 mg/kg diet**
Vitamin E	**59 mg/kg diet**
Vitamin K	**1 mg/kg diet**
Thiamin	**5.2 mg/kg diet**
Riboflavin	**8.1 mg/kg diet**
Vitamin B_6	**6.5 mg/kg diet**
Pantothenic acid	**23 mg/kg diet**
Niacin	**40 mg/kg diet**
Biotin	**0.5 mg/kg diet**
Vitamin B_{12}	**0.04 mg/kg diet**
Folate	**2.7 mg/kg diet**
Choline	**1455 mg/kg diet**
Myoinositol	**390 mg/kg diet**
Vitamin C	**88 mg/kg diet**

Segedi (1977) found that vitamin C toughens cellular membranes making them less susceptible to infections. This provides more resistance to parasitic attacks and promotes rapid healing.

Blasiola (1988) clearly demonstrated the effect vitamin C can play in the pigmentation of surgeonfish, *Paracanthurus hepatus*. Using a flake food deficient in vitamin C, test fish started to show depigmentation of the head region in just 21 days. Whereas the controls, fed green algae and a diet with vitamin C, were normal. Within 10 days of restoration of vitamin C in the diet for the test fish, color was starting to reappear. Vitamin A and E are fat-soluble and are essential as antioxidants of fat.

Essential Fatty Acids

I have devoted considerable discussion to fats within diets. This is deliberate to help demonstrate how important fat content, rancidity, and diets are in rearing juvenile clownfish and no doubt many other species. The majority of this information is data from in-house reports generated by Vickie Blazer, who was a consultant to IOH in 1979.

Fat levels in a diet are important to obtain optimal growth in young fish. The National Research Council (1977) reports that warm water fish are able to utilize saturated fats and that dietary lipid levels up to 15% show growth enhancing effects. Takeuchi et.al. (1978) found that the optimum ratio of protein to lipid for rainbow trout was 35% protein to 15-20% lipid which is considerably higher than most foods available to the aquaculturist or aquarist. Initially, larval fish require a 15-20% lipid and 20-30% protein diet which changes to 5% lipid and 35-45% protein in juvenile stages. Table 50 shows fatty acid composition of foods utilized at IOH compared to wild and cultured clownfish.

Fish contain high levels of poly unsaturated fatty acids, predominantly of the linolenic acid (omega 3) types, while land animals contain more saturated fats with unsaturated fats of the linoleic series (omega 6). Fish have been shown to have a requirement of the omega 3 fatty acids which are often lacking in commercial prepared diets. Freshwater fish require 18:3 omega 3 whereas 10:5 omega 3 is more suitable for marine fish. Within the literature these are referred to as "essential fatty acids" or EFA. Most studies on EFA have been on trout and salmon, but no doubt have a bearing on all fish, including clownfish. Yu and Sinnhuber (1979) found that the optimum level of dietary omega 3 fatty acids within a diet was 1 to 2.5%. Takeuchi and Watanabe (1978) found certain marine invertebrate oils, like squid oil, contained growth-enhancing components. These were identified as 20:5 omega 3 and 22:6 omega 3 and satisfied the EFA requirement at 0.5% of the diet. Other enhancing oils commonly used are menhaden and cod liver oil.

Deficiency of EFA within a diet may produce varied results with very similar symptoms of most diseases. The main internal indicator in all organisms is the biosynthesis and deposition of 20:3 omega 9 fatty acids (Sinnhuber et al. 1972). Fatty livers have been reported in a number of cases (Sinnhuber, et al 1972, Castell, et al, 1972 and Farkas, et al, 1978). Impaired pigmentation was reported in chinook salmon (Nicolaides and Woodall, 1962). Shock syndromes were observed in rainbow trout deficient in linolenic acids (Castell, et al, 1972). Goldstein (1990) noted shock syndrome in juvenile clownfish and indicated that diet may be responsible. Farkes, et al, (1978) showed that there was over production and deposition of certain fatty acids in carp fed less than 1% linolenic acid. This increase of unneeded fats coupled with unchanged or decreased rate of removal of fatty acids from the blood results in fat accumulations in the liver and other parts of the body including between the external fin rays.

Blazer investigated a juvenile clownfish growth problem that we referred to as "wasting". Clownfish which were about 60 days old often started to show symptoms of "wasting" but not all fish within a tank reacted the same or at the same time, which is more typical of a protozoan or bacterial disease. Younger fish usually did not show the typical symptoms. Typically, early stage, sick fish would reduce feeding, hide in corners, breath rapidly, and the gill operculum (outside flap) would become flared. Gills appeared fragile with deterioration at the filament bases and scattered aneurysms. More advanced cases showed severe "creased" or flared operculums, abnormal swimming behavior, pinched faces, and occasionally locked jaws (mouths always open). External microscopic examination showed, round globules of "fat" being deposited along fin ray insertions of the dorsal and anal fin. In more advanced cases, these globules were evident in tissue between fin rays. Internally, tissues were full of oil droplets when smashed. The liver was pale with heavy fat and ceroid deposits within macrophages (yellow/brown deposits). These deposits were also evident in the kidney and spleen. The spleen was enlarged and extensive muscle damage was evident in many places.

Diet at this time, consisted of lobster eggs two times per day and a gelatin mix also containing some lobster eggs which was fed two times per day. Experiments conducted on diets with and without lobster eggs dramatically demonstrated that "wasting" occurred mainly in those fed extra lobster eggs. Yet, growth rates were better in those being fed lobster eggs. Basically, we found we had two distinct problems occurring at once, not enough EFA or total fat content in the mix diet, and rancid fats in the lobster eggs.

The questionable lobster eggs were obtained from a cold storage facility in mass quantities, but it was not clear how long they had been stored or how they were handled. They were excellent color and growth enhancers. Obviously, we felt they would provide essential fatty acid (EFA) requirements since they were of marine origin.

Improperly stored products containing marine fats and oils become oxidized or begin to become rancid when exposed to atmospheric oxygen and/or heat. They are very susceptible to auto-oxidation, a self-propagating reaction, which when started, proceeds rapidly. It continues at low freezing temperatures, especially if the product had been exposed to a higher temperature

Table 50 - Research conducted for IOH. Fatty acid composition (% weight) of samples of food items used at IOH verses cultured and wild clownfish. Clownfish tested were A. ocellaris.

Lipids	FDP	DMD	GBM	MR	LE	CMA	CJ	WMA
12:0	0.85	0.04	0.02	0.01	0.03	0.01	0.02	-----
14:0	5.48	1.86	1.76	1.93	1.48	1.75	3.89	1.45
15:0	0.96	0.69	0.89	0.74	1.57	1.78	0.84	0.33
16:0	22.70	25.56	15.14	7.86	15.65	27.25	24.62	31.40
16:1w7	12.13	3.16	13.35	25.84	13.72	18.84	12.59	23.88
16:2w4	2.40	-----	1.99	2.83	1.06	0.70	.075	-----
18:0	2.15	7.72	4.60	2.75	5.27	4.38	4.69	5.33
18:1w9	15.15	38.14	27.20	10.64	21.47	18.60	19.60	14.57
18:2w6	1.46	16.83	13.51	3.18	1.34	4.08	6.84	1.99
18:3w3	1.48	0.94	1.40	1.71[3]	trace	-----	2.01	1.05
18:4w3	3.72	-----	0.99	2.91	-----	-----	-----	3.99
20:1	-----	1.22	-----	-----	2.91	2.92	5.64	-----
20:2w6	0.04	0.04	0.25	1.30	0.50	0.62	0.25	-----
20:3w6	-----	-----	0.01	0.92	-----	0.02	0.04	-----
20:4w6	1.64	0.70	2.74	3.35	5.38	2.40	2.14	1.59
20:4w3	-----	-----	0.17	1.94	-----	0.40	4.41	2.36
20:5w3	18.91	0.78	4.75	10.06	10.16	2.11	4.14	1.59
22:1	-----	0.78	-----	-----	-----	-----	-----	-----
22:5w3	-----	trace	2.36	5.89	-----	4.11	1.77	1.97
22:6w3	9.18	1.39	8.38	9.53	17.34[4]	7.68	5.33	6.76[5]
22:4w6	1.40	0.15	0.49	5.18	0.76	2.35	0.48	1.08
% Lipid	**22.10**	**8.06**	**3.63**	**15.83**	**6.71**	**10.06**	**9.75**	**3.56**

FDP = Freeze Dried Plankton, **MD** = Dry Minnow Diet, **GBD** = Gelatin Based Diet, **MR** = Mullet Roe, **LE** = Lobster Eggs, **CMA** = Cultured Male Adult, **CJ** = Culture Juvenile, **WMA** = Wild Male Adult at some time during processing. Vitamin E functions as an biologic antioxidant within processed food or within the body. If levels of vtamin E are deficient, oxidation of polyunsaturated fats may eventually occur in the lipoprotein membranes (membranes containing fats) of a body, releasing toxic by-products that cause problems in other parts of the body (Recknagel and Glende, 1978). Woodall et al (1964) found that chinook salmon fed diets containing up to 5% herring oil and less than 3 mg of vitamin E per 100 gm of food prevented fatty acid deficiency symptoms. Other researchers found that adding 25 mg of vitamin E per Kg (2.2 lb) of diet along with ethoxyquin prevents the onset of muscular degeneration in channel catfish when their diets contained high levels of oils. Without additions of ethoxyquin to the food, vitamin E could be destroyed by oxidation (Blasiola 1991).

Large ceroid deposits observed in IOH clownfish appeared to be a result of lipid breakdown within the lobster eggs and within the body tissues of the fish. South African rock lobster eggs are full of gonadal oils including phospholipids and polyunsaturated fatty acids. Certain enzymes are essential to the biosynthesis (making) fatty acids within a fish resulting in the synthesis of long chain fatty acids. Hyper-production and deposition of these fatty acids depends on EFA intake. Farkas, et al (1979) showed overproduction and deposition of fatty acids within the body tissues occurred in carp fed less than 1% linolenic acid.

In conclusion, it is essential that the food fed to clownfish contain good levels of EFA and especially fats derived from marine origin. Storage and handling of all foods must be strictly controlled.

Syacium papillosum (Linnaeus) Female

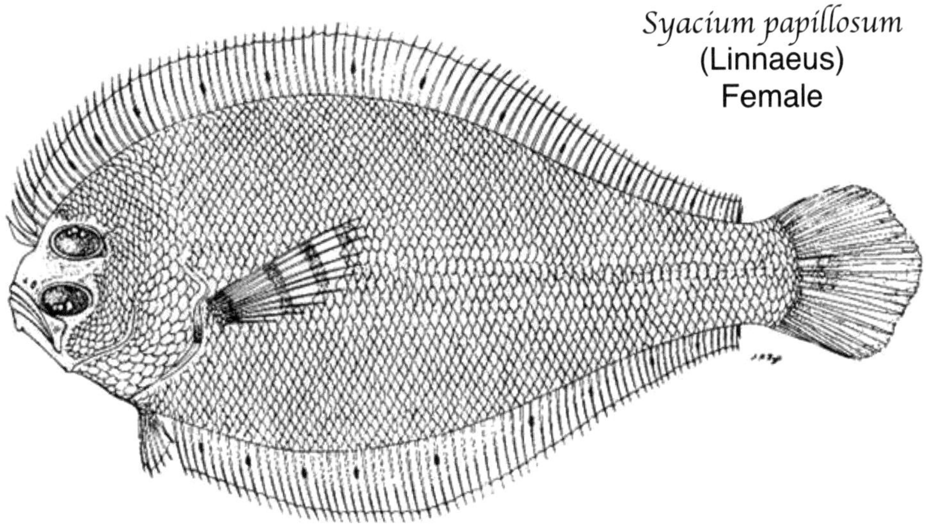

Drawings By F. H. Hoff
Memoirs of the Hourglass Cruses
Flatfishes (Pleuronectiformes)
By Robert W. Topp & Frank H. Hoff, 1972

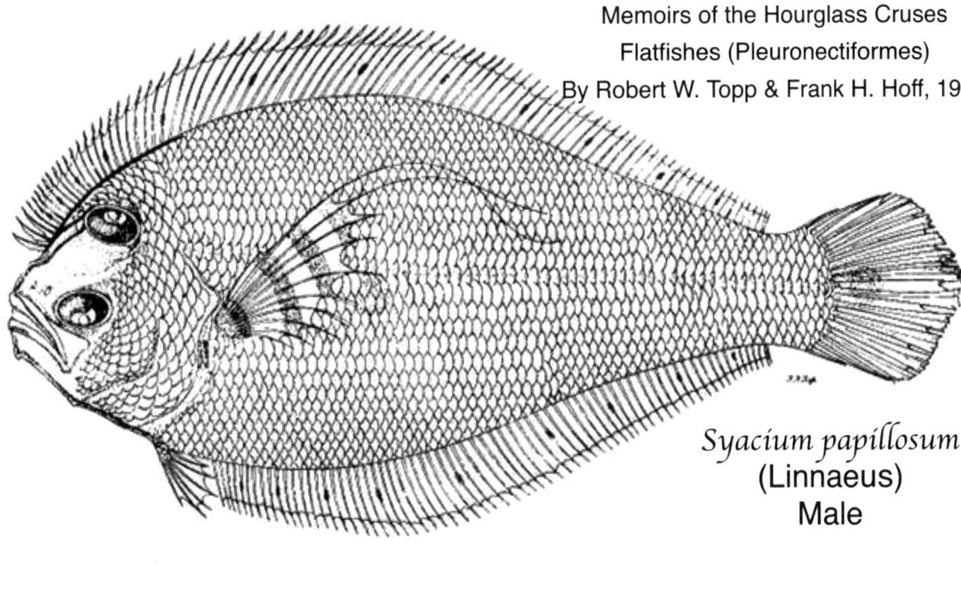

Syacium papillosum (Linnaeus) Male

COMMON JUVENILE DISEASES

The diseases listed are the most common ones found in rearing clownfish juveniles. They are listed and discussed by order of importance. Certain diseases are often seasonal and can be routinely expected. In commercial operations, routine prophylactic measures are recommended to prevent outbreaks. Again, I want to strongly stress that disease problems are closely associated with poor water quality, unclean tanks, overloaded biological filters, overloaded tanks, and improperly stored foods. Most disease causing organisms are present at all times and can be triggered by prolonged internal or external stress. There are those that think we can provide a virtually sterile environment and this is good for the fish. Wrong, This is wrong, the first time a fish is exposed to an invader, antibodies are formed which will provide protection from future infections by the same organism. Exposure to sublethal concentrations of pathogens is extremely important for fish to develop a competent immune system (Francis-Floyd, 1991).

Treatment Agents

The usual approach for treatment has been to reach for the antibacterial or therapeutic agents, yet in a closed recirculating system you must be very careful. Hawke (1991) reviewed common treatments and their effect on nitrification in recirculating systems. Failure of a biological filter is serious and more toxic than many diseases you want to treat. Table 51 lists many papers that have evaluated that effect. Long term effects of selected therapeutic agents on ammonia oxidation can continue even though treatment has ceased. No reduction in ammonia was noted after water exchanges in tanks treated with methylene blue and copper sulfate when monitored over 21 days after treatment. Of the compounds tested by Bower and Turner (1982), only neomycin and copper sulfate caused significant increases in nitrite. The impact of erythromycin on ammonia oxidizing bacteria and the delayed affect on nitrite bacteria is significant even after 24 days post treatment (Collins et al. 1976).

Calculating Treatments

Normally the entire water volume is treated although many antibiotics are now being ad-

Table 51 - Effects of antibacterial and therapeutic agents on closed recirculating systems.

Compound	Dose	Effect	Source	Compound	Dose	Effect	Source
Chloramine T	0.00 mg/l	N	Muir and Roberts (1988)	Copper Sulfate	1.0 mg/l	N	Collins et al (1976)
Chloramphenicol	13.3 mg/l	S	Bower & Turner (1982)	Copper Sulfate	0.3 mg/l	M	Bower & Turner (1982)
Chloramphenicol	50.0 mg/l	E	Levine & Meade (1976)	Copper Sulfate	5.0 mg/l	N	Levine & Meade (1976)
Chloramphenicol	50.0 mg/l	N	Collins et al (1976)	Formalin	10.0 mg/l	N	Muir & Roberts (1988)
Chloratracycline	10.0 mg/l	E	Levine & Meade (1976)	Formalin	25.0 mg/l	N	Collins et al (1976)
Erthromycin	50.0 mg/l	E	Collins et al (1976)	Formalin	15.0 mg/l	S	Levine & Meade (1976)
Gentamycin	5.0 mg/l	N	Bower & Turner (1982)	Malachite Green	0.1 mg/l	S	Muir & Roberts (1988)
Neomycin	66.0 mg/l	M	Bower & Turner (1982)	Malachite Green	0.1 mg/l	N	Collins et al (1976)
Nifurpirinol	1.0 mg/l	N	Collins el al (1976)	Malachite Green	0.5 mg/l	S	Levine & Meade (1976)
Nifurpirinol	0.1 mg/l	S	Levine & Meade (1976)	Methylene Blue	5.0 mg/l	E	Levine & Meade (1976)
Nifurpirinol	0.1 mg/l	N	Bower & Turner (1982)	Methylene Blue	8.0 mg/l	E	Bower & Turner (1982)
Oxytetracyline	50.0 mg/l	N	Collins el al (1976)	Methylene Blue	5.0 mg/l	E	Collins et al (1976)
Sulfamerazine	50.0 mg/l	N	Collins el al (1976)	Pot Permanganate	4.0 mg/l	N	Collins et al (1976)
Sulfanilamide	25.0 mg/l	M	Levine & Meade (1976)	Pot Permanganate	1.0 mg/l	E	Levine & Meade (1976)
Trimethoprim	10.0 mg/l	S	Muir & Roberts (1988)	Quinacrine	12.0 mg/l	N	Bower & Turner (1982)
				Salt (NaCl)	5000 mg/l	N	Collins et al (1976)
				Salt (NaCl)	30000 mg/l	E	Spotte (1979)

E = Extreme effects on nitrification
N = None, S = Slight, M = Moderate

ministered orally. Johnson (1990) has provided a fairly easy method to calculate treatments for a specific static water volume.

First calculate the weight of the water.

Water weight (WW) = L x W x D x 62.4 (L = length, W = width, D = depth, 62.4 lbs/cu ft.))

Next calculate the weight of chemical needed by comparing the known ratio and the unknown ratio.

1 ppm = 1/1,000,000

1/1,000,000 = (N) pounds divided by WW

By cross multiplication the following is obtained.

1,000,000 x (N) = 1 x WW

N = 1 x WW/1,000,000 = pounds of chemical needed.

To convert to grams there are 454 gms/16 ounces (1 lb)

N pounds into grams is : N x 454 = grams of chemical needed

N pounds into ounces is: N x 16 = ounces of chemical needed

Often in small tanks, the amount of chemical needed is small, hence hard to determine and administer. The best thing to do is weight ten times the amount calculated, add to 500 ml, mix and dissolve, then add enough distilled or deionized water to make a liter (1000 ml). Add 10% of the volume (100 ml) to the tank. The purpose is to allow for the best accuracy of the drug being administered.

Usually metric units are used when working with small amounts of chemicals in small volumes of water, e.g. grams per gallon, cubic centimeters per cubic foot, and milligrams per liter. With large volumes, such as acre-feet, it is more convenient to use a large unit of weight such as pounds or kilograms.

The following information is probably the best description of calculating treatment levels that I have ever found. Therefore, I have elected to transmit this information as written by Welborn (1987). "In treating fish, it is common practice to add a chemical to a specific volume of water to produce a known concentration of the chemical. The desired concentration is usually expressed as parts per million (ppm). Parts per million can only be used in a weight-to-weight relationship; a weight-to-volume relationship cannot be used direct because various chemicals have a different weight per unit of volume. Using only the weight of chemicals or the appropriate conversion factor with the weight of the water will avoid confusion in calculating the amount of chemical needed to give the desired concentration in parts per million"

Table 52 -Weight of chemical needed to a unit volume of water to yield 1 ppm

1 ppm = pounds per acre foot
1 ppm = 1,233 grams per acre foot
1 ppm = 0.0283 grams per cubic foot
1 ppm = 0.0000624 pounds per cubic foot
1 ppm = 0.0038 grams per gallon
1 ppm = 0.0584 grains per gallon
1 ppm = 1 milligram per liter
1 ppm = 0.001 gram per liter
1 ppm = 8.34 pounds per million gallons of water

Table 53 - Conversion factors%- ppm

0.1	1:10,000,000	0.000010
0.25	1:4,000,000	0.000025
1.0	1:1,000,000	0.0001
2.0	1:500,000	0.0002
3.0	1:333,333	0.0003
4.0	1:250,000	0.0004
5.0	1:200,000	0.0005
8.4	1:119,047	0.00084
10.0	1:100,000	0.001
15.0	1:66,667	0.0015
20.0	1:50,000	0.002
25.0	1:40,000	0.0025
50.0	1:20,000	0.005
100.0	1:10,000	0.01
150.0	1:6,667	0.015
167.0	1:6,000	0.0167
200.0	1:5,000	0.02
250.0	1:4,000	0.025

"One part per million (ppm) refers to 1 pound of chemical to 999,999 pounds of water or 1 gram to 999,999 grams of water to give a total weight of 1 million pounds or grams of solution. This means that there is 1 pound or 1 gram of substance in 1 million pounds or grams of solution or mixture, thus 1 part per million (ppm). To avoid needless calculations in determining the appropriate conversion factors for different units of volume, the weight of chemical needed to give 1 part per million in each of the standard units of volume is given in Table 52

"Some chemicals are in liquid form and contain a stated weight of active ingredient per gallon of liquid, e.g., 4 pounds per gallon. In this case, simply calculate the number of pounds of active chemical needed and then divide by the weight of active chemical per gallon of liquid to get the number of gallons needed. For example, suppose you need 18 pounds of a chemical to get 1 ppm. The container has 4 pounds active material per gallon. Divide 18 by 4 (18/4) to get the number of gallons of chemical needed which is 4.5 gallons."

"Other chemicals are liquid in their pure form. In this case it is necessary to know how much the liquid chemical weighs per unit volume in order to calculate how much to use. If the chemical is heavier than water, less will be needed for the desired weight, and if it is lighter than water, more will be needed for the desired weight of chemical. To determine the volume of this type of chemical needed to give the desired concentration, you must know the specific gravity (SG) of the chemical. This is calculated by dividing the weight of one unit of volume of chemical by the weight of the same unit volume of water. For example, one gallon of 37-40 percent formaldehyde (formalin) is 9 ÷ 8.34 = 1.08."

'If the amount of liquid needed is calculated in grams, it must be divided by the specific gravity to convert it to the number of cc required to give the desired co¡ncentration. However, if the amount of liquid needed is calculated in pounds, it must be divided by the weight of 1 gallon of water (8.34) times the specific gravity of the liquid (8.34 lb x SG) to convert to the number of gallons of liquid needed to give the desired concentration."

Treatment rates are sometimes given in percentages of chemicals to use in proportions. It is easier to convert these to ppm to do your calculations. Table 53 provides a quick conversion from percent to ppm.

As stated, you must know the volume of the unit being treated. To calculate the correct amount of chemical needed use the following formula:

Weight of Chemical = V x CF x ppm x (100 ÷%AI)

V = volume of water being treated

CF = conversion factor, weight of chemical used to give one ppm in one unit volume of water. (see Table 52). ppm = concentration

100 ÷ %AI = 100 divided by the percent active ingredient of the chemical to be used.

Example: we determine how much copper sulfate is needed to treat a tank 2'x4'x2' (16 cu ft) deep with a concentration of 0.5 ppm active ingredient Copper sulfate is 100% active ingredien.

Weight of Chemical = V x CF x ppm x (100 ÷ AI)
Volume = 16 cu ft
Weight of Chemical = 16 cu ft x 0.0283 gm/cu ft (Table 52) x 0.5 ppm x (100 ÷ 100)
Weight of Chemical = 0.23 gm (230 mg)

Example: we determine how much Masoten (80% active) is needed to treat a tank that is 10 meters x 3 meters x 3 meters deep with 0.25 ppm active ingredient.

Weight of Chemical = V x CF x ppm x (100 ÷ AI)
Volume = 90 cu meters = 90 x 1000 = 90,000 liters
Weight of Chemical = 90,000 L x 0.001 gms/l (Table 52) x 0.25 ppm x (100 ÷ 80)
Weight of Chemical = 28.13 grams

Example: we determine how much potassium permanganate is needed to treat a holding tank that measures 36" x 24" x 10" deep with 2 ppm active ingredient. Potassium permanganate is 100% active ingredient.

Weight of Chemical = V x CF x ppm x (100 ÷ AI)
Volume = 8640 cu in = 8640 ÷ 1728 (in/cu ft) = 5 cu ft
Weight of Chemical = 5 cu ft x 0.0283 gms/cu ft (Table 52) x 2 ppm x (100 ÷ 100)
Weight of chemical = 0.28 grams (280 mg)

Example: we determine how much formalin is needed to treat a round tank that is 8' in diameter and water depth of 2' with 250 ppm active ingredient. Formalin is considered 100% active even though it is a dilution of formaldehyde.

Weight of Chemical = V x CF x ppm x (100 ÷ AI)
Volume = π x R² x D = 3.14 x (4x4) x 2 = 100.5 cu ft
Weight of Chemical = 100.5 x 0.0283 gms/cu ft (Table 52) x 250 x (100 ÷ 100)
Weight of Chemical = 711 grams
Formalin is a liquid, therefore we must converted it to a volume by dividing the calculated grams by the specific gravity of the fluid.
711 grams ÷ 1.08 SG = 658 cubic centimeters (cc or ml) of formalin

Oodinium ocellatum

Commonly called "velvet disease", this is the most significant disease problem. *Oodinium ocellatum (Amyloodinum)* is a severe cyclic problem characterized by dust-like white spots or haze on the skin or fins. *Oodinium* is usually seasonal, being more prevalent in the warmer months. It is a single cell dinoflagellate protozoan similar to those that create toxic red-tide along warm water coastal regions. As opposed to the dinoflagellate that causes red-tide, *Oodinium* does not release toxic substances but rather attacks gill membranes in such numbers that it disrupts normal transfer of oxygen and ion exchanges. Basically it is external parasite with most of the infestation concentrated around the head region.

The life cycle consists of three stages: the parasitic stage called the trophont, the reproductive stage called the tomont, and the free-swimming stage called the tomite or dinospore. Examination of the gill membranes will reveal smaller pear-shaped cells to larger roundish type cells attached to the gill filaments. Size of these cells are 12.4 x 9.9 to 103.7 x 80.5 microns with a mean of 61.1 x 50.1 microns (Spotte 1992). Heavily infected fish can have 200 to 500 trophonts attached to a single gill filament. As the trophont matures (3 to 7 days) it detaches, sinks to the bottom of the aquarium, and is then referred to as an encysted tomont. The tomont then undergoes sporulation (a form of cloning) to a point where as many as 256 free swimming tomites can develop. Development takes 3 to 5 days at 72° to 77°F (22°-25°C). Swimming tomites (dual flagellated dinospores) are small, 12 to 15 microns, with a red eyespot and single flagellum. One end of the flagellum is wrapped around the middle of the tomite forming a groved girdle. When it finds a host fish the dinospore attaches and transforms into a trophont to begin a new cycle. Theoretically, a single *Amyloodinium* can engender more than four billion others in less than a month (Bower 1987). The encysted stage (tomont) is the hardest to eradicate. Under normal aquarium temperatures 75°F to 80°F (24°C to 26°C) all dinospores should hatch and die within 7 to 9 days providing no host is present. However, Bower (1987) found evidence of tomont survival even after 37 days and speculated that low oxygen, or bacterial toxins found in filter impacted beds may help inhibit normal hatching.

Early detection is hard since the gills are attacked first. Yet, clownfish have a tendency to form tight balls or bunch together in the back corner of the aquarium. Continuous heavy breathing is another clue. Skin of heavily infected fish appears dull, dusty, patchy or velvet-like. The mucus layer may appear pinkish. General behavior may include heavy breathing, congregating at surface, scratching, shaking or shuddering while swimming, sluggishness, and bunching together in tight balls. Later stages the fish may lie on their side on the bottom of the tank for periods of time. The period of infestation may be as quick as 3 to 4 days or may be slower due to efficiency of the filtration , temperature and tank population (Moe 1982).

Frakes (1987) states that in low density, noncommercial situations, infected fish stocked at 1 inch of fish per 16 gallons become symptom free in 3 weeks without therapy. Fish maintained in reef tanks with an abundance of filter and tactile feeding invertebrates rarely are infected with *Amyloodinium*.

Treatment For Oodinium

Initially, quinine and similar synthetic chemicals like atabrine, quinacrine, chlorquine, primaquine, etc. have been used to remove attached tomites and trophonts. Quinine sulfate or quinine hydrochloride (HCl) administered at 250 mg/38 liters (10 gal.) will force the parasitic stages to form cysts and drop off in 2 to 3 hours. However, this does not kill the parasites so a follow-up treatment of copper sulfate is needed. Removal of quinine can be accomplished by passing water through carbon (Frakes 1986).

Highly infected fish can be removed from the infected tank and given a 2 to 3 minute freshwater dip and then placed in a clean tank. When using a freshwater dip, adjust the pH to around 8.0 using sodium bicarbonate (baking soda). Temperature should be the same during the dip. This will remove tomonts and dinospores but does not guarantee that the fish will not become reinfected. Freshwater will not kill the encysted tomonts. They can survive longer than 6 weeks in freshwater and can resume reproduction when placed back in saltwater (Bower 1987).

Prolonged treatment with copper sulfate has been shown to be the most effective. Dosage levels of 0.3 to 0.5 mg/L (3 to 5 ppm) over a 10 to 14 day period are best. Copper treatment is more effective in bare tanks without carbonate substrate since the copper ion is attracted to the surface of the substrate and becomes ineffective.

During the copper treatment, aeration should be high. Water quality should be maintained with water exchanges if needed and retreatment to maintain copper ions at the prescribed level. Copper may alter the efficiency of the biological filter within the tank. Daily additions of liquid solutions of *Nitrosomonas* and *Nitrobacter* bacteria supplements may help maintain water quality. In order to completely break the reproductive cycle it is important to continue treatment for the full 10-14 days. Copper can be reduced or removed by passing the water through activated carbon or carbonate substrate.

Stock solutions of copper sulfate can be made by adding 4.46 grams of copper sulfate (pentahydrate) and 3.00 grams of citric acid to one liter (33.82 fluid ounces) of distilled or deionized water. This will produce a 0.3 ppm concentration at a dose of one ml per 3.75 liters (20-25 drops per gallon). In commercial hatcheries where many tanks are involved, treatment can be quickly administered using a spray bottle calibrated at the proper concentration per squirt. Be sure to know the exact volume of your tank since overdoses can cause permanent injury to the gills. A good quality copper test kit or meter is highly recommended. Copper activity is influenced by total alkalinity (ppm). Toxicity to fish increases as total alkalinity decreases. Although this is not normally a problem with regard to saltwater situations, a word of caution should be noted if you are using a freshwater dip with copper. The rule is to divide total alkalinity (ppm) by 100 to determine the copper sulfate (crystal or powder form) needed in ppm.

In commercial hatcheries, prophylactic applications of copper sulfate for a 10-14 day period every six weeks will greatly reduce possible infestations of *Oodinium* . Continuous use of a low level of copper is not recommended since it leads to slower growth and permanent disfigurement of the operculum and gill filaments. In addition, utilization of ultraviolet radiation and/or diatomaceous earth filtration helps control the spread of the disease.

There is the question of whether total eradication is possible. Basically you are faced with the problem of destroying the tough encysted capsule. Bower (1987) suggests: 1) leave the tank unstocked for 6 to 8 weeks. This process can be shortened to 2 weeks by increasing the temperature to 35°C (95°F) or higher; 2) add no less than 4 ml/liter (3 tsp/gal) of chlorine bleach to the tank, recirculate for 24 hours, flush with freshwater and restart; 3) drain tank and dry everything since encysted *Amyloodinium* cannot survive desiccation.

Photo 25 - Cryptocaryon on gill filiments and caudal fin rays.

Cryptocaryon irritans

Cryptocaryon disease (Photo 25) is called "white spot" or "saltwater ich.". It seems to be triggered by seasonal temperature fluctuations which usually occur in the spring and fall months. It is not a common problem during the summer when temperatures are elevated above 85°F (29° to 30°C). At these temperatures reproduction is limited or ceases. As mentioned, well adjusted adult broodstock clownfish are basically immune to the effects of this disease but juveniles may experience only temporary immunity after being infected. *Cryptocaryon* disease is caused by a ciliated protozoan, *Cryptocaryon irritans.* In this case, the trophont (parasitic stage) lives embedded in the skin, gills, and sometimes the eyes of the fish. It is considerably larger than *Oodinium* trophonts, measuring 48 x 27 to 450 x 350 microns (Spotte 1992). The average length of time that *Cryptocaryon* spends as a parasite varies with temperature. At 21° to 24°C (70° to 75°F) they survive for 4 days. At the optimal reproductive temperature of around 21° to 27°C (70° to 80°F) the period is reduced to 2 days. But, at 15°C (60°F) they remain as a parasite on the body for 14 days. During this period they are able to move about the skin, ingesting tissues and body fluids. Eventually they develop into tomonts (reproductive stage) which can be quite large measuring 440 x 250 microns which can be easily distinguished as pinpoint size, white spots on the skin. The tomonts may or may not drop off the host fish. Within 3 to 28 days as many as 200 tomites (swimming stage) can emerge from a single tomont. These are short-lived and must find a host within 48 hours. Newly hatched tomites are 56.5 x 35 microns in size.

Detection is relatively easy in the early morning hours since the large white tomonts are quite evident on the dorsal, pectoral and caudal fins and dorsal crest of the fish. They are about the size of a pinhead. Later in the day most of the tomonts will drop off and the infestation is not as evident. Rapid breathing and scratching are common symptoms. Signs of advanced infestation include opaque eyes, excess mucus, and high respiration. Patchy loss of color or pigmentation can occur as the disease progresses. Blindness can occur when the eyes are invaded.

Treatment For Cryptocaryon

Cryptocaryon irritans tomonts are susceptible to salinities below 16 ppt (1.0114 sg) or greater than 55 ppt (1.0420 sg). At elevated temperatures of 30° to 32°C (85° to 90°F) reproduction ceases.

Clownfish easily tolerate lower salinities and higher temperatures provided the disease is not advanced. But, they may not be able to adjust to higher salinities. Five minute dips in 55 ppt salinity kills or inactivates trophonts but could create delayed mortality of the host (Spotte, 1992). Moe (1982) suggests removing the fish and giving them a 2 to 3 minute freshwater dip but states that this is not as effective as with *Oodinium* since many *Cryptocaryon* trophonts are embedded under the skin. He further suggests a series of three 1/2 hour formalin dips, one teaspoon to 5 gallons (1 ml per 3.75 L), spaced every other day. The aquaria they are housed in can be treated with copper, 0.4 mg/L, or the tank can be broken down and cleaned. In commercial operations, the latter method is preferred.

At IOH, we used quinacrine hydrochloride at a dosage of 4 to 6 mg/L (4 to 6 ppm) dosed twice over a two day period. Spotte (1992) further expanded this treatment to include initially raising the salinity to 45 to 48 ppt (1.0340 to 1.0364 sg) and maintained for 20 minutes. Then reduce the salinity to 36 ppt (1.0268 sg). With the flow off, add a combination of quinine HCL at 3.96 mg/L (4 ppm) and chloroquine at 15.90 mg/L (16 ppm) for two days. The Shedd Aquarium suggests using two-day treatment, consisting of lowering the salinity (1.017 specific gravity or about 3.5 ppt) and formalin (2 drop/gallon) for the first day, followed by 1 drop/gallon the second day. In recent years formalin has been a good economical choice for treating many ectoparasites.

Bacterial Infections

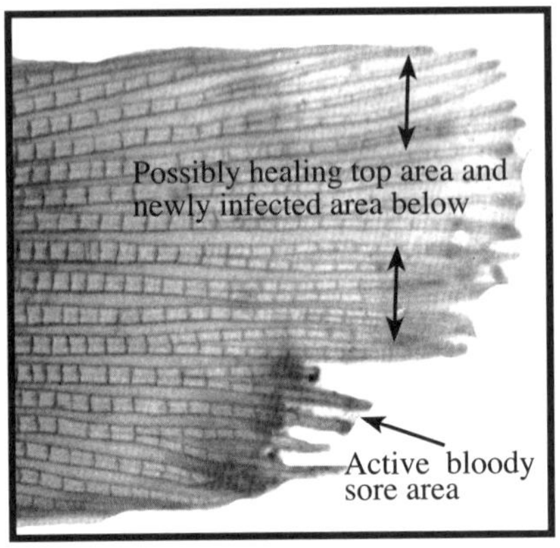

Photo 25a - Tail and fin rot which is usually caused by a bacterial infection.

Bacterial infections are not usually a primary disease, but secondary in effect, and are results of induced stress or injury. Most disease causing bacteria in marine fish are gram-negative. Bacterial infections are usually enhanced by elevated temperature and poor water quality. High densities of bacteria turn normally clear water to a light, milky white, color. Care should be taken because high bacteria levels are usually associated with high ammonia levels. Typically a few weaker individuals in a tank become infected and if the stress continues all can become infected. Death is not usually rapid and may take one to two weeks.

Infected fish often exhibit frayed pectoral and/or caudal fins with bloody tips (Photo 25a). Open body sores that do not quickly heal and/or have a reddish bloody colored edge, indicate bacteria. Note, healing body sores and fins have a whitish colored edge. These symptoms are accompanied by the usual rapid breathing, along with other signs such as a graying opaqueness of the eyes and bloody scales at the base of the fins. Blindness may also be caused by bacterial infection. At times, the mucus pores around the head and lateral line region may be clearly visible and appear to be dry and open. The overall appearance is that the skin looks dry with mucus pores look like a series of obvious holes around the head region.

Treatment For Bacteria

Antibiotics are usually applied. By technical definition, an antibiotic is a natural chemical produced by microorganisms, such as bacterium or fungus, that is used to kill other microorganisms. It is basically a form of chemical warfare (Dixon 1992). As usual, infected tanks should be isolated from the system and treated separately. Moe (1982) suggests several drugs such as neomycin at 250 mg/3.75 L (66 ppm), chloramphenicol sodium succinate (chloromycetin) at 50 mg/3.75 (13 ppm), erythromycin at 40 to 50 mg/3.75 L (11 to 13 ppm) or tetracycline at 50 mg/3.75 L (13 ppm). Positive results should be noted in three to four days. Use of bacterial drugs will have an adverse effect on your biological filters, so water quality must be carefully watched. Daily additions of *Nitrosomonas* and *Nitrobacter* concentrates may help. At IOH, we found the use of nitrofurazone (Furanace) 5 to 6 capsules per 40 gallons may be helpful without strong adverse effects on biological filters.

The best treatment is to clean the water and tanks and not use antibiotics. Utilization of ultra violet sterilizers, ozone and/or diatomaceous earth filters will effectively reduce bacteria levels. Check your UV, make sure it is working and the bulb is not older then one year. Turn up the flow and aeration, clean the bottom and filter. Bacterial diseases are triggered by poorly enriched water conditions. Extreme levels of dissolved organics contribute, so increase the protein skimmer output or intermittently use activated carbon. Antibiotics should be considered only as a last resort.

Fish Tuberculosis (Mycobacterium)

This is a widespread aquatic related disease caused by a gram-positive bacteria of the genus *Mycobacterium* . Prevalence of mycobacteriosis may run as high as 15% in some fish populations (Parisot & Wood 1970). The most frequently isolated species from fish are *M. fortuitum* and *M. marinum*. Infections have been described in over 150 species of fish representing 40 families (Nigrelli & Vogel 1963). In Israel, 399 strains of *M. simiae* were isolated from humans and most are considered to have come from environmental water sources.

Commonly, human infections are diagnosed as "fish handlers disease". People who handle raw fish are at times infected on the hands and arms. Aquarists and aquaculturists are also susceptible. Sites of entry of the infection are through open cuts and abrasions, therefore use of rubber gloves are recommended when cleaning filters and tanks. Lesions on humans appear only on cooler surface skins and look like a wart but do not respond to normal wart medicine. They can become sore and enlarge if not treated. The bacteria is susceptible to high heat applied externally or long term (2 to 4 weeks) and internal treatment of minocycline. Routine application of iodine directly on the sore may also help. In fish culture, this disease is also referred to as "wasting disease."

Mycobacteriosis is generally characterized as a systemic disease affecting several organ systems and in advanced cases granulomas can be found in the viscera (Lansdell et. al. 1993). Within fish, the kidney is a primary site of infection and it is thought that the bacteria is released into the water through the urine, then ingested through contaminated food particles. Development of fish TB is slow and may stay in remission for years like human TB.

Initially, infected fish become lethargic, hide in corners, loose their appetite, and respire rapidly. Infected young fish may appear stunted and grow slowly and develop a slightly concave stomach region. Loss of scales, depigmentation, or hyperpigmentation is common (Noga, et al 1990). In salmon the skin color becomes brighter whereas in other fish it becomes darker. In neon tetras the bright red line becomes yellowish. As the disease spreads the eyes may become clouded and exothalmus (popeye) usually develops. The skin may have blotches or lighter colored areas. Raised scales may appear in certain areas. Death is slow and the fish just "wastes" away.

Treatment For Mycobacterium

Mycobacteria are widespread throughout aquatic environments and cannot be easily eradicated. They are triggered by poor husbandry practices. Routine removal of uneaten food and fecal material is imperative. Feeding of raw fish and shellfish is believed to be a source and should be considered. Utilization of organic loaded, dirty, clogged, sub-gravel filtration is considered a focal point for the incubation of mycobacteria.

Visibly infected fish should be destroyed and the tank sterilized with a chlorine based oxidant. An actively infected fish is a carrier and can infect others, so removal of infected fish is best. Internal treatment is most effective but is hard to administer if fish are to the point where they will not feed. External treatment of isoniazid or streptomycin at 40 mg/3.75 L (gallon) to the aquarium can be used. Change water every three days and remedicate for a period of 10 days. Moe (1992) recommends soaking the food in a streptomycin solution or adding rifampin at 6 mg/100 gm of food. This can become a serious disease problem in crowded hatchery conditions, and when detected, a vigorous effort should be made to clean and sterilize the tanks, increase the filtration efficiency and destroy all infected individuals. At IOH, results were obtained using Minocycline mixed into the gelatin-based food.

General Alternative Treatments

Less toxic, more natural treatments have been developed in many fish industries. Many come from the food fish industry since all treatments must be FDA approved for human consumption and non adverse impact on the environment. Since the treatment can often be as detrimental as the disease, consideration of alternative medication might be important.

The following compounds have undergone review of the Food and Drug Administration and have been determined to be new animal drugs of low regulatory priority (Trosclair 1994).

Acetic Acid: 1000 to 2000 ppm dip for 1-10 minutes as a parasiticide.

Calcium Chloride: Used to increase calcium concentration to insure proper egg hardening, 10-20 ppm $CaCO_3$ in freshwater. Sustained levels up to 150 ppm in freshwater to 300-400 ppm in saltwater to maintain osmotic balance.

Calcium Oxide: Use as an external protozoacide for fingerlings and adult fish at a concentration of 2000 mg/L (2000 ppm) for 5 seconds.

Carbon Dioxide Gas: For anesthetic purposes for cold or warm water fish.

Fuller's Earth: Reduce adhesiveness of fish eggs to improve hatchability.

Garlic (whole form) Used for control of helminth and sea lice infections in marine salmonids at all life stages.

Hydrogen Peroxide: Used at 250-500 mg/L (250-500 ppm) to control fungi on fish and eggs.

Ice: Reduce metabolic rate during transport.

Magnesium Sulfate: Treat external monogenic trematodes and other external crustacean infestations in juveniles and adults. Dip in 30,000 mg/L of $MgSO_4$ and 7000 mg NaCl/L solutions for 5 to 10 minutes.

Onion (whole form): Use whole onion to treat external crustacean parasites, and to deter sea lice from infestation.

Papain: Use a 0.2% solution to remove gelatinous matrix of fish egg masses in order to improve hatchability and decrease incidence of disease.

Potassium Chloride: To aid osmoregulation, relieves stress and prevents shock. Dosages necessary to increase chloride ion concentration to 10-2000 mg/L.

Povidone Iodine: Use as a egg disinfectant before and after water hardening, 100 ppm solution for 10 minutes. Use as a direct treatment on open sores.

Sodium Bicarbonate: A means of introducing carbon dioxide into the water to anesthetize fish, 142 to 642 ppm for 5 minutes.

Sodium Chloride: Reduce stress and aid osmoregulation, a sustained 0.5% to 1.0% (5 to 10 ppt) solution for freshwater fish and 20 to 25 ppt for saltwater fish. As a parasiticide, 3% (30 ppt) for 10 to 30 minutes for freshwater and 5% (50 ppt) or 0.1% to 0.3% (1 to 3 ppt) for saltwater fish.

Sodium Sulfite: Treat eggs to improve hatchability, 15% solution for 5 to 8 minutes.

Urea and Tannic Acid: Used to denature egg adhesiveness, 15 g urea and 20 g sodium chloride per 5 liters water for 6 minutes, followed by a separate solution of 0.75 g tannic acid per 5 liters for additional 6 minutes. Good for 400,000 eggs.

Probiotics

A new and exciting alternative treatment process uses aggressive microbial species to combat or compete against pathogenic species. This process offers a new field of treatment without residual effects associated with antibiotics. Treatment consists of introducing aggressive non-pathogenic bacteria strains that actually compete with pathogenic bacteria strains like *Vibrio*, *Aeromonas* and *E. coli*. Pathogenic species are not totally eliminated, but reduced to non effective levels. Probiotics are added routinely, creating a dominating population that compete for the same environment and subduing common pathogenic bacteria. This process is not unique to aquaculture but is also being applied in human medicine and farming practices. Use of yogurt is a common example. Yogurt is actually a live culture of bacteria which aids digestion and is known to help eliminate yeast infections in women. Generally, probiotics are obtained in freeze-dried form and are mixed with a culture media. Once mixed, the media must be applied within 72 hours since probiotics grow quickly and consume the nutrients provided. Probiotics are used to reduce turbidity, reduce algae concentrations, and remove ammonia. There is no doubt that products based on natural processes will become the preferred treatment in the future.

Douillet (1996) presented results of using pure and combined strains of bacteria to feed rotifers as an enhancement food. He found that the growth rate and production were consistently improved in all experimental trials. Also the coefficients of variation of production were significantly reduced.

Larvae of red fish, *Sciaenops ocellatus*, were cultured for 10 days in recirculating systems with internal biological filters. The culture tanks were seeded twice during the culture period with singular and muti strains of probiotic bacteria. Control tanks were not seeded. There were no differences in larval growth between treatments, but there was an improvement in larval survival with both treatments and especially multi strains. Similar work on shrimp, *Penaeus vannamei*, juveniles showed better growth rates but no difference on singular or muti strains. This is a new field and considerable work must be done to really evaluate the true value of using positive strains of bacteria.

Neon Gobie - *Elacatinus oceanops*

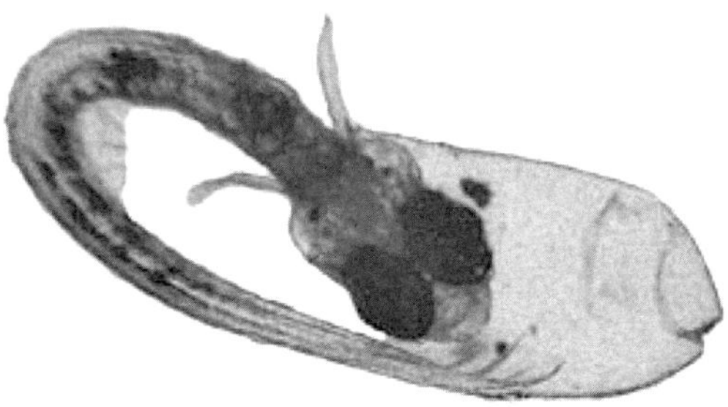

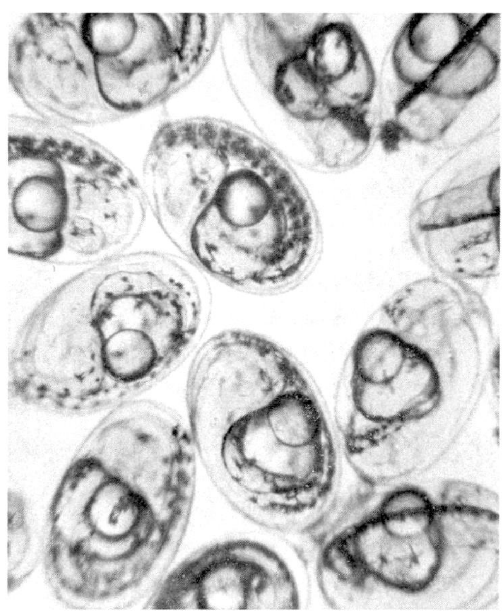

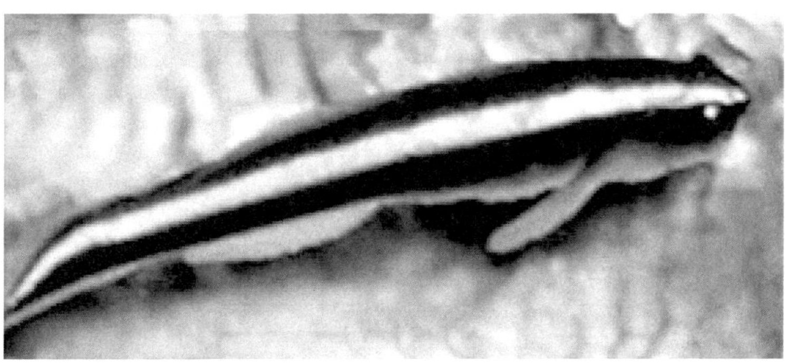

WATER QUALITY STRESS

Our biggest investment and highest risk is in juvenile rearing. Ability to recognize problems and make quick positive decisions will result in higher survivals. Masser *et al* (1992) stated that in order to commercially rear fish in a profitable manner, intense stocking is necessary. Food fish must be reared intensively at levels of 0.5 lb per gallon or greater for recirculating systems to be cost effective. As an analogy, a 20-gallon home aquarium, which is a miniature recirculating system, would have to produce at least 10 pounds of fish to be cost effective. Just think about trying to maintain 10 pounds of fish in 20 gallons of water. Basically this is the problem you are faced with as you increase your production. In tropical fish this is not as serious since 300 to 400 clownfish weigh about 1 pound.

Oxygen is an important controlling aspect of water quality. Without backup systems, intense culture systems can change drastically in a very short time. For example, if a power failure occurred in a warm-water system (84°F = 29°C) at saturated oxygen concentrations containing one-half pound fish at a density of one-quarter pound of fish per gallon of water, it will only take 16 minutes until the oxygen concentration is down to 3 ppm and fish start to stress. The same system with one-pound fish stocked at a pound of fish per gallon would take less than 6 minutes to plunge to an oxygen concentration of 3 ppm (Masser *et al* 1992).

As mentioned, biological filters can fail due to senescence, chemical treatments such as antibiotics, tea seed, residual chlorine, detergents, etc. or anoxia (lack of oxygen). Nitrifying bacteria are basically slow growing and can easily be inhibited by high concentrations of dissolved organic matter greater than 50 ppm, high ammonia and nitrite concentrations greater than 2 ppm, temperatures over 82.5°F (28°C) and low dissolved oxygen levels below 4 ppm. In addition, limited attachment sites, poor circulation, excessive exchange rates, and poor compatibility with fast growing, gram negative, heterotrophic, bacteria effects nitrifiers. Since the filter is the heart, it is very important to take care and make sure it continues to beat.

Carbon dioxide is produced by respiration of fish and bacteria in the system. Fish begin to stress at carbon dioxide levels above 20 ppm because it interferes with oxygen uptake. Ranges of pH from 6 to 9.5 can usually be tolerated by fish although rapid changes of 2 units or more is harmful, especially to fry. In recirculating systems, pH declines as bacteria produce acids and carbon dioxide is generated by the fish and biofilter. Carbon dioxide reacts with water to form carbonic acid which drives the pH downward. Below a pH of 6.8 nitrifying bacteria are inhibited and do not remove toxic nitrogen wastes. Utilization of sodium bicarbonate, calcium carbonate, calcium hydroxide, calcium oxide, and/or sodium hydroxide should be used to maintain alkalinity buffers.

An estimated 2.2 pounds of ammonia is produced from each 100 pounds of feed. If the bacteria levels are not adequate at peak feeding periods, problems can occur. If the biofilter is disrupted, quickly reduce or stop feeding. Feeding, even after a complete water exchange, can cause ammonia or nitrite concentrations to rise to stressful levels in a matter of hours if the biofilter is not functioning properly.

Nitrite toxicity varies with species. Scaled fish and those in saltwater containing chloride ions are generally more tolerant of high nitrite. Usually 5 to 6 parts of chloride protect fish from 1 part nitrite. Increasing nitrite is a signal that biofiltration is failing or not large enough to control the amount of waste.

Temperature stress, particularly cold, can completely halt activity of the "killer cells" in the immune system (Francis-Floyd 1991). Fish can usually tolerate a diurnal ±5°F (±2°C) temperature change without significant problems. Temperature changes of more than this should be controlled to about 1°F (0.4°C) every 20 to 30 minutes (Masser *et al* 1992). To be on the safe side, it is advisable that temperature changes beyond the ±5°F range be extended on a daily basis per degree.

Figure 62 illustrates the overall affect of continued stress, nutritional disorders, and/or excessive pathogen levels. All contribute, but when out of balance for a period of time, disease will often result. Table 54, and 55 offer water management suggestions and possible solutions to prevent continued stress.

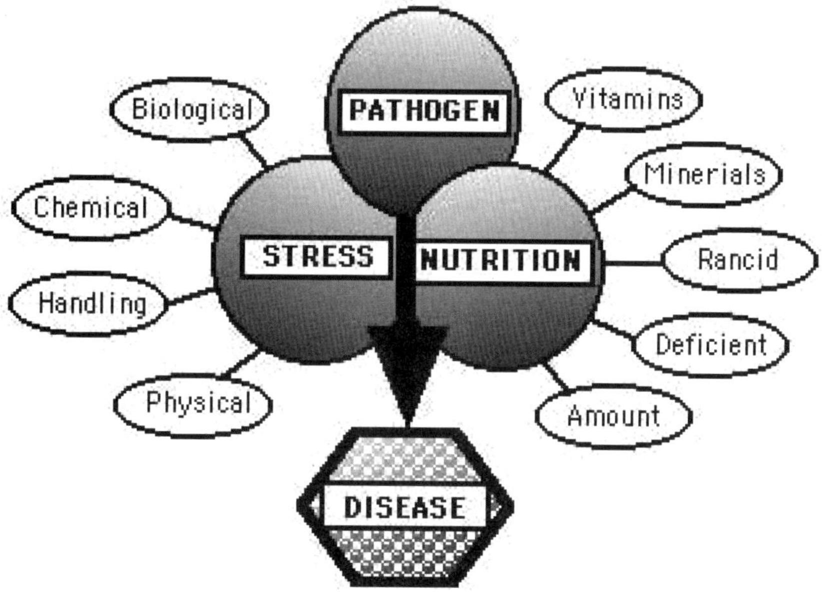

Figure 62 - Stress contributing external and internal, nutritional and pathogenic factors, which when in excess and/or sustained too long result in the host becoming diseased.

Table 54 - Possible management options based on water quality and feed observations (Masser et al 1992).

Observation	Possible Management
Low dissolved oxygen (less than 5 ppm)	Increase aeration Stop feeding until corrected Watch for symptoms of new parasite/disease
High carbon dioxide (above 20 ppm)	Add air stripping column Increase aeration Stop feeding until corrected Watch for symptoms of new parasite/disease
Low pH (less than 8.0 saltwater) (less than 6.8 freshwater)	Add alkaline buffers (sodium bicarbonate etc.) Reduce feeding rate Check ammonia and nitrite concentrations
High ammonia (above 0.05 ppm as unionized)	Exchange system water Reduce feeding rate Check biofilter, pH, alkalinity, hardness, and dissolved oxygen in the biofilter Watch for symptoms of new parasite/disease
High nitrite (above 0.5 ppm)	Exchange system water Reduce feeding rate Add 5 to 6 ppm chloride per 1 ppm nitrite
Low alkalinity	Add alkaline buffers
Low hardness	Add calcium carbonate or calcium chloride
Discolored/clumped feed	Purchase new feed, discard old feed Watch for symptoms of new parasite/disease

Table 55 - Possible options in managing a recirculating tank system based on observations of the fish (Masser et al, 1992).

Observation	Possible Cause	Possible Management
Excitable/darting/ erratic swimming	Excess of intense sounds or lights	Reduce sound levels, pad sides of tank, Reduce light intensity Examine fish for symptoms Check ammonia concentration Check carbon dioxide level
Flashing/whirling	Parasite	Fully examine fish with symptom
Discolorations/sores	Parasite/bacteria	Fully examine fish with symptom
Bloated/eyes bulging out	Virus or bacteria	Fully examine fish with symptom
	Gas bubble disease	Check for supersaturation and examine fish with symptom
Laying at surface or not swimming off	Parasite	Fully examine fish with symptom
	Low oxygen	Check dissolved oxygen in tank
	High ammonia or nitrite	Check ammonia and nitrite concentrations
	Bad feed	Check feed for discoloration, clumping, smell, and fungus. Discoloration of fish blood
Crowding around water inflow and/or aerators	Low oxygen	Check dissolved oxygen in tank
	Parasite/disease	Fully examine fish with symptom
	High ammonia or nitrite	Check ammonia and nitrite concentrations
	Bad feed	Check feed for discoloration, clumping, smell, fungus and the blood of the fish
Gulping at surface	Low oxygen	Check dissolved oxygen in tank
	Parasite/disease	Fully examine fish with symptom
	High ammonia or nitrite	Check ammonia and nitrite concentrations
	High carbon dioxide	Check carbon dioxide level
	Bad feed	Check feed for discoloration, clumping, smell, fungus and the blood of the fish
Reduced feeding	Low oxygen	Check dissolved oxygen in tank
	Parasite/disease	Fully examine fish with symptom
	High ammonia or nitrite	Check ammonia and nitrite concentrations
	Bad feed	Check feed for discoloration, clumping, smell, fungus and the blood of the fish
Stopped feeding	Low oxygen	Check dissolved oxygen in tank
	Parasite/disease	Fully examine fish with symptom
	High ammonia or nitrite	Check ammonia and nitrite concentrations
Discolored brown blood	High nitrite	Examine fish with symptom; add 5 to 6 ppm chloride for each 1ppm nitrite; purchase new feed and discard old feed
Clear (no blood)	Vitamin deficiency	Examine fish with symptom; check feed for discoloration/clumping; purchase new feed and discard old feed
Broken back or "S"	Vitamin deficiency	Examine fish with symptom; purchase new feed and discard old feed

SHIPPING AND RECEIVING

Now we have successfully reared fish to a salable size. How do we safely transport them to friends, distributors and dealers. Even the healthiest individuals can be seriously stressed and damaged during shipping. Water quality changes within the small confined shipping bag can be dramatic without taking preventative measures. In aqueous solutions, ammonia exists in two forms: ionized (NH_4^+) and unionized (NH_3). Of the two, NH_3 generally is regarded as the toxic form. Concentrations of unionized ammonia (NH_3) increases by a factor of 10 with each increase of a pH or 1.0. Marine fish should be the most susceptible to ammonia toxicity due to high pH levels. However, in high salinity marine fish, especially clownfish, ammonia and nitrite are not as toxic as they are to brackish or freshwater fish (see figure 59, page 152). This is only true if pH remains high 8.0 to 8.3, carbon dioxide levels remain low and alkalinity remains high. Clearly, it should be understood that use of antibiotics, copper and other therapeutics are not recommended. During shipping is not the time to treat fish.

Turner and Bower (1982) conducted a series of tests to see if the use of preconditioned biological substrates added to shipping bags would effectively control ammonia and nitrite. Substrates tested were activated carbon, crushed oyster shell and polyurethane pads which were used alone, and with a buffer, Trizma™ -8.3 (Sigma Chemicals) at a rate of 6.77 gm/L. In addition, the buffer was tested without substrate. Four tests with three replicates were run including controls which received no media. Into each bag 500 ml of Instant Ocean (30 ppt) was added and a single 6.8 gm fish, equivalent to 25 or 13.6 gm/L of salable clownfish. Each bag was filled with pure oxygen at a 5:1 ratio and sealed. Preconditioned substrates were added at 50 cm^3 per bag. Bags were maintained in styrofoam coolers for 24 hours.

Results showed a drop of pH of 1.0 in unbuffered tests and <0.2 in buffered tests with or without substrates. The pH values in unbuffered seawater containing nitrifying bacteria on all three substrates was more pronounced in bags containing activated carbon and polyurethane foam then those containing crushed oyster shell. Ammonia (NH_4-N), regardless of substrate used or whether buffered or not, was 93% to 95% lower than those in the control bags. Buffered or unbuffered tests showed no significant differences indicating that use of Trizma™-8.3 does not inhibit ammonia-oxidizing bacteria. Concentrations of unionized ammonia were 10 times higher in buffered tests however, the fish were not affected. Experiments conducted at IOH showed concentrations as high as 1.0 ppm total ammonia and 1.4 ppm nitrite were not toxic to clownfish (page 151-152). As expected, nitrate increased significantly in bags containing nitrifying bacteria as compared to controls. Dissolved oxygen levels in all tests exceeded saturation levels.

Based on their data the addition of preconditioned substrates is a practical and effective method of maintaining low concentrations of ammonia during transport. Of the substrates tested they felt the polyurethane pads might be best since it is not selective for specific dissolved media like buffers or stress conditioners.

Tests conducted at IOH used *A. ephippium* at (25 fish/bag), 1.31 L water containing NovaAqua™ (4 ml to 16 L of water) and 1 ppm copper solution. Results in 24 hours showed a significant drop in pH from 7.86 to 6.75 which we felt is critical to marine fish.

Newly made Instant Ocean and other trade salts are not suitable for shipping fish. However, if recirculated through a calcium carbonate substrate for 24 hours or longer they stabilize and can be safely used. Fish should be culled the night before and placed in plastic mesh baskets (Figure 35) within the system they came from to help reduce handling stress Packing water (5 gallons at a time) should be drawn from a special highly aerated recirculating system contain ing adult pairs. Treat with Trizma™ -8.2 and StressGuard™ (SeaChem), aerate as recommended by the manufacturer, add 1 L per quarter bag and place 20-25 fish in each, add pretreated aged polyurethane pad, a squirt of nitrifying bacteria (EnviroReps), fill with pure oxygen, double band and place in a full square bottom bag. Seal quickly in styrofoam cooler and store out of the sun.

FUTURE OF MARINE ORNAMENTAL FISH CULTURE

Pet Sales

This chapter is presented to assess the current demand and direction of future marine tropical fish culture endeavors. The discussion is in part from a speech I presented to a special session of the 1993 World Aquaculture Society Meeting in Tourmalines, Spain.

Ornamental freshwater tropical fish culture in Florida dates back to the 1930's. Within the last 20 years a major expansion of the existing industry has occurred. Farms in Florida now produce over 700 varieties of ornamental fish. Current farm gate value of this crop within Florida alone is $33.7 million (1989). Aquatic plants represent an additional $11 million.

Marketing potential of dry goods is primarily motivated by appeal, innovation, demand, price, and availability. Whereas, marketing potential for aquacultured products is chiefly motivated by biological knowledge, ease of culture, growth rate, and price. Tropical fish are often marketed on an individual farm basis in an attempt to obtain the highest price possible for their products. Farm-gate prices are controlled by wholesalers, retailers, and by supply and demand. Extreme direct competition from the Far East, local "garage hatcheries", and lastly wild caught fish has forced the freshwater industry to consistently accept low farm-gate prices with little change over the last ten years. Typically a $0.15 cent fish to a Florida producer may sell for $2.89 in a northern pet shop. Competition in the USA freshwater market results in few new farms entering the field and a general condensing of smaller farms into conglomerates or coops.

Cultured marine tropicals, on the other hand, virtually represent an unlimited, untapped, field coupled to a constantly diminishing wild-caught supply. Estimated price per pound of freshwater tropical fish is $35 to $60 whereas marine tropical fish are estimated at $400 to $500. Of the few marine tropicals being reared, average farm gate price is around $1.65 with a wholesale price of $3.50 to $4.50 and retail price of $9.00 to $20.00.

Currently 54 million households (58%) in the USA own at least one pet of which 10% have freshwater fish and 0.8% have saltwater fish. Within the USA it is estimated that 89 million freshwater fish are maintained in 12.1 million tanks while 5.6 million marine tropical fish are maintained in 1.1 million tanks. Currently 80% of the freshwater fish are cultured from farms in Florida and Southeast Asia. Whereas, virtually 100% of the marine fish, invertebrates, and algae are collected from the wild.

Worldwide it has been estimated that the aquarium industry is a $4 to $15 billion industry. Pet sales (all animals) and pet related products in the USA were a $12 to $13 billion industry during 1994. Nine billion dollars represent dog and cat food sales of which 60% were sales in supermarkets. Individual independent pet store sales in the USA in 1991 was $3.5 billion rising to $4 billion in 1994 The average pet store inventory, including livestock and dry goods was $71,751 in 1991 dropping to $58,443 in 1994. Approximately $56,883 (1992) and $48,133 (1994) of the inventory consists of dry goods while $14,868 (1992) and $10,310 (1994) is in livestock.

Of the livestock, 26% consists of fish and related products in 1991 and 21% in 1994, second only to dogs at 37% and 34% respectively. Total inventory in 1991 indicated that it had doubled over past surveys but remained the same or lower in 1994. Yet, this overall net increase clearly indicates where the public interest is going. Of the fish and fish product sales, 15% were products such as food, aquarium supplies, chemicals and drugs in 1991 and 11% in 1994. Aquarium livestock was 8% freshwater fish in 1991 and 7% in 1994, 2% saltwater fish in 1991 and 3% in 1994. Reef products such as "live-rock", plants, and invertebrates may represent an additional 1%. In 1991, shops that sell only aquarium products were 4% and rose to 6% in 1994. Shops selling some level of fish or fish related products is 68% while only 30% were full line shops including fish, dogs, cats, etc. in 1991 and 44% in 1994. Ironically, 32% of the shops carry no fish at all (Pet Dealer, 1992 & 1995).

A survey of 1991 inventory changes showed a 42% increase in inventory with emphasis directed towards increasing dogs and dog items. Freshwater fish showed a net increase of 5% where saltwater fish showed a net increase of 0% and reef livestock showed a 3% decrease (Pet Dealer, 1992 & 1994). Although there has been about a 1% increase in saltwater fish sales over the last four years, growth is still poor compared to overall interest. Slow growth and decreases in saltwater livestock are no doubt due to the following:

1) Diminishing, consistent, suppliers of saltwater livestock.
2) Diminishing quality of livestock.
3) High cost of saltwater livestock verses handling losses.
4) Diminishing sources and supply of saltwater livestock due to government fishery restrictions.
5) Inability and/or general lack of knowledge of store owners with regard to the care of saltwater livestock.

One has to question whether interest in marine tropical livestock actually exists if less and less shops are carrying livestock and related products? This can probably be easily answered by the virtual surge of advertising, articles, and new publications on saltwater livestock and products. Freshwater and Marine Aquarium Magazine (FAMA) is one of the leading pet magazines in the USA. Over the last 6 years there has been a significant shift from mostly freshwater articles to mostly saltwater articles. Advertising has increased 10 fold and leans heavily towards reef systems and related components. Tropical Fish Hobbyist, the oldest existing aquarium magazine, was virtually a freshwater fish magazine with absolutely little interest in marine articles. Within 6 years, they now dedicate at least one third of the magazine to saltwater livestock. Marine Fish Monthly probably has the third highest circulation and is the newest publication totally dedicated to saltwater fish and invertebrates. Publications respond to demand and it should be very clear where the interest in fish keeping is going.

With increasing world-wide fishery restrictions, virtual closing of reef areas in the Florida Keys, import and export embargoes in Europe, and public environmental concern, the "showdown" is soon to be eminent. Supply and demand will play a key roll in further expansion of this important hobby. You simply can not have a saltwater aquarium without saltwater fish, invertebrates and/or plants. The feasibility of marine culture has never been higher in priority than now.

Past And Current Endeavors

Biologically, significant data and information exist on the culture of clownfish (*Amphiprion sp.*.). For years only two significant facilities existed in the USA which produce tank-raised clownfish: C-Quest, Puerto Rico and Reef Propagations, Chicago Illinois. Over the last three years Aqualife Research, Walkers Cay Bahama; Desert Fisheries; Nevada, and Dynasty Marine, Marathon Florida have stopped production. However within the last 12 years Ocean Reefs and Aquarium (ORA) in Fort Pierce Florida has become the worlds largest and most profitable facility. Recently several small new entities have emerged. Like most products in the marine ornamental industry, supply often does not meet demand on a consistent basis. Past facilities, Aqualife Research founded by Martin Moe, Sea World Aquaculture founded by Chris Turk, and Instant Ocean Hatcheries founded by Frank Hoff, made great strides in commercially producing clownfish. Unfortunately, like most pioneering aquaculture facilities, expenses for continued research and development exceed projected allocations. Coupled with long-term limited returns, the sustaining power and interest of the investor diminishes and the business usually ends in failure. Up to this year (1996), only Aqualife Research existed. Yet, even after 13 years, Instant Ocean Hatcheries was probably the most intensive, furthest advanced, most unusual facility that maintained detailed records and data.

On January 27, 1981, Martin Moe, Chris Turk and Frank Hoff conducted a joint conference at IOH to compare and hopefully solve mutual biological problems. This unprecedented conference between competitors clearly illustrates our dedication and desire to keep one or all businesses alive. Table 56 is a comparison of the three facilities and production procedures which resulted from this conference. The table is broken down into three distinct areas of broodstock management, larval rearing and juvenile rearing. Table 57 illustrates the major differences between the three hatcheries. In addition, we conducted comparative histological studies of juvenile clownfish (*A. ocellaris*) from the three facilities. Of the fish studied during this test, those from Aqualife Research were considered the healthiest (Table 58).

Accomplishments of IOH were significant. Eight species of clownfish, *Amphiprion* were successfully reared in mass. Third generation broodstock fish were being utilized. Three cross-hybrids were developed and one species of neon goby were reared.

Table 56

Comparison of facilities and production at Instant Ocean Hatcheries (I.O.H.), Sea World (S.W.), and Aqualife Research (A.R.). This is unpublished information from a joint conference at I.O.H. between Martin Moe, Chris Turk, and Frank Hoff on January 27, 1981.

COMPARISON OF BROODSTOCKS			
QUESTION	I. O. H.	S. W.	A. R.
Number of Tanks	174	25 - 60	100
Tank size, gal	25 to 40	25 to 60	30 to 75
Tank construction	plywood w glass front	all glass	all glass
System or static	system	system	static
Tank exchange rate	4x per hr	1x per 2 hrs	1x per week
Tank filter	sub-biofilter	sub-biofilter	sub-biofilter
Filter substrate	coral hash	sand	sand & oyster shell
Use anemones	Usually not	always	usually
Anemone types	Atlantic mainly	Pacific	mostly Pacific
Spawning substrate	flower pots	flat rocks	flat slate rocks
Avg. spawns / month	2.1	about 2.0	2.0
Feedings per day	4x	3x	3x
Food types	roe + gel mix with roe, shrimp, chicken hearts, clams & kelp	frozen mix	frozen mix with conch, lobsters and mainly shrimp
Broodstock	70% wild	100% wild	50% wild
Number of pairs	160	40	55
Number active pairs	105	10	25
System size	4,000 gal	2,000 gal.	no systems
System exchange	20% per month	100% per 3 days	50% to 75% /wk/tank
Pairs per system	58	all in one	no systems
Airlifts per tank	1	2	2 or 3
Lighting	solar plus indirect florescent lighting	florescent plus some indirect solar light	total florescent
Lighting period	14 to 16 hours depending on season	not noted	14 hours
Auxiliary lighting	dim indirect "moon light" all night	none	none
Salinity	28 ppt	34 ppt	35 ppt
pH	7.9 to 8.3	8.0 to 8.1	7.8 to 8.0
Desired temperature	25 to 29 C	27 C	27 C
Algae growth in tank	brown & bluegreen	red & purple	mainly green & red
Nests are hatched	in larval area	in tank	in larval area
Egg bound pairs	yes - usually occurs in spring	yes - when cod liver oil is fed	yes - when too much shrimp is fed
Egg attachment problems	no	yes	yes
Diseases	lymphocystis in maroons and "bloat"	none mentioned	none mentioned
Initial pair size	sub adults	adults	adults

Table 56 Continued - COMPARISON OF LARVAL REARING

QUESTION	I. O. H.	S. W.	A. R.
Number of tanks	72	?	30
Tank size	40 gallons	60 gallons	60 to 250 gallons
Tank depth	20"	not noted	18" to 36"
Tank width	24"	not noted	15" rectangular ones 125-250" round dia.
Tank length	24"	not noted	30"
Construction	all glass	all glass	glass, fiberglass, plywood with black plastic liner
Outer cover	black plastic	black plastic	black plastic
Airstones per tank	one in corner	two opposite corners	two middle
Air flow	strong	moderate	moderate
Other obstruction	center standpipe	corner drain	center standpipe
Water flow	none for 7 to 9 days	none all the time	none for 4 to 5 days
Water exchanges	30% first 7-9 days	50% per day	50% first 4 -5 days
Tanks cleaned	3 times per week	1-2 times per day	1 time per day first 4-5 days then 2X/wk.
Water loss/cleaning	5 to 10%	20%	20%
Feedings	6x per day	3x per day	3x per day
Live food per day	2x per day	yes but no number	2x per day
Live food type	rotifers & artemia	rotifers & plankton	rotifers & artemia
Dry food fed	4x per day	none fed	1x per day
Dry food type	flakes & salmon mash	none fed	flakes
Size of dry food	80 & 50 mesh	none fed	80 mesh
Frozen food	none fed	copepods	sometimes copepods
Stocking range	2,000 to 3,000	not determined	not determined
Average harvest	250 avg.	500 avg.	400 to 500 avg.
Highest harvest	2,200 per 40 gal.	800-900 per 60 gal	967 per 75 gal
Metamorphosis	6 to 9 days	8 to 10 days	6 to 10 days
Temperature range	25 to 29 C	26 to 27 C	27 to 29 C
Capture method	net	net	net
Shock evident	yes	yes	yes
Reason for shock	Artemia, dinoflagellates	artemia	lower temperatures
Dark larvae indicate	stress	stress	stress
Clear larvae indicate	healthy larvae	healthy larvae	healthy larvae
Pre-metamorphosis	dark coloration, change in body shape	dark coloration	dark coloration and mouth change
Microalgae in tank	no	no	no
Water colored	yes-green	yes-green	no
Source of color	spirulina	green light filters	none
Natural algae growth	mainly brown	brown and red	mainly red
Macro or microalgae	attached microalgae	attached microalgae	attached microalgae
Lighting	solar + indirect two fluorescent bulbs	fluorescent	direct fluorescent 4 bulbs
Lighting period	14 to 16 hours	not noted	24 hours
Bottom of tank	white	not noted	black
Best production	spring and summer	not noted	spring and fall
Highest production	30,000 per month	not noted	20,000
Treatments use	Furan & neomycin	Furan	
Bacteria counts made	no	yes	no
Tanks sterilized	yes, chlorine	no	yes chlorine and sodium thiosulfate
Nests treated	formalin, 1 ml/gal dip/1 minute	no	no
Hatch irregularities	yes, mainly due to adult feeding	no	mainly tomatoes in summer months
Days when larvae ussally need treatment	hatch, day 4 to day 14 prior and after metamorphosis	not noted	hatch and day 8 to 10 especially after feeding artemia
Symptoms for treatment	shock and starvation	not noted	not noted
Water treatment	UV, DE and biofilter	ozone, sand filter	none

Table 56 Continued - COMPARISON OF JUVENILE REARING

QUESTION	I. O. H.	S. W.	A. R.
Number of tanks	280	not noted	50
Capacity of tanks	40 gallons	100 gallons	100-500 avg 300
Construction	plywood with single front window	fiberglass cone with internal rearing basket no window	plywood, black plastic liner, no window
Tank water exchange	4x per hour	0.50x per hour	4x per hour
On system or static	system	system	static
Open or closed water exchange	mainly closed system	basically open system	open tanks, no system
Filter in tank	sub gravel biofilter	none	none
Filtration	DE, UV, & biofilter	sand filter and ozone	sand filter
Sys. water exchange	100% per 4 months	100% per 3 days	100% per 6 hours
System size	4,000 gallons	9,000 gallons	tank size only
Stocking size	3/8" to 1/2"	same	same
Stocking age	15 to 20 days	not noted	21 to 42 days
Stocking density/tank	8.5 per gallon	1.5 to 2.5 per gallon	1.6 per gallon
Stocking per system	19,600 or 4.9 /gal.	not noted	1.6 per gallon
Number of feedings	4x	5x	3x
Food types fed	2x gelatin mix, roe, freeze dried plankton, salmon mash	frozen mix	1x frozen mix, 2x flakes #80 mesh
How administered	mix pushed through sieve, or is pre-grated	not noted	mix fed through food grater
Vitamins used	special mix in gelatin mix full bodied dry	dry pet types	A, B, C and E
Feeding time	0800-11-14-1700	not noted	0930-13-1630
% algae in diet	10%	30%	5%
Algae type	dry kelp	fresh kelp & ulva	hypnea 20% & ulva
Lighting	ambient solar	like adult broodstock	solar 9 months only rest on lights
Algae growth in tanks	bluegreen & brown	micro green & red	macro enteromorpha
Timed lighting	no	like broodstock	not noted
Harvest after hatch	3 to 5 months	3 months	4 months
Average survival	50% from stocking	60%	70%
% salable at harvest	20%	30%	45%
Mortalities monitored	daily	daily	none
Market size (length)	1 to 1.25 TL	1" TL	1 to 1.5" TL
Prophylactic used	none	copper sulfate	copper sulfate

Table 57 - Major Differences Between IOH, Sea World, and Aqualife Research

Item	I.O.H.	S.W.	A.R.
Location	Central Florida	San Diego Calif.	Marathon Florida
Water source	artificial	natural	natural
Water exchange	25% per year	100% per 3 days	65% per week
Water volume	110,000 gallons	approx. 12,000	approx 10,000
Salinity	28 ppt	34 ppt	35 ppt
Nitrate	20 to 40 ppm	< 5 ppm	< 5 ppm
pH	7.9-8.3	8.0-8.1	7.8-8.0
Phosphate	<10 ppm	< 1 ppm	< 1 ppm
Number of pairs	105	10	23
Post larval production.	20,000/mo	3,000/mo	8,300/mo
Post larval survival	50%+	50%	50%
Juvenile stocking density per tank	8.5 per gallon	1.5 to 2.5 per gal.	1.6 per gallon
Juvenile survival	50%	60%	70%
Salable juveniles	20%	30%	45%
Species reared	11	4	6
Main food	Gelatin mix	Frozen mix	Frozen mix
Main Ingredients	Chicken hearts, shrimp	not noted	Shrimp
Algae in diet	10% kelp	30% ulva	5% hypnea

Table 58 - Comparative histological study on clownfish from Instant Ocean Hatcheries, Sea World and Aqualife Research. (unpublished data from conference between Frank Hoff, Martin Moe and Chris Turk , January 27, 1981).

Comparison of juvenile *A. ocellaris*

ORGAN	I.O.H.	S.W.	A.R.
Gills	NSL	NSL	NSL
Brain	NSL	NSL	NSL
Liver	ceroid deposits	many fatty deposits	few fatty deposits
Pancreas	NSL	ceroid deposits	NSL
Kidney	MM centers present	few MM present	NSL
Spleen	MM centers present	not noted	NSL
Heart	ceroid deposits	not noted	NSL
Intestine	NSL	NSL	NSL
Stomach	NSL	NSL	NSL
Fin rays	many fat deposits	some fat deposits	few fat deposits

Comparison of juvenile *A. frenatus*

ORGAN	I.O.H.	S.W.	A.R.
Gills	NSL	NSL	NSL
Brain	NSL	NSL	NSL
Liver	NSL	few fatty deposits	NSL
Pancreas	few ceroid deposits	few ceroid deposits	few ceroid deposits
Kidney	many MM centers	some MM centers	few MM centers
Spleen	some MM centers	some MM centers	NSL
Heart	macrophages present	macrophages present	NSL
Intestine	NSL	some ceroid deposits	NSL
Stomach	NSL	NSL	NSL
Fin rays	fatty deposits	fatty deposits	few fatty deposits

General histological comments:

1. Of the fish examined, *A. ocellaris* seem to have more problems with lipids than *A. frenatus*. *A. ocellaris* consistently have more fat deposits under the skin and between the fin rays.
2. *A. ocellaris* also seem to have more problems handling fat since they generally have fatter livers, inflammation of the fat that is deposited, and more ceroid deposits. Possibly this species requires more vitamin E or different essential fatty acids.
3. Histologically the fish from Aqualife Research appeared the healthiest. Both Instant Ocean Hatcheries and Sea World have significant deposits and inflammation of the fat.

Table 59 - Post larval production at Instant Ocean Hatcheries from 1981 to 1982

1981

Item	Jan	Feb	Mar	Apr	May	Jun	Jul	Aug	Sep	Oct	Nov	Dec
Pairs used	79	67	89	86	85	85	85	85	85	83	79	79
Post larvae	8439	15540	23320	19796	20569	21339	24129	18373	17701	17738	14680	14880
PL/pair	107	232	262	230	242	250	284	216	208	214	98	188
Nests taken	104	96	146	119	167	160	170	158	169	171	150	160
No hatch	6	1	0	4	3	4	4	5	1	0	3	1
Avg. nest	418	404	396	468	372	345	343	363	381	397	321	403
Total eggs	43445	38819	57814	55678	62067	55243	58401	57298	64439	67903	48179	64569
Dead eggs	18563	10932	15648	18212	20280	17435	16759	16667	23783	27681	10875	13044
Larvae	24911	27887	42166	37466	41787	37808	41642	40631	40656	40222	37304	51525
% hatch	57%	72%	72%	67%	67%	68%	71%	71%	63%	59%	77%	80%
% survival	34%	56%	55%	53%	49%	56%	58%	45%	43%	44%	39%	29%

1982

Item	Jan	Feb	Mar	Apr	May	Jun	Jul	Aug	Sep	Oct	Nov	Dec
Pairs used	68	69	66	70	81	79	66	80				
Post larvae	11215	22458	22063	27160	34059	27103	22161	19787				
PL/pair	165	326	334	388	420	343	336	247				
Nests taken	110	117	110	133	132	124	110	169				
No hatch	3	3	2	0	0	23	9	0				
Avg. nest	429	446	432	435	409	486	495	331				
Total eggs	47144	52166	47484	57900	61988	60351	54525	55980				
Dead eggs	11401	9688	6291	7364	6810	6170	10860	6280				
Larvae	35743	42478	41193	50536	55178	54181	43665	49700				
% hatch	76%	81%	87%	87%	89%	90%	80%	89%				
% survival	31%	53%	54%	54%	62%	50%	52%	40%				

Survival And Sales At IOH

Table 59 provides detailed postlarval production data for the last two years of operation. During this period about 840,000 larvae were produced with an average survival to post-larvae (20 to 30 day old) of 47.9 %. This is a very good survival percentage when considering this work was conducted over 13 years ago. Yet, the real problem at IOH was not in broodstock or larval rearing, but in juvenile growout.

Table 60 summarizes five years of postlarval production and percent survival after 3 to 4 months growout. As shown, our average survival was 43.6% during juvenile growout but only

Table 60 - Juvenile percent survival verses percent marketable at Instant Ocean Hatcheries

Year	Post Larvae Produced	% Survival	% Marketable	Average Price
1978	144,303	50%	37%	$0.95
1979	185,140	40%	22%	$1.05
1980	263,321	55%	22%	$1.07
1981	216,394	39%	19%	$1.21
1982	190,240 8 months data	34% 8 months data	20% 8 months data	$1.29 8 months data
5 years	**999,398**	**43.6%**	**24%**	**$1.15**

Table 61 - Main problems at Instant Ocean Hatcheries.

1. Poor juvenile survival coupled with significant loss of marketable juvenile fish after 3 to 4 months growout.
2. Investment over 9 years totaled around $2 million. with 1/8 monetary return. (Basically this price for this type of research over 9 years is low.)
3. Inability to convince wholesalers and dealers to pay higher prices for tank-raised fish.
4. Inability to compete against prices of wild-caught fish from the Philippines
5. Inadequate water quality with regard to the types of filtration verses bioloads.
6. Inability to clearly outline dietary needs at all levels of the hatchery but more specifically in the juveniles.
7. Unwillingness of owners to scale down overhead to meet production and sales.
8. Top heavy in biologists and chiefs in later years.
9. Initial unwillingness of owners to accept consultants to deal with specific problems until later years.
10. Retention of information slowed progress, and prevented more open exchange with others.
11. Expansion of product line to augment sales came into effect too late.

Suggestions for future endeavors

1. Start slowly and build on your progress.
2. Keep initial investment minimal but not to the point where it hampers advancement and production.
3. Be open minded, share information with others that might be able to help you. Free advice is everywhere and you have no competition.
4. Keep focused on your goal, yet be open to alternatives ideas that may save your investment or augment your income without greatly hampering your initial goals.
5. Used paid consultants and governmental agencies for difficult problems if you cannot solve them within a reasonable time and progress is hampered.
6. Address dietary needs very heavily. Be aware that problems with the broodstock are passed directly via the egg to the larvae, etc.
7. I strongly feel that closed or semi-closed (<10% / day) systems are much more advantageous than open systems in rearing fresh or saltwater tropical fish and invertebrates.

24% (239,855) fish were considered to be in marketable condition. Significant losses of quality salable fish, after 3 to 4 months growout, greatly hampered our efforts to cover costs. Adding insult to injury, we were forced to accept an average farm gate price of $1.15 each ($275,833) due to the cheap prices of imported wild stocks. Had our production margin of salable fish been greatly reduced, IOH would have started breaking even in 1982 or 1983.

Table 62 is a summation of production and sales at IOH. Average cost to produce a salable fish started at $3.87 each in 1978 and dropped to $1.92 per fish in 1983. Investment over the nine years amounted to approximately two million dollars ($222,222/year). By today's aquaculture standards, IOH can be considered an exceptionally cheap investment into virtually unexplored fields. Other cost cutting measures could have been incorporated into our expenses, however, the investors elected not to enact these changes. Prior to closing IOH in 1984, it was determined that this large facility could be efficiently run with 3.5 people, half the current staff, providing we catered to a small, but very lucrative, customer base of wholesalers.

Table 62 - Summation of production and sales at Instant Ocean Hatcheries.

Year	1978	1979	1980	1981	1982	1983
Avg post larvae./mo	12,032 12 mo	15,453 12 mo	21,943 12 mo	18,042 12 mo	23,250 12 mo	3 mo
Avg. juven. sold./mo	4,384	3,364	3,351	3,371	3,189	4,000
Avg price per fish	$0.95	$1.05	$1.07	$1.21	$1.29	$1.27
Proposed expenses	$220,000	$194,000	$203,000	$180,000	$201,000	$54,00 3 mo
Actual expenses	$201,000	$180,000	$173,000	$189,000	$193,000	$48,000 3 mo
Total revenue	$52,000	$45,000	$46,000	*$56,000	*$72,000	*$25,000 3 mo
Loss/year	$149,000	$135,000	$127,000	$133,000	$121,000	$23,000 3 mo
Cost to produce salable fish	$3.87 ea	$4.00 ea	$3.76 ea	$3.38 ea	$2.68 ea	$1.92 ea

*In 1981 we started to sell related items that were being used at IOH These included macroalgae, special fertilizers, our gelatin mix and small items from our sister company Aquarium Systems. These items drastically reduced our costs to produce fish In 1981 related items represented 18% of our gross sales, 1982 = 31% and for the first 3 months of 1983 = 39%.

Final Analysis of Operation Procedures At IOH

Although IOH accomplished more than anyone else has ever done, we recognized problem areas that needed attention and research. Some required considerable attention while most needed refinement and tuning to iron out problems. None of the problems were significant enough to deter production or stop sales. Table 63 outlines the level of expertise we had obtained. Projects with three asterisks (***) were essentially completed and could be easily controlled by non-research level personnel. These included items that had established symptoms and patterns which could be easily recognized by trained technicians. Likewise, controls and preventative measures were well established and could be administered by hatchery personnel. Those with two asterisks (**) were problems, somewhat under control, yet we were not positive of patterns and/or lacked positive control measures. In some cases, patterns had been established but set rules had not been implemented therefore, non-research personnel had not taken over completely. Projects with one (*) asterisks were incomplete, clearly lacking defined patterns and guidelines of control and were still in full control of the research staff.

At the closing of Instant Ocean Hatcheries our obvious area of concern was the juvenile

TABLE 63

FINAL ANALYSIS OF OPERATION PROCEDURES

At Instant Ocean Hatcheries

ADULT BROODSTOCK		
***	1.	Constant control of nest production (approximately 2/month/pair).
***	2.	Could produce enough quality nests and eggs to meet larval rearing facility demands.
***	3.	Could establish and maintain high hatching rates.
***	4.	Could maintain consistent suitable water quality.
***	5.	Could routinely establish pairs as needed.
***	6.	Could develop new species or crosses.
***	7.	Could control diseases.
***	8.	Could control environment necessary for good quality spawns.
LARVAL REARING		
***	1.	Established a successful feeding regime that met dietary needs.
***	2.	Maintained minimum production of 20,000 post larvae per month.
**	3.	Control high early mortality after hatch (first three days).
**	4.	Control metamorphosis (change from larval fish to true juvenile).
***	5.	Control periods of non-feeding.
***	6.	Control shock and nervous disorders.
**	7.	Control water quality.
**	8.	Control production of specific species as needed.
***	9.	Control the physical environment including light, temperature, and circulation.
***	10.	Control of hatching techniques.
JUVENILE REARING		
**	1.	Establish feeding regimes and meet dietary needs.
*	2.	Control necessary water quality.
*	3.	Control Oodinium.
**	4.	Control nutritional disorders that cause gradual deterioration of juveniles.
*	5.	Control mycobacterium.
**	6.	Control Cryptocaryon.
*	7.	Control of growth rates.
*	8.	Increase survival rates.
*	9.	Increase the number of marketable fish.
**	10.	Control over-feeding.
***	11.	Obtain desired coloration.
PLANKTON CULTURE AND FOOD PRODUCTION		
***	1.	Control of rotifer requirements per day (20-30 million/day).
***	2.	Control microalgae production.
***	3.	Consistent supply of gelatin based foods for adults and juveniles.
***	4.	Formulate a suitable vitamin package for all stages.
**	5.	Formulate a suitable mineral package for all stages.
*	6.	Develop suitable method for mass rearing adult Artemia.
*	7.	Develop alternative live or processed foods for juveniles.
CRITERIA FOR SALES		
***	1.	Determine best packing density.
***	2.	Determine best water condition for shipping.
***	3.	Establish standards of quality control for salable fish.
**	4.	Regulate production according to demand.
**	5.	Finding markets for products.
*	6.	Develop effective advertising for product that will bring you the highest return.

growout systems. We could easily produce all the 20 to 30 day old post-larvae needed to stock our growout tanks. Best growth and survivals were obtained when stocked at 350 to 400 fish (about 22,000 to 24,000 fish per system). Yet, after two to three months growout we were only able to sell 20% of what was stocked. This does not imply we lost 80% of the fish but rather the quality was not considered suitable for sale. Overall survival was actually 43.6% (Table 59) therefore, we were discarding 23% of what survived. However, quite frequently tanks with 50% to 80% salable fish and 80% to 98% survival occurred while others in the same system had 10% to 20% salable. Ironically, both came from the same larval rearing tank and parents, indicating problems occurred on an individual juvenile tank basis. This phenomenon is not clearly understood, but seems to correlate well with: 1) Overfeeding of dry feeds, resulting in eventual collapse of the biological filter in the rearing tank. (often referred as the "toxic tank syndrome"); 2) Delayed detection of disease or nutritional problems resulting in stunting of growth and deformation; 3) Failure in our ability to control water quality problems in growout systems especially after three months use; 4) Inadequate water treatment system for high density fish culture. Table 63 outlines what I consider the main problems of IOH with regard to continued survival.

Future Considerations

Recently, there has been a resurgence of interest in the culture of marine tropicals and more specifically, clownfish. Many are conservative home projects with only a small possibility of commercialization. Yet, of significant note is a small basement size, family owned operation (Reef Propagations) founded by Joe Lichtenbert in April 1990 and located in the Chicago, Illinois area. By October 1990, this small 400 square foot operation produced their first salable fish. To date (1996), 50,000 have been sold for a total of $150,000 ($3.00/fish USD). Average sales have increased from 70 per week for the first two years to 250 fish per week at present. The ultimate goal of this small operation was 450 fish sold per week with gross sales of $71,000 per year and a anticipated operating expense of $10,000 to $15,000. Total cash outlay to date is $53,000 of which about $10,000 was allocated towards for research and development. Currently operating revenue is now positive and all capital was repaid by mid 1994. Average work week is 45 hours for 1.5 people.

Future considerations should start slowly and build upon progress and returns. Keep investment minimal, but not to the point where it hampers advancement and production. Be open minded, willing to share information with others that might be able to help you. Free advice is everywhere, and since you virtually have no competition to worry about, you should take advantage of any help. Use paid consultants and government agencies for difficult problems that you cannot solve within a reasonable amount of time. Be careful to listen, do not get carried away expounding on your achievements. Everyone is aware of your goals and achievements and are proud just to be able to help you in some manner or form.

Address dietary needs heavily. Be aware that problems begin with the broodstock and are passed via the egg to the larvae, etc. I strongly feel that closed or semi-closed (<10% per day exchange) are much more advantageous than open systems in rearing fresh or saltwater tropical fish and invertebrates. Concentrate heavily on filtration and water rejuvenation processes. Keep focused on your goal, but, be open to related alternatives that might save your operation or augment sales without seriously disrupting your ultimate goal.

Devotion

Aquaculture is agriculture! Either you like what you are doing or just don't get started. Successful fish farming is accomplished by dedication, working two jobs, and spending 365 days a year, working whatever time it takes to get the job done. Farming of live animal products, whether terrestrial or aquatic, places heavy year-round demands on both the farmer and his or her family. If you and your family are not willing to meet these demands, then aquaculture is not the field for you.

Farming is like any new small business and is subject to the same attrition rates and failures. Most do not survive over two to three years. Generally, investment, dedication, and preplaning is not sufficient to sustain losses for over two years and the doors must be closed. Working more than one job to make farming work is common. Farming is full of feast or famine situations. Knowing how to manage money, people, and product becomes a fine line between success and failure. I would rather discourage possible commercialization of aquaculture of saltwater fish rather than see another investor blindly enter this field based on projected possible quick monetary returns.

Large investors should be especially careful since they seem to be more vulnerable to hiring "bio-prostitutes". "Biologists?" of this special classification consist of "yes men" who promise you anything you want to hear, but actually have very little hands-on practical experience. Three years later you find they are basically experimenting and learning at your expense and have accomplished very little. The field of aquaculture has been greatly tarnished by "bio-prostitutes" and poor investment propositions. Make sure the persons you select to work with have indeed accomplished what they expound upon. Check with others in the field before you select a biologist to work with. Make sure they provide references. Have they published papers regarding their expertise? If you have limited funds and sustaining power, select people with direct or close experience. You do not want to become a training facility.

Small investors on the other hand are less prone to hiring anyone with any experience since they cannot afford them. Small investors normally get more involved into their investment, are tenacious, but often become involved to the point where it may actually hamper or be detrimental to future growth. The old saying; "a little knowledge can be a dangerous thing" certainly applies to small investors with little biological knowledge in aquaculture.

Keeping focused is a problem of large and small investors alike. You only see obstacles when you take your eyes off the goals. Set realistic goals and at all times keep focused on what you set out to accomplish. Do not get side tracked just because something looks good. Sidetracking into other related ventures should be a last resort idea based on observations and facts you have gathered while trying to accomplish your main goal. If these side ventures can provide quick positive money flow to help support your ultimate goal, they should be considered. However, they must not require more time than you are capable of providing. Doing a good job on one or two projects is significantly better than doing a poor job on ten.

Location

Reef Propagations and IOH clearly demonstrate that a location with a pristine saltwater supply or a particular water type is not, and should not be, the main priority consideration. Water management in closed systems is the key and no other arrangement can compete once you have mastered your specific problems. Although artificial seawater is initially a high investment, by being conservative, and utilizing high tech recirculating filtration systems, yearly costs can be minimal. As I mentioned before, I feel closed systems located right next to pristine water and operating on a maximum of 5% to 10% exchange per day are more advantageous with regard to rearing fish, than an open flow-through system. Access to pristine open ocean waters is a nice option but is usually detrimental with regards to farm-gate price, market potential, and storm damage.

Unfortunately, said waters are usually located in remote areas which greatly hampers transportation and dictate higher end product pricing. Shipping live animals from remote areas is expensive and provides a decided advantage for future competitors that locate closer to more lucrative markets. Aquaculturists in remote situations who sell live products are always scrambling to lower production costs and accept lower farm-gate prices to compensate for ever rising freight costs to the consumer.

Knowledge on water recirculation is quickly maturing. Probably more important than finding a pristine water source, is locating a suitable facility with regard to future markets. Live products should be located close to large live markets and/or within an hour drive of a large airport. Locating near large cities allows for easy access to local pet markets and future national and international markets. Local large metropolitan markets initially allow you to receive maximum farm-gate prices especially when production is low, irregular and your cash-flow is limited. As production increases, access to airports, and other overnight shipping will become a priority.

Fortunately, overnight shipping by UPS, DHL, and Federal Express has relaxed some of the priorities with regard to location. Currently, many live fish, invertebrates and plants are being sent by these means with door-to-door delivery within a 24-hour period. Mail-order live products is certainly a real possibility and would allow even more flexibility with regards to location and facility costs. To date, none of these carriers will insure live product against loss or injury. This risk factor must be considered. Yet, based on 12 years experience using UPS, I have found they are very efficient except around Christmas and a month after. I might add that this same problem exists for commercial airlines and other carriers. Losses are usually the result of packages that don't get delivered due to severe weather conditions or are not received at the specified location. Fortunately, all packages sent by UPS can be traced and if the receiver is at fault the loss is his or hers. It is highly advisable that all sales the prepaid, especially in the beginning.

Pilot Scale Farms

Think big, start small! I highly recommend you first get your hands wet before immersing your entire body. Unless you are totally knowledgeable about what you are doing and have a well rounded support team, do not invest heavily. Assemble considerable knowledge and hands-on expertise first. Initially, I feel the investor should be a part of the operation in order to gain insight into the scope of problems. Your initial experimental facility can be transformed into a pilot facility yet, consist of just a garage or basement as clearly illustrated by Reef Propagations. As you gain expertise and marketing experience, consideration of full commercialization can begin. I feel it takes a minimum of two years of pilot-scale operation to fully address future considerations. Income data presented for Reef Propagations shows what can be accomplished in just 400 square feet of space. Niche marketing directly to local shops and wholesalers has provided this facility maximum return with minimal initial investment and minimal shipping loss or problems. How much further this operation can grow without expansion is becoming a factor to consider.

There are those who may wish to subsidize their income or consider what they are capable of doing alone or as a family unit. That is fine, and I highly encourage this type of development, especially in marine tropical fish businesses. After your initial investment in equipment, the business will be much more stable if allowed to grow upon itself. In other words, it should be managed as money-in/money-out business only. Do not spend what you do not have except for the initial investment in equipment and stock. Be careful not to "bleed" your company of needed cash too quickly. You may indeed have to maintain two jobs to keep it going. Yet, when the day comes that you can devote all your time, you will owe no one. Try to pay back your investment and loans as soon as possible, since this seems to build confidence to go on. Using this method normally results in slower initial growth, but you will be able to maintain the business longer through the hard, trial and error, learning periods.

Success and future expansion of small business brings more problems, especially with regard to trust and faith in those you are forced to employ. Key people are absolutely essential in the first phases of expansion. I recommend that they are hired prior to expansion and be involved in the design, purchasing, and construction. Devotion by these key people to your business should be immediately enhanced by incentives such as: goals coupled with reward, profit sharing, future ownership, retirement programs and security. Keep in mind, you cannot "buy people" so there is no need to throw good money at a possible good employee. Strategic incentives based on realistic goals built into your employment program, will do wonders with motivated people. Good people become better because of their interest in your business and your willingness to release responsibility to them.

People who start small, learn the business, and nurture it from conception are often less prone to turn over responsibility to another person. Intern, they can actually hamper future growth and intimidate employees by constantly being involved and not providing the trust they need to gain your respect. Stay involved, but not to the point where you are always calling the shots. I feel that selection of "the person or persons" who will initially take over parts of your expanding business is a very, very, important step in the success or failure of your business. A good business plan with realistic goals coupled with enough time to accomplish these goals should be setup before you expand. If the person(s) you hire cannot meet these goals, first review your objectives, make sure they are realistic, then adjust your plan and/or release this employee. Do not waste valuable time on people who say they can do the job but never accomplish your goals in a reasonable time and/or never have time to stay and complete a job. By being initially involved in your business, you know what can be accomplished in a given period of time. Be a hard, but fair owner. You have too much too lose in trying to train people over and over. Your initial selection of those who have good experience and/or are totally devoted like you is imperative.

Moving from a pilot to commercial scale is supposed to be just an expansion of methods you have been doing before, right. Wrong, expect considerable problems. For some reason, things that are accomplished in test tubes within labs just do not work out as well in a larger scale. Ideally, your pilot scale operation should mimic your commercial farm at least for the last year. But, this is usually not the case since you are usually typically experimenting with this tank or that filter all through a pilot scale operation. Do not expand until you are confident that you have:

1. enough time to devote to your new operation,
2. mastered the biological techniques,
3. can produce sufficient numbers of salable juveniles,
4. established some marketing expertise,
5. selected people who can really help you, and
6. carefully considered the design of your new facility.

Facility Design

Centralized, simple, and modular are key factors in your design. Do not spend money on buildings you do not intend to occupy within a year or two. This just ties up necessary capitol. Design your facility so it can be easily expanded from at least two sides in a modular fashion. Centralization of key components such as offices, food producing areas, shipping and receiving saves considerable time with regard to daily activities. The facility should flow from beginning broodstock to the finished product (salable juveniles). Yet, these should be distinctly separate entities within the design (Figures 64,65 & 66). Filtration, pumps, valves, heating, and lights should be easily accessible and not hidden. All plumbing should be under boardwalks or in open, grid covered, floor drains. You should not have to walk over any pipe or through any standing water. If using boardwalks, (Figure 17) untreated wood can be used in saltwater facilities since salt will preserve wood. In freshwater facilities treated wood must be used. The cement floor should be tilted slightly towards the center or sides of the building (Figure 63) to facilitate cleaning, draining, and help keep the floor dry. When laying the cement floor, it should receive a heavy brush finish in the direction of the tilt. This will eliminate future, costly, employee injuries due to wet, slick floors.

If starting with a new building, I suggest a greenhouse type structure that can be interconnected so new buildings can be easily added. Ceiling height should be 8 feet if no boardwalks are used or 10 feet otherwise. Maximize natural light initially in your design since light levels can be easily reduced with shade cloth or paint, but it is hard and costly to increase light. Depending on where you are located, ventilation will be a serious problem. Summer heat levels can easily destroy your animals and plankton cultures. In warm tropical climates being able to lower or open the side walls will help. Selecting greenhouses which have top ventilation systems is

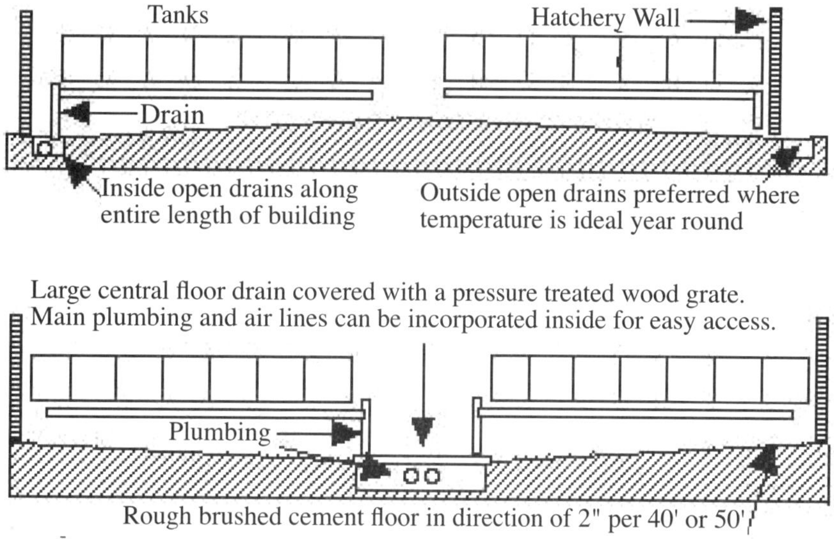

Figure 63 - Sloping floors and drains for hatcheries.

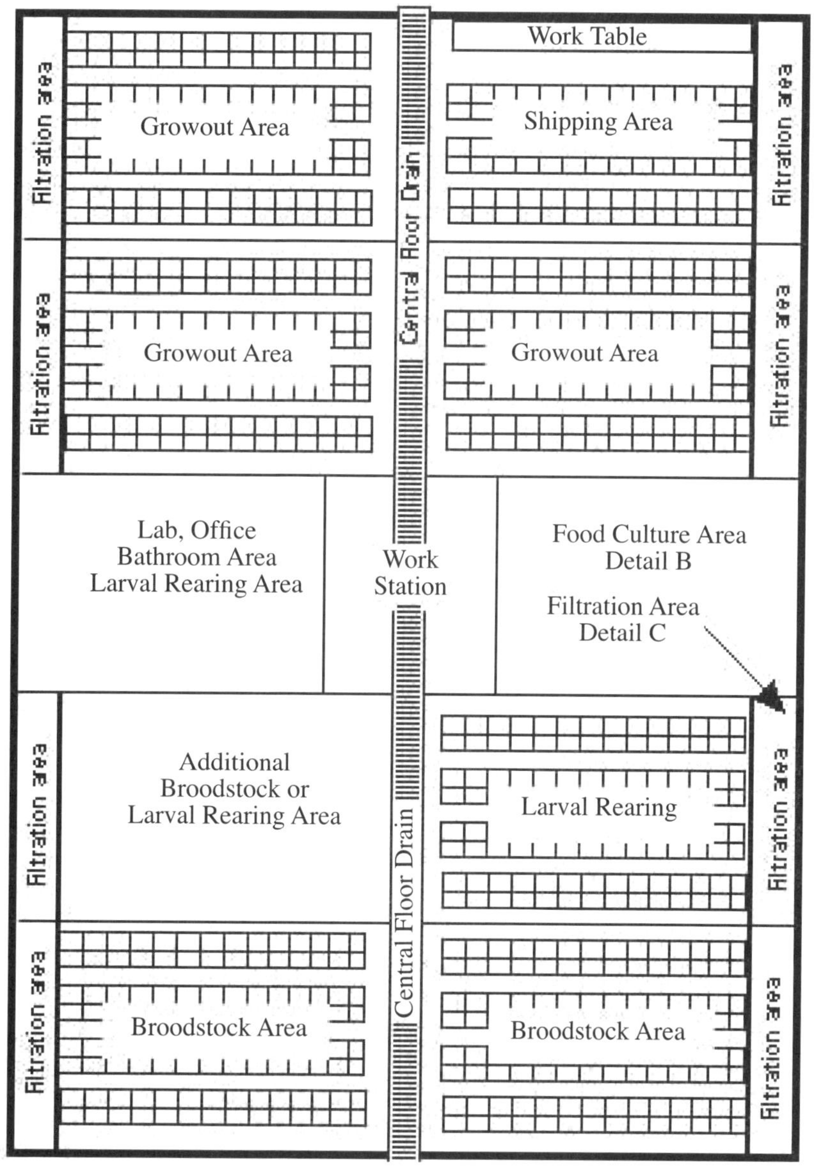

Figure 64 - Tropical fish hatchery designed based on centralization of primary work areas and stock isolation.

Detail A
Office Area

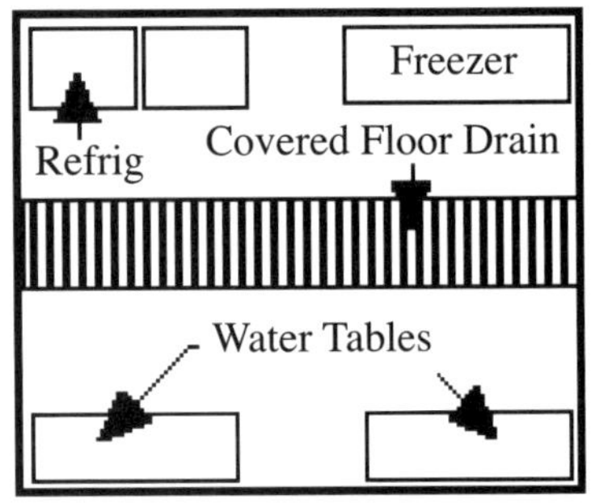

Detail B
Work Station

Figure 65 - Detail "A" and "B" of fish hatchery design based on C-Quest in Puerto Rico.

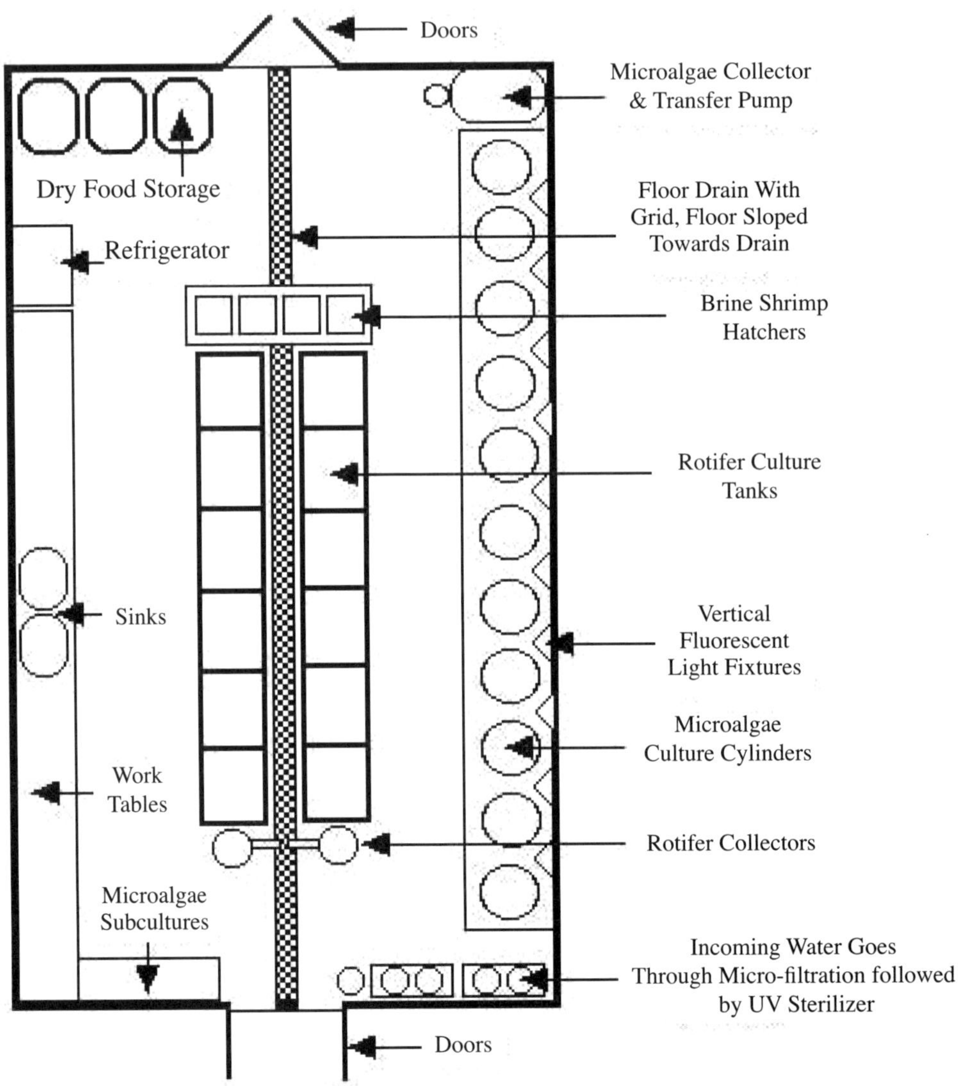

Figure 66 - Detail "C" culture room facilities based on C-Quest in Puerto Rico.

very are beneficial. In semitropical and warm temperate areas, solid insulated walls which can be opened up during the summer months is beneficial. Large thermostatically controlled ventilation fans located on the ends or on one side of the buildings can maintain comfortable greenhouse temperatures. Selecting sites with southern exposure and protection from north winds will help reduce heating costs. In northern areas building selection becomes a real problem. Utilization of natural light may not be efficient with regard to the costs of trying to maintain heat. Utilization of double layers of fiberglass roof panels creating an air pocket helps reduce quick temperature drops in the evening. Double layered poly greenhouses with air forced between the layers can also help reduce quick heat loss.

Heating will depend on the type of building. In general, heating water is more efficient than heating a building. If using a central recirculating system, submersible heating is best. Insulating tanks with one inch foam on at least three sides and the bottom will help conserve heat. Pipes can also be insulated. The larger the volume of water you have in the facility, the less chance you have of quick temperature changes. Hot air or individual immersion heating can be used in areas that have static tanks such as the larval rearing and culture areas.

Modular facility design currently used by C-Quest in Puerto Rico is detailed in Figures 64, 65 & 66. Ideally this type of building should be located centrally within the property so it can be expanded in any direction. The floors are tilted toward troughs located along the edges of the buildings.

Budget

Accountants and investors always want to know where the money will be spent. Below (Table 67) is the average percent of the budget for running IOH in 1981. This was based on 7 full-time employees. Since this was basically a research and development facility, the salaries may

Table 67 - Average percent of monthly expenses for maintaining IOH from 1980-81

New Broodstock	0.52%
Medications	0.71%
Farm supplies	7.13%
Building Repairs	2.10%
Food Costs	3.34%
Freight costs in & out	0.78%
Research & Development	0.83%
Office Supplies	1.22%
Library Costs	0.16%
Transportation Costs	0.66%
Truck Repairs	0.43%
Mortgage payments	9.03%
Advertising	0.64%
Telephone	1.00%
Utilities (electric, oil)	4.51%
Salaries	53.9%
Payroll Taxes	3.50%
Health Insurance	2.54%
Travel & Entertainment	0.13%
Consulting Fees	1.22%
Insurance	3.74%
Fees, Dues & License	0.39%
Other Taxes	1.26%
Bank Charges	0.13%
Total expenses per month	99.87%

REFERENCES

Allen, G.R. 1972. The Anemonefishes. TFH Publications, Inc. Ltd. Surray England. 288 p.

Allen, G.R. 1991. Damselfishes of the World. Mergus Publishers Hans A. Baensch, Melle, Germany. 271 pp.

Arthur, J.W. & D.I. Mount. 1975. Toxicity of a disinfected effluent to aquatic life. First International Symposium on Ozone in Water and Wastewater Treatment, Intern. Ozone Inst., Syracuse, NY.

Baragi, V. & R.T. Lovell. 1986. Digestive enzyme activities in striped bass from first feeding through larval development. Trans. Am. Fish. Soc. 115:478-484.

Bardach, J.E. 1968. The status and potential of aquaculture, particularly fish culture. Vol 2, Part III, Fish Culture. 22-29.

Belsare, D.K. 1976. Studies on the development of the endocrine glands in fishes. V. Development of the thyroid gland in two species of teleosts. Morphologisches Jahrbuch, 122: 118-122.

Bertschy, A.. 1979a. Unpublished data presented to the author for review and use.

Bertschy, A. 1979b. Essais de reproduction en aquarium d'Amphiprion ocellaris Cuv. (Pomacentrides). Rev. Fr. Aquariol., 6 (Sept 79), 3:91-94.

Blasiola, G.C. 1991. Recent discoveries are being used to improve foods for freshwater fishes. Pet Age Mag. May Issue. pp 36-37.

Blasiola, G.C. 1995. Toxic shock. Pet Age Mag. October issue. pp 53-56.

Blaxter, J.H.S. 1962. Herring Rearing - IV Rearing beyond the yolk-sac stage. Mar. Res. Scot. 1962(1):18 pp.

Blaxter, J.H.S. 1968. Rearing herring larvae to metamorphosis and beyond. J. Mar. Biol. Ass. U.K. 48:17-28.

Blaxter, J.H.S. 1969. Experimental rearing of pilchard larvae, *Sardina pilchardus*. J. Mar. Biol. Ass. U.K. 49:557-575.

Blazer, V. 1979. Diet experiments vs wasting disease at Instant Ocean Hatcheries. Unpublished, in-house consulting report.

Bovendeur, J., A.B. Zwaga, B.G.J. Lobee & J.H. Blom. 1990. Fixed-biofilm reactors in aquacultural water recycle systems: effect of organic matter elimination on nitrification kinetics. Wat. Res. Vol. 24. pp 207-213.

Bower, C.E. & D.T. Turner. 1982. Effects of seven chemotherapeutic agents on nitrification in closed seawater culture systems. Aquaculture 29:331-345.

Bower, C.E. 1987. Update on *Amyloodinium ocellatum.* Sea Scope, Published by Aquarium Systems. Vol. 4.

Boyd, C.E. & B.J. Watten. 1989. Aeration systems in aquaculture. CRC Critical Reviews in Aquatic Sciences 1:425-472.

Bremer, H. 1980. Some investigations on the relationship between the feeding behavior and the secretory mode of exocrine pancreas of the hybrid sturgeon (Huso huso x Ancipenser ruthenus) (in German). Zool. Jahrb. Anat. 104:69-76.

Bryan, P.G. & B.B. Madraisau, 1977. Larval rearing and development of Siganus lineatus (Pisces: Siganidae) from hatching through metamorphosis. Aquaculture 10:243-252.

Bryant, P.L. & A.J. Matty. 1980. Optimisation of *Artemia* feeding rate for carp (*Cyprinus carpio* L.) larvae. Aquaculture 21:203-212.

Bryant, P., 19??. How to take uncertainty out of spawning. Fish Farmer Vol. 4(2):22-23

Browman, H.I. & W.L. O'Brien. 1992a. Foraging and prey search behavior of golden shinners (*Notemigonus crysoleucas*) larvae. Can. J. Fish Aquat. Sci. Vol 49:813-819.

Browman, H.I. & W.L. O'Brien. 1992b. The ontogeny of search behavior in the white crappie, *Pomoxis annularis*. Environ. Biol. Fish., in press.

Brown, J.A, & P.W. Colgan. 1984. The ontogeny if feeding behavior in four species of centrachid fish. Behav. Proc., Vol 9:395-411.

Brown, J.A. & P.W. Colgan. 1985. Interspecific differences in the ontogeny of feeding behavior in two species of centrachid fish. Z. Tierpsychol. Vol 70:70-80.

Brune, D.E. & D.C. Gunther. 1981. The design of a new high rate nitrification filter for aquaculture water reuse. J. World Maricul. Soc. 12:20-31.

Brune, D.E. & R.H. Piedrahita. 1983. Operation of a retained biomass nitrification system for treating aquaculture water for reuse. Proc. First Inter. Conf. on Fixed-film Biological Processes. pp 845-869.

Castell, J.D., R.O. Sinnhuber, J.H. Wales and D.J. Lee. 1972. Essential fatty acids in the diet of rainbow trout (*Salmo gairdneri*) : Growth, feed conversion and some gross deficiency symptoms. J. Nutr. 102:77-86.

Charlon, N. & P. Bergot. 1984. Rearing system for feeding fish larvae on dry diets. Trial with carp (Cyprinus carpio). Aquaculture 54:83-88.

Charlon, N., H. Drante, A.M. Escaffre & P. Bergot. 1986. Alimentation artificielle des larves de carpe (Cyprinus carpio). Aquaculture 41:1-9.

Chen, S., D.E. Coffin & R.F. Malone. 1991. Sludge management for recirculating aquaculture systems. Design of High Density Recirculating Aquaculture Systems. Workshop Proceedings, Louisiana Sea Grant, Baton Rouge. pp 36-43.

Collie, N.L. & J.J. Stevens. 1985. Hormonal effects on L-proline transport in coho salmon (*Oncorhynchus kisutch*) intestine. Gen. Comp. Endocrinology 59:399-409.

Collins, M.T., J.B. Gratzek, D.L. Dawe & T.G. Nemetz. 1975. Effects of parasiticides on nitrification. J. Fish. Res. Board Can. 32:2033-2037.

Collins, M.T., J.B. Gratzek, D.L. Dawe & T.G. Nemetz. 1976. Effects of antibacterial agents on nitrification in an aquatic recirculating system. J. Fish. Res. Board Can. 33:215-218.

Colt, J. 1986. An introduction to water quality management in intensive aquaculture. In oxygen supplementation, a new technology in fish culture. U.S. Dept. of the Interior, Fish and Wildlife Serv. Region 6, Information Bulletin #1. Denver Colorado.

Colt, J. & B. Watten. 1988. Applications of pure oxygen in fish culture. Aquacultural Engineering 7:397-441.

Conde, B., 1986. Longevity of marine tropicals at the Nancy Aquarium. Seascope, Aquarium Systems Volume 3, summer issue.ton Research Vol. 15:117-123

Coughlin, D.J. 1993. Prey location by clownfish (*Amphiprion perideraion*) larvae feeding on rotifers (Brachionus plicatilis). Jour. Plankton Research, Vol 15 #2:117-123.

Coughlin, D.J., J.R. Strickler & B. Sanderson. 1992. Swimming and search behaviour in clownfish, Amphiprion perideraion, larvae. Animal Behaviour, 44:427-440.

Crozier, G.F. 1969. Effects of controlled diet on the morphological color change of a marine teleost. Journal of Experimental Marine Biology & Ecology. 4:1-8.

Crozier, G.F. 1970. Tissue carotenoids in prespawning and spawning sockeye salmon (*Oncorhynchus nerka*). Jour. Fish. Res. Bd. Canada. 27:973-975.

Curtis, B. 1949. The life story of fish, his manners and morals. Dover Publications USA. 284 pp.

Czeczuga, B. 1975. Carotenoids in fish. IV. Salmonidae and Thymallidae from Polish waters. Hydrobiologia 46:223-239.

Czeczuga, B. 1979a. Carotenoids in fish. XIX. Carotenoids in the eggs of *Oncorhynchus keta* (Walbaum). Hydrobiologia 63:45-47.

Czeczuga, B. 1979b. Carotenoids in fish. XX. Carotenoids in *Salmo gairdnri* Rich. and *Salmo trutta morpha fario* L. Hydrobiologia 64:251-259.

Dabrowski K. 1984a. The feeding of fish larvae. Present state of the art and perspectives. Reprod. Nutr. Develop. 24:807-833.

Dabrowski K. 1984b. Influence of initial weight during the change from live to compound feed on the survival and growth of four cyprinids. Aquaculture 40:27-40.

Dabrowski K. & Bardega, 1984. Mouth-size and recommendation of feed size preferences in three cyprinid fish. Aquaculture 40:41-46.

Delmonte, P.J., I. Rubinoff & R.W. Rubinoff. 1968. Laboratory rearing through metamorphosis of some Panamanian gobies. Copeia 1968:411-412.

Delbeek, C. & J. Sprung. 1994. The Reef Aquarium, A comprehensive guide to the identification and care or tropical marine invertebrates. Ricordea Publishing, Coconut Grove Florida. 544 pages.

Donaldson, E.M. & U.H.M. Fagerlund. 1970. Effect of sexual maturation and gonadectomy at sexual maturity on cortisol secretion rate in sockeye salmon (*Oncorhynchus nerka*). Jour. Fish Res. Bd. Canada. 27:2287-2296.

Douillet, P. 1996. Use of single and multiple probiotics in hatcheries. Presentation at the World Aquaculture Assoc. Meeting, Bangkok, Thailand.

Ekman, S. 1953. Zoogeography of the Sea. London, Sidwick & Jackson, Ltd., 417 p.

EPA. 1975. Process Design Manual for Suspended Solids Removal. U.S. EPA Technology Transfer Publication.

Eriksen, B.F. 1991. The right touch for efficiency in feeding. Fish Farming International, May issue, pp 38-39.

Fahey, W.E. 1964. A temperature controlled salt water circulating apparatus for developing fish eggs and larvae. J. Cons. 28:364-384.

Farkas, T., I. Csengeri, F. Majoros & J. Olah. 1978. Metabolism of fatty acids in fish. II. Biosynthesis of fatty acids in relation to diet in the carp, *Cyprinus carpio* L. 1758. Aquaculture 14: 57-65.

Fishelson, L., 1965. Observations and experiments on the Red Sea anemones and their symbiotic fish, *Amphiprion bicinctus*. Bull. Sea Res, Station, Haifa, 39:1-14.

Fowler, H.W. 1944. Results of the fifth George Vanderbilt Expedition. The Academy of Natural Sciences of Philadelphia. Wickersham Printing Company, Lancaster PA., 583 pp.

Frakes, T. & F. Hoff. 1982. Effect of high nitrate-N on the growth and survival of juvenile and larval anemonefish, *Amphiprion ocellaris*. Aquaculture, 29:155-158.

Frakes, T., F. Hoff & W. Hoff. 1983. Delayed metamorphosis of larval anemonefish (*Amphiprion ocellaris*) due to iodine deficiency. In: Proceedings of the warmwater fish culture workshop. R.R. Strickney & S.P. Meyers (eds.) Spec. Pub. No. 3, World Mariculture Society, Baton Rouge LA. pp. 117-123.

Frakes, T. & F. Hoff. 1983. Mass propagation techniques currently being used in the spawning and rearing of marine fish species. Zoo Biology 2:225-234.

Frakes, T. 1986. Control of Marine Parasites. Sea Scope. Published by Aquarium Systems Vol. 3.

Frakes, T. 1987. Update on *Amyloodinium ocellatum*. Sea Scope, Published by Aquarium Systems. Vol. 4.

Francis-Floyd, R. 1991. Stress-its role in fish disease. Florida Coop. Extension Ser. Inst of Food and Ag. Univ. of FL, Circular 919.

Freiden, E., 1962. The chemistry of amphibian metamorphosis. Scientific American 209:110-118.

Fujita, S. & C. Kitajima. 1978. Occurance of the ladotic deformity in larva and juveniles of red seabream. Kaiyo-kagaku (Marine Science Monthly), 107:721-727 (In Japanese)

Gohar, H.A.F., 1934. Partnership between fish and anemone. (With a description of *Amphiprion bicinctus* Ruppell) Fouad I. Uuiv. Publ. Mar Biol. Sta. Ghardaqa (Red Sea), 6: 35-44.

Goldstein, R.J. 1989. Breeding marine clownfish. Aquarium Fish Magazine. June Issue, pp 18-23.

Goldstein, R.J. 1990. Tomato clownfish raised in the home aquarium. Aquarium Fish Magazine. Dec. Issue, pp 56-59.

Grady, C.P.L. & Lim, H.C. 1980. Biological Wastewater Treatment. Theory and Applications. Marcel Dekker, Inc. New York 963 pp.

Graefe, G. 1964. Zur Anemonen-Fisch-Symbiose, nach Freilandunter- suchungen bei Eilat/ Rotes Meer. Z. f. Tierpsychol., 21(4): 468-485.

Gross, J. & P. Budowski. 1966. Conversion of carotenoids into vitamin A in two species of fresh-water fish. Biochem. 101:747-754.

Halver, J.E. 1985. Nutrition and feeding of fish. IN, pp 415-429.

Harremoes, P. 1982. Critera for nitrofication in fixed film reactors. Wat. Sci. Technology. 14:167-187.

Hartmann, M., F.G. Medem, R. Kulm & H.J. Bielig. 1947. Untersuchungen uber die befruch-tungsstoffe der regenbogenforelle. Z. Naturf. B., 2:330-343.

Hata, M. & M. Hata. 1972. Carotenoid pigments in goldfish. 4. Carotenoid metabolism. Bull. Japan Soc. Sci. Fish. 38:331-338.

Hata, M. & M. Hata. 1976. Carotenoid metabolism in fancy red carp, Cyprinus carpio, 2. Metabolism of 14C-zeaxanthin. Bull. Japan Soc. Sci. Fish. 42:203-205.

Hawke, J.P. 1991. Potential Effects of Therapeutic Agents on Biological Filtration in Closed Aquaculture Systems. Design of High Density Recirculating Aquaculture Systems. Workshop Proc., Louisiana Sea Grant, Baton Rouge. pp 56-60.

Herre, A.W. & H.R. Montalban 1927. The Philippine butterfly fishes and their allies. The Philippine Jour. of Sci. Vol 2, Articles 31-59 (1927-1936).

Hwewig, N. 1980. Disease prevention and control, seemingly inexplicable cases of blindness in fish. Freshwater and Marine Magazine, Vol. 3(2): 21-33.

Hoff, F., C. Rowell & T. Pulver. 1972. Artifically induced spawning of the Florida Pompano under controlled conditions. Proc. World Mariculture Soc. 3:53-64.

Hoff, F., J. Mountain & T. Frakes. 1973. Induced spawning of pompano and seabass . FAO Aquaculture Bulletin. October 6 (1):4.

Hoff, F. 1976. Masters Thesis, Univ. South Fla. Library.

Hoff, F., T. Pulver, & J. Mountain. 1978a. Conditioning Florida pompano (*Trachinotus carolinus*) for continuous spawning. Proc. World Mariculture Soc. 9:299-309.

Hoff, F., T. Frakes & K. Halscott. 1977. Larval rearing of the Florida pompano and southern sea bass. Fl. Dept Nat Resources, Contribution report 310, unpublished data.

Hoff, F., J. Mountain, T. Frakes & K. Halscott. 1978b. Spawning, oocyte development and larval rearing of the Florida pompano (*Trachinotus carolinus*). Proc. World Mariculture Soc. 9:279-297.

Hoff, F. H. 1981. Unpublished data from a conference between Aqualife Research, Sea World Research and Instant Ocean Hatcheries, January 27, 1981.

Hoff, F. & T. Snell. 1993. Plankton Culture Manual, Third Edition. Publish. Florida Aqua Farms Inc. 136 p.

Honn, K.V. & W. Chavin. 1976 Utility of ozone treatment in the maintenance of water quality in a closed marine system. Marine Biol. 34:201-209.

Hopkins, T.A. & W. E. Manci. 1989. Feed Conversion, Waste and Sustainable Aquaculture, the Fate of the Feed. Aquaculture Magazine, 15(2):30, 32-36.

Houde, E.D. 1974. Effects of temperature and delayed feeding on growth and survival of larvae of three species of subtroical fishes. Marine Biology 26:271-285.

Houde, E.D. & B.J. Palko. 1970. Laboratory rearing of clupeid fish *Harengula pensacolae* from fertilized eggs. Marine Biology 5:354-358.

Hsiao, S.I.C. 1969. Life history and iodine nutrition of the marine brown alga *Petalonia fascia* (O.F. Mull) Kuntze. Canad. J. Botany 47:1611-1616.

Hunter, J.R. 1972. Swimming and feeding behavior of larval anchovy, *Engraulis mordax*. Fishery Bulletin U.S. 70:821-838.

Hunter, J.R. 1980. The feeding behavior and ecology of marine fish larvae. In: Fish Behavior and its Use in the Capture and Culture of Fishes. J.E. Bardach, J.J. Magnuson, R.C. May & J.M. Reinhart (eds), ICLARM Conference Proceedings 5:287-330.

Iseda, H. 1982. Prevention of demonstration in the juveniles of red seabream. *Pagrus major*, reared in Ponds-III. Relationship between the initial conditions of rearing environment and gas content in air bladders. Bull. Kummanoto Pref. Inst. Fish., 2:25-45 (In Japanese).`

Iwama, G.K. 1991. Interactions Between Aquaculture and the Environment. Critical Reviews in Environmental Control. 21(2):177-216.

Johnson, S.K. 1990. Calculating treatments for fish vats, hauling tanks, and ponds. In: Inland Aqaculture Handbook, Pub. Texas Aquaculture Assoc., Austin TX. pp A1107-A1108

Joseph, E.B. & V.P. Saksena. 1966. Determination of salinity tolerances in mummichog (*Fundulus heteroclitus*) larvae obtained from hormone-induced spawning. Chesapeake Science (now Estuaries) 7:193-197.

Kaiser G.E. & F.W. Wheaton. 1983. Nitrification filters for aquatic culture systems: state of the art. J. World Mariculture Soc. 14:302-324.

Kashahara S.R. Hirano & Y. Oshima. 1960. A study on the growth and rearing methods of black porgy, Mylio macrocephalus (Basilwesky). Bul. Jap. Soc. Sci. Fish. 26:239-244.

Kaushik, S. & K. Dabrowski. 1983. Nitrogen and energy utilization in juvenile carp (*Cyprinus carpio*) fed casein amino acid or protein-free diet. Reprod. Nutr. Dev. 23:741-754.

Kincheloe, J.W., G.A. Wedemeyer & D.L. Koch. 1979. Tolerance of developing salmonid eggs and fry to nitrate exposure. Bull. Environ. Contamination. Toxicology, 23:575-578.

Kitajima, C. 1983. The present status of marine fish seed producing techniques in Japan. In: Advances and Perspectives in Aquaculture, H.R. Fuentes, J.G. Castillo, and L.H. Disalvo editors. Universidad del Norte, Coquimbo, Chile, pp. 375-390.

Kitajima, C., Y. Tsukashima, S. Fujita, T. Watanabe, & Y. Yone. 1981. Relationship between uninflated swim bladders and lordotic deformity in hatchery reared red seabream, *Pagrus major*. Bull. Japan. Soc. Sci. Fish., 47(10):1289-1294. (English summary).

Korringa P. 1947. The mood and periodicity in breeding of marine animals. Ecol. Monographs 17(3): 349-381.

Kramer, D. & J.R. Zweifel. 1970. Rearing and growth of anchovy larvae (*Engraulis modax* Girard) as influenced by temperature. Calif. Coop. Oceanic Fish. Invest Rep. 14:84-87.

Kunihiko, F. 1985. Status of marine larval culture in Japan. In: Reproduction and Culture of Milkfish. (ed. Cheng-Sheng Lee & I-Chiu Liao), Pub. by Oceanic Institute Hawaii & Tungkang Marine Laboratory, Taiwan. pp 127-140.

Lam, T.J., 1980. Thyroxine enhances larval development and survival in *Sarotherodon* (Tilapia) *mossombious* Ruppel. Aquaculture 21:287-291.

Lansdell, W., B. Dixon, N. Smith, & L. Benjamin. 1993. Isolation of several Mycobacterium species from fish. Journal of Aquatic Animal Health. 5:73-76.

Lasker, R., H.M. Feder, G.H. Theilacker & R.C. May. 1970. Feeding, growth and survival of *Engraulis mordax* larvae reared in the laboratory. Marine Biology. 5(4):345-353.

Lauff, M. & R. Hofer. 1984. Development of proteolytic enzymes in fish and the importance of dietary enzymes. Aquaculture 37:335-346.

Lawson, T.B. & G. Merry. 1991. Ozone use in recirculating systems comparisons with other disinfection options. Design of High-Density Recirculating Aquaculture Systems. Workshop Proceedings Lousiana Sea Grant, Lousiana State Univ., Baton Rouge. pp 80-88.

Lefkowitz, R.J. 1995. The nose knows love; so do sperm. Report on work being conducted at Duke University, Tampa Tribune.

Levine, G. & T.L. Meade. 1976. The effects of disease treatment on nitrification in closed system aquaculture. Proc 7th Annual Meeting World Aquaculture. Soc. pp 483-493.

Li, W.H & W. Brocksen. 1977. Approaches to the analysis of energetic costs of intraspecific competition for space in rainbow trout (Salmo gairdneri). Journal Fishery Biology 11:329-341.

Liaaen-Jensen, S. 1978. Marine carotenoids. Marine Natural Products: Chemical and Biological Perspectives, Vol. 2:1-73. Schever, P.J. ed.
Academic Press, New York.

Lichtenbert, J. 1992. Presentation In: Marine Fish Culture Session. Florida Aquaculture Association 8th Annual Meeting.

Linder, R. 1986. Some notes on breeding clownfish. Freshwater and Marine Aquarium Magazine, June Issue, pp 15-51.

Losordo, T.M., M.P. Masser & J.E. Rakocy. 1992a. Recirculating aquaculture tank production systems an overview of critical considerations. N.C. Coop. Extension Ser. SRAC Pub No.451.

Losordo, T.M., J.E. Rakocy & M.P. Masser. 1992b. Recirculating aquaculture tank production systems component options. N.C. Coop. Extension Ser. SRAC Pub No.453.

Lovell, T. 1979. Effects of diet on reproduction of brood catfish. Aquaculture magazine . Volume 5, 6:28.

Lovell, R.T. et al. 1993. Nutrient Requirements of Fish. National Academy Press, Washington D.C. 113 pp.

Lu, Z. & R. Piedrahita. 1993. Nitrifying characteristics of a high rate packed column. Techniques for Modern Aquaculture. Am. Sco. of Agri. Eng. pp 345-351.

Lubbock, R. 1980. Why are clownfishes not stung by sea anemones? Proc. Royal. Society. London. B 207:35-61.

Malone, R.L. & D.G. Burden. 1988. Design of recirculating Blue Crab shedding systems. Louisiana State University, Sea Grant Program. 76 pages.

Mariscal, R.N. 1966. The symbiosis between tropical sea anemones and fishes: a review In: R.I. Bowman (ed), The Galapagos: 157-171. Berkeley, Univ. Calif. Press.

Mariscal, R.N. 1970. The nature of the symbiosis between Indo-Pacific anemone fishes and sea anemones. Mar. Biol., 6(1): 58-65.

Masser, M.P., J. Rakocy & T.M. Losordo. 1992. Recirculating aquaculture tank production systems management of recirculating systems. N.C Coop Ext Ser. SRAC Pub. No 452.

Metcalf & Eddy, Inc. 1979. Wastewater Engineering Treatment/Disposal/Reuse. Second Edition, Mc Graw Hill, 920 pp.

Meyers, S.P. & H.M Chen. 1983. Astaxanthin and its role in fish culture. Proc. Warmwater Fish Culture Workshop, Spec. Pub. 3 World Mariculture Assoc. pp 153-165.

Mikulin, A.Y. & S.G. Soin. 1975. The functional significance of carotenoids in the embryonic development of teleosts. Jour. Ichthyology (Engl. Transl. Vopr. Ikhitol.) 15:749-759.

Miller, G.E. & G.S. Libey. 1984. Evaluation of three biological filters suitable for aquacultural applications. J. World Mariculture Sco. 16:158-168.

Moe, A.M. 1981. Unpublished data from a conference between Aqualife Research, Sea World Research and Instant Ocean Hatcheries, January 27, 1981.

Moe, A.M. 1982. The Marine Aquarium Handbook, Beginner to Breeder. Green Turtle Publications. 319 p.

Moe, A.M. 1991. Personal communication.

Morris, R.W. 1956. Some aspects of the problem of rearing marine fishes. Bull. Inst. Oceanog., Monaco 1082, 61 pp.

Mouw, K.W. 1990. Ozone generators, the need for standards and specifications. Jour. Water conditioning and Purification, August: 46-48.

Moyer, J.T. 1976. Geographical variation and social dominance in Japanese populations of the anemonefish, *Amphiprion clarkii*. Japanese Journal of Ichthyology. 23(1):12-22

Moyer, J.T. 1980. Influence of temperature waters on the behavior of the tropical anemonefish *Amphiprion clarkii* at Miyake-Jima, Japan. Bull. Mar. Sci. 30:261-272.

Moyer, J.T. & L.J. Bell. 1976. Reproductive behavior of the anemonefish Amphiprion clarkii at Miyake-jima, Japan. Japan Jour. Icthyol. 23(1):23-32.

Muir, J.F. & R.J. Roberts. 1988. Recirculated water systems in aquaculture. Recent Advances in Aquaculture Vol. 3 Westview Press Boulder CO.

Muller, R.K., K. Bernhard, F. Kienzle, H. Mayer & A. Ruttimann. 1980. Some recent advances in the synthesis of natural carotenoids. Food Chem. 5:15-45.

Murphy, J.P. & R.I. Lipper. 1970. BOD production of Channel Catfish. Progressive Fish Culturist, 32:195-198.

National Research Council. 1977. Nutrient requirements of warmwater fishes. National Academy of Sciences, Wash. D.C. pp. 13-16.

Nicolaides, N. & A.N. Woodall. 1962. Impaired pigmentation in chinook salmon fed diets deficient in essential fatty acids. J. Nutr. 78:431.

Nigrelli, R.F., & H. Vogel. 1963. Spontaneous tuberculosis in fishes and other cold-blooded vertebrates with special reference to *Mycobacterium fortuitum* Cruz from fish and human lesions. Zoologica (New York) 48:131-144.

Nikolsky, G.V. 1963. The ecology of Fishes. Original by Academic Press, London 1963. Reprinted by T.F.H. Publications in 1978, Neptune City, N.J. pp 352.

Noga, E.J., M.J. Dykstra & J.F. Wright. 1989. Chronic inflammatory cells with epithelial characteristics in teleost fishes. Veterinary Pathology 26:429-437.

O'Kelly, J.C. 1974. Inorganic nutrients. In W.P.D. Stewart, ed., Algal Physiology and Biochemistry. Univ. Calif. Press, Berkeley, pp 610-635.

Okey, R.W. & O.E. Alberston. 1989a. Diffusion's role in regulating and masking temperature effects in fixed-film nitrification. J. Water Pollution Control Fed. 61:500-509.

Okey, R.W. & O.E. Alberston. 1989b. Evidence of oxygen-limiting conditons during tertiary fixed-film nitrification. J. Water Pollution Control Fed. 61:510-519.

Owsley, D.E. 1988. Ozone for disinfecting hatchery rearing water. Presented at the Bio-Engineering Symposium, Amer. Fish. Soc., Portland OR, Oct. pp 24-28.

Parisot, T.J. & J.N. Wood. 1970. Fish mycobacteriosis (tuberculosis). U.S. Fish and Wildlife Service Fish Disease Leaflet 7.

Parker, D.S., M.P. Lutz, and A.M. Pratt. 1990. New Trickling Filter Applications in the U.S.A. Water Science Tech. 22:215-226.

Qasim, S.Z. 1955. Rearing experiments on marine teleost larvae and evidence of their need for sleep. Nature. Vol. 175:217-218.

Randall, J.E. 1955. Fishes of the Gilbert Islands. Atoll Research Bulletin. No. 47:243p.

Randall, J.E. 1961. A contribution to the biology of the convict surgeon fish of the Hawaiian Islands, *Acanthurus triostegus sandvicensis*. Pac. Sci. 25(2): 215-272.

Rice, B.J. & J. Overton eds. 1993. Nutrient requirements of fish. National Academy Press, Wash. DC. 114 pp.

Richards, T. & D. Reinhart. 1989. Evaluation of plastic media in trickling filters. Journal WPCF, Vol. 58(7):774-783.

Richards, W.J. & B.J. Palko. 1969. Methods used to rear thread herring, *Opisthonema oglinum*, from fertilized eggs. Trans. Amer. Fish. Soc. 98:527-529.

Roberts E.R. 1987. Photoperiod / temperature control in the commercial production of red drum (*Sciaenops ocellatus*) eggs. Manual on Red Drum Aquaculture Texas Ag. Ext Ser., Texas A&M Univ. Setion II:10-26.

Rodriguez, J. 1993. Considerations for the use of ultraviolet in fish culture. In: Techniques for Modern Aquaculture. Proc. Aqua. Engineering Conf., Pub. American Soc. Agricultural Engineers. St. Joseph Mich. pp 481-485

Roselund, B. 1974. Disinfection of hatchery effluent by ozonation and the effect of ozonated water on rainbow trout. In: W.J. Biogoslawski and R.G. Rice, eds. Aquatic Applications of Ozone, International Ozone Inst., Syracuse, 226 pp.

Rosenthal, H. & G. Otte. 1979. Adaptation of a recirculating fish culture system to salinity change. Biologische Anstalt Helgoland, Zentrale, Hamburg. pp 203-206.

Ross. R.M. 1977. Reproductive behavior of the anemonefish, *Amphiprion melanopus* on Guam. Marine Laboratory, Univ. of Guam. Exact journal title unknown. pp 103-107.

Rottmann, R.W. & J.V. Shireman. 1990. Hatchery manual for grass carp and other riverine cyprinids. Coop. Extensive Service, Univ. of Florida, Inst of Food and Agricultural Sci. 27 pages.

Schumann, G.O. 1969. Rearing marine larvae. Salt Water Aquarium magazine. 5 (5):131-141.

Shelbourne, J.E. 1964. The artificial propagation of marine fish. Advances in Marine Biology 2:1-83.

Shepherd C.J. & N.R. Bromage. 1988. Intensive Fish Farming. Blackwell Scientific, BSP Professional Books, London, pp 404.

Siegrist, H. & Gujer, W. 1987. Demonstration of mass transfer and pH effects in a nitrifying biofilm. Water Res. 21:1481-1487.

Siddall, S.E. 1979. The culture of marine fish larvae. Freshwater and Marine Magazine. Nov. Issue.

Simpson, K.L. & Kamata, T. 1979. Use of carotinoids in fish feeds. Finfish Nutrition and Finfish Technology, Vol. 2 pp 415-424. Berlin, Satz and
Druck: H. Heeneman GmbH.

Sinnhuber, R.O., J.D. Oastell & D.J. Lee. 1972. Essential fatty acid requirements of the rainbow trout, *Salmo gairdneri*. Fed. Proc. 31:1436-1441.

Spotte, S., 1979. Fish and Invertebrate Culture: Water management in closed systems. 2nd edition. Wiley, New York, 2nd ed. 179 pp.

Spotte, S. 1992. Captive Seawater Fishes, Science and Technology. Wiley-Interscience, John Wiley & Sons Inc. 942 p.

Spotte, S. 1979. Seawater Aquariums, The Captive Environment. John Wiley & Sons Inc. New York, N.Y. 413 p.

Steven, D.M. 1949. Carotenoid in the reproductive cycle of the brown trout. Jour. Exp. Biol. 26:295-303.

Sutterlin, A.M., C.Y. Couterier & T. Devereaux. 1984. A recirculation system using ozone for the culture of Atlantic salmon. Prog. Fish Cult. 46:239-244.

Tacon, A.G.J. 1981. Speculative review of possible carotenoid function in fish. Progressive Fish Culturist. 43:205-208.

Tanaka Y. 1978. Comparative biochemical studies on carotenoids in aquatic animals. Mem. Fac. Fish., Kagoshima Univ. 27:255-422.

Takeuchi, K. 1960. The behavior of carotenoid and distribution of xanthophores during development of the medaka (*Oryzias latipes*). Embriologia. 5:100-177.

Takeuchi, K. 1968. Specificity of carotenoid transfer in the larval medaka, (*Oryzias latipes*). Jour. Cell Physiology. 72:43-48.

Takeuchi, T. & T. Watanabe. 1978. Growth enhancing effect of cuttlefish liver oil and short-necked clam oil on rainbow trout and their effective components. Bull. Jap. Soc. Sci. Fish 44:733-738.

Taniguchi, A. 1978. Reproduction and life histories of the tintinnid ciliates (Review). Bull. Plankton Soc. Japan. 25(2): 123-134. Thune, R.L. & T.E. Schwedler. 1991. Fish Health Management in Recirculating Systems. Design of High Density Recirculating Aquaculture Systems. Workshop Proceedings, Louisiana Sea Grant, Baton Rouge. pp 48-54.

Tave, D. 1993. Genetics for fish hatchery managers. Second Edition. Pub. Van Nostrand Reinhold, New York. pp 415.

Torrissen, O.J., R.W. Hardy & K.D. Shearer. 1989. Pigmentation of salmonids-carotenoid deposition and metabolism. Rev. Aquat, Sci. 1:209-225.

Trosclair, C., 1994. Industry, feds and states differ on approaches to fish transportation. Water Farming Journal Vol. 8:7 pp 5-10.

Tullock, J. 1992. Clownfish Host Anemones. Marine Fish Monthly Vol. 7, Issues 1,2 &3.

Turk, C.T. 1981. Unpublished data from a conference between Aqualife Research, Sea World Research and Instant Ocean Hatcheries, January 27, 1981.

Turk, C.T. 1987. Feeding marine fish, good nutrition is critical for health and color. Pet Age Magazine, July issue. pp 32-42.

Young, F.A. 1984. Raising clownfish for the hobbyist. Freshwater and Marine Aquarium magazine. Sept. 9:70-72.

Yu, T.C. & R.O. Sinnhuber. 1979. Effect of dietary *w*3 and *w*6 fatty acids on growth and feed conversion efficiency of coho salmon (*Oncorhynchus kisutch*).

Valenti, R.J. 1975. Semi-closed system culture of the northern puffer, *Sphoeroides maculatus*. Proc. World Mariculture Soc. 6:479-485.

Verwey, J. 1930. Coral reef studies. I. The symbiosis between damselfishes and sea anemones in Batavia Bay. Treubia, 12(3-4): 303-366.

Wedemeyer, G.A., N.C. Nelson & W.T. Yasutaka. 1979. Potentials and limits for the use of ozone as a fish disease control agent. Ozone: Science and Engineering 1:295-318.

Wellborn, T.L. 1987. Catfish Farmers Handbook. Cooperative Extension Service, Mississippi State University. Report 134, 35 pp.

Wheaton, F.W. 1977. Aquacultural Engineering. John Wiley and Sons, New York, 708 pp.

Wimberly, D.M. 1990. Development and Evaluation of a Low-Density Media Biofiltration Unit for Use in Recirculating Fish Culture Systems. Masters Thesis, Louisiana State University, Baton Rouge, Louisiana.

Ziegler, A.C. 1960. Annotated list of Pycogonida collected near Bolinas, California. The Veliger, 3(1): 19-22.

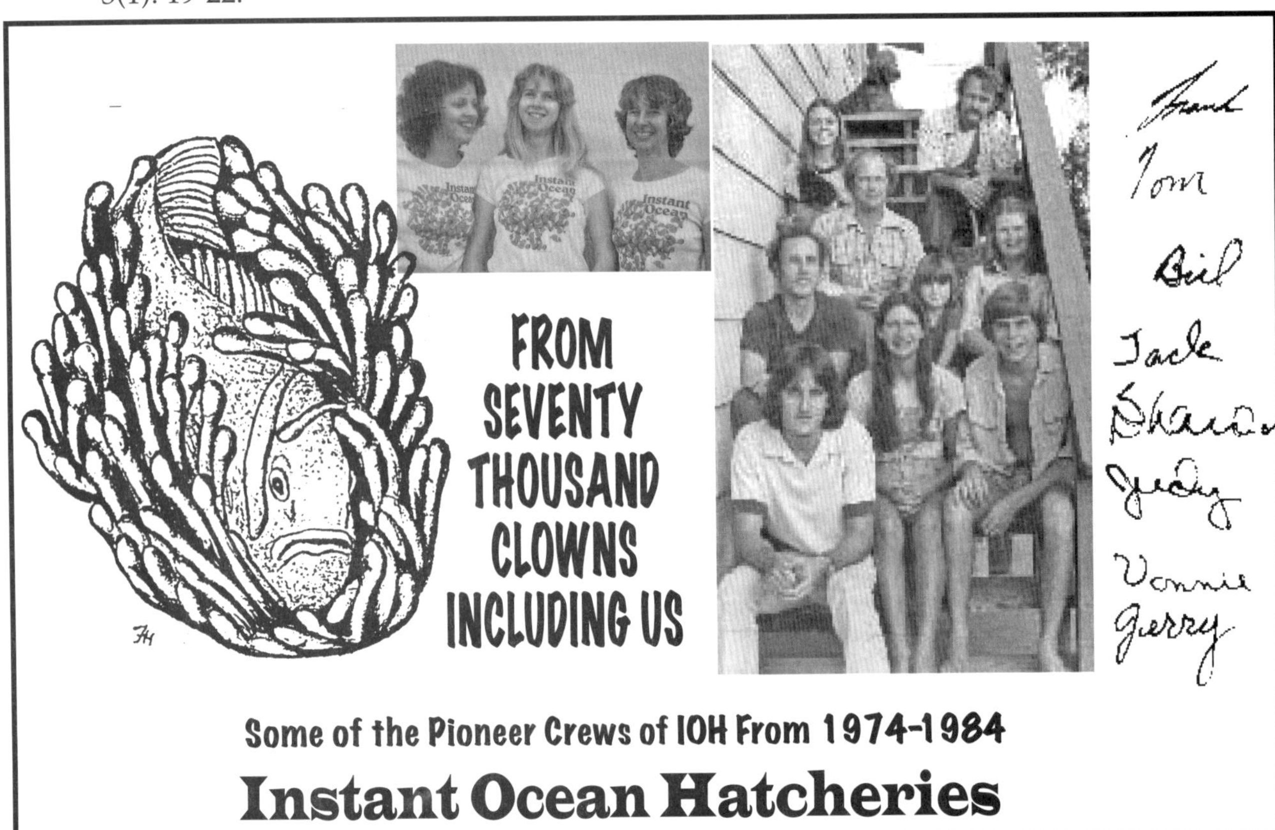

Girls on left Rose Torintino, Judy Johnson and Nicki. Top deck Frank's dog Sassy, Right Frank Hoff, left Marsha Madill, middle William Hoff, below left Tom Frakes, right Sharon Autman and her daughter, middle right against rail Jack Madill and left Tim Barfield. Not shown is Vonnie DeRosure, Lin Corella, King Wen Sun, Bob Presley and Big Jerry.

Conversion Tables

Conversion of Salinities in ppt To Densities

Salinity	Density	Salinity	Density
1.1 ppt	1,0000	23.0 ppt	1.0168
2.0	1.0007	24.1	1.0176
3.0	1.0015	25.0	1.0183
4.1	1.0023	26.0	1.0191
5.0	1.0030	27.1	1.0199
6.0	1.0038	28.0	1.0206
7.1	1.0046	29.0	1.0214
8.0	1.0053	30.1	1.0222
9.0	1.0061	31.0	1.0229
10.1	1.0069	32.0	1.0237
11.0	1.0076	33.1	1.0245
12.0	1.0084	34.0	1.0252
13.1	1.0092	35.0	1.0260
14.0	1.0099	36.0	1.0268
15.0	1.0107	37.1	1.0276
16.0	1.0114	38.0	1.0283
17.0	1.0122	39.0	1.0291
18.0	1.0130	40.1	1.0299
19.0	1.0137	41.0	1.0306
20.0	1.0145	42.0	1.0316
21.0	1.0153	43.0	1.0330
22.0	1.0160		

Major Ions in Seawater at 35ppt

ION	Chemical Name	g/kg^{-1}
Cl	Chloride	19.354
SO^2	Sulfate	2.712
Br	Bromine	0.0673
F	Iron	0.0013
B	Boran	0.0045
Na	Sodium	10.77
Mg^2	Mangnesium	1.290
Ca^2	Calcium	0.4121
K	Potassium	0.399
Sr^2	Strontium	0.0079

Ionic Composition of Seawater at 35ppt

Element	Chemical Species	Name	Molar	$\mu g/L$
H	H2O	Water	55	$1.1x10^8$
He	He (gas)	Helium	$1.7x10^{-9}$	$6.8x10^{-3}$
Li	Li^+	Lithium	$2.6x10^{-5}$	180
Be	$BeOH^+$	Berium	$6.3x10^{-5}$	$5.6x10^{-3}$
B	$B(OH)_3$, $B(OH)_4$	Boron	$4.1x10^{-4}$	4440
C	HCO_3, CO_3, CO_2	Carbon	$2.3x10^{-3}$	$2.8x10^4$
N	N^2, NO_3 NO_2, NH_4	Nitrogen	$1.07x10^{-2}$	1.5x
O	H2O, O2	Oxygen	55	8.8x108
F	F-, MgF+	Flourine	6.8x10-5	1.3x103
Ne	Ne (gas)	Neon	7x10-9	1.2x10-1
Na	Na+	Sodium	4.68x10-1	10.77x106
Mg	Mg2+	Magnesium	5.32x10-2	12.9x105
Al	Al(OH)-4	Aluminum	7.4x10-8	2
Si	Si(OH)4	Silica	7.1x10-5	2x106
P	HPO24-, PO34-, H2PO3	Potassium	2x10-6	60
S	SO24-, NaSO-4	Sulphur	2.82x10-2	9.05x105
Cl	Cl-	Chlorine	5.46x10-1	18.8x106
Ar	Ar (gas)	Argon	1.1x10-7	4.3
K	K+	Potassium	1.02x10-2	3.8x105
Ca	Ca2+	Calcium	1.02x10-2	4.12x105
Sc	Sc(OH)3		1.3x10-11	6x10-4
Ti	Ti(OH)4		2x10-8	1
V	H2VO4-, HVO24		5x10-8	2.5
Cr	Cr(OH)3, CrO24-		5.7x10-9	0.3

Metric Conversions

Cubic Foot (ft^3)
- 28.32 L
- 28.317 ml or cc
- 7.48 gallons
- 62.43 lb of water
- 957.5 fl oz

Cup
- 8 fl oz
- 1/2 pt
- 237 ml

Gallon (gal)
- 3.75 L
- 3.785.4 ml or cc
- 128 fl oz
- 0.13 ft^3
- 133.52 oz of water
- 8.35 lb of water

Gram (g)
- 0.035 oz
- 1 ml or cc of water
- 1,000 mg
- 1,000,000 μg

Kilogram (kg)
- 2.20 lb
- 35.27 oz
- 1.0 L of water

Liter (L)
- 33.82 fluid oz.
- 1.057 qt
- 0.26 gal
- 1 kg of water
- 2.20 lb of water
- 1000 ml or cc

Teaspoon (tsp)
- 4.93 ml or cc
- 1/3 tbsp
- 1/6 fl oz

Tablespoon (tbsp)
- 14.79 ml or cc
- 3 tsp
- 1/2 fl oz

Ounce (weight)
- 28.35 gm
- 0.063 lb
- 0.96 fl oz of water

Ounce (fluid)
- 29.57 gm of water
- 29.57 ml or cc
- 1.043 oz of water
- 1/8 cup
- 6 tsp
- 2 tbsp

Part per million (ppm)
- 1 mg/L of water
- 3.78 mg/gal
- 3.78 g/1,000 gal of water
- 0.13 oz/1,000 gal of water
- 1 oz/1,000 ft^3

Pint (pt)
- 473.17 ml or cc
- 1/2 qt
- 16 fl oz
- 1/8 gal
- 1.04 lb of water

Quart (qt)
- 946.34 ml or cc
- 32 fl oz
- 4 cups
- 2 pt
- 1/4 gal
- 2.09 lb

Meter (m)
- 39.37 in
- 100 cm
- 1000 mm
- 1,000,000 μm

Cubic Meter (m^3)
- 35.31 ft^3
- 1.308 yd^3
- 264.2 gallons
- 1000 L
- 1234 acre-feet

Abbreviations in table above:

Cubic centimeter= cc	Meter= m
Cubic inch= in^3	Milligram= mg
Cubic yard= yd^3	Milliliter= ml
Fluid Ounce= fl oz	Square meter= m^2
Pound = lb	Square yard= yd^2
Ounce = oz	Square foot = ft^2

Liquid conversions, TSB/TSP to ML & Dram (DM) to fluid ounce.

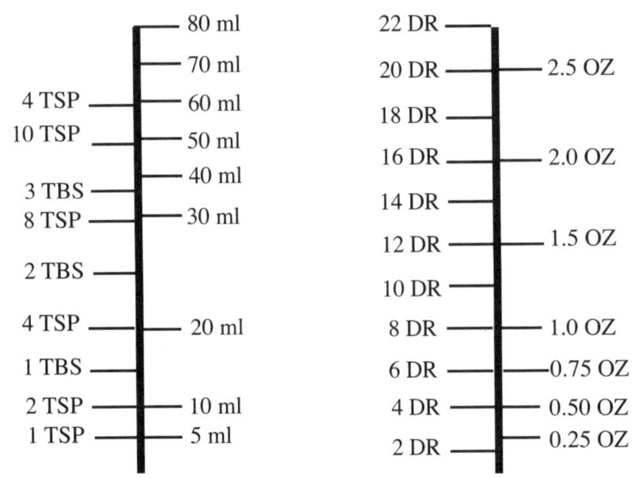

Temperature Conversions

°C	°F
0	32.0
1	33.8
2	35.6
3	37.4
4	39.2
5	41.0
6	42.8
7	44.6
8	46.4
9	48.2
10	50.0
11	51.8
12	53.6
13	55.4
14	57.2
15	59.0
16	60.8
17	62.6
18	64.4
19	66.2
20	68.0
21	69.8
22	71.6
23	73.4
24	75.2
25	77.0
26	78.8
27	80.6
28	82.5
29	84.3
30	86.0
31	87.8
32	89.6
33	91.4
34	93.2
35	95.0
36	96.8
37	98.6
38	100.4
39	102.2
40	104
41	105.8
42	107.6

$$°C = \frac{°F - 32}{1.8}$$

$$°F = 1.8 \times °C + 32$$

Calculating Pond Volumes

Need for calculating pond volumes

1.) Knowing your pond volume will help to determine how many fish you can keep.
2.) Help to determine partial water exchanges.
3.) Essential in determining levels of conditioners, and medication treatments.
4) Determination of flow rates and size of UV filters.
5.) Determination of flow rates and size of biological filters.
6.) Determination of flow rates and size of particle filters.

Ways to measure water volume while filling a pond

1.) The easiest method is use of a water meter while pond is being filled.
2.) Another easy way is to calculate the time it takes to fill a 5 gallon (20L) bucket with a garden hose with a constant flow rate. If you are on a well system water pressure will vary so it would be best to take 5-8 samples at the various pressures and average them. Use this time figure as a standard rate for 5 gallons and record the time in seconds that it takes to fill your pond using the same garden hose. Use the following formula to determine the total volume of your pond.

Volume of pond (gallons or liters) = Time (seconds) to fill pond X Volume (gal or L) of bucket / Time required to fill bucket in seconds

Example: Volume = Total time to fill tank 770 seconds X 5 gal divided by 20 seconds
Volume = 192.5 gallons

Ways to measure water volume of filled ponds

1.) Square and rectangular ponds with the same depth over the entire pond.
Length(L) X Width(W) X Depth(D) (inch/feet or cm/meters) X 0.785 = Vol. of Pond

Example: 9'(2.74 m) L X 6'(1.82 m) W X 2'(0.6 m) = 108 cubic feet or 3 cubic meters

2.) Square and rectangular ponds with irregular depth over the entire pond. In this case at least 10 samples of depth should be taken and averaged.
Length X Width X Average Depth (inch/feet or cm/meters) X 0.785 = Volume

3.) Round ponds with the same diameter from top to bottom and the same depth over the entire pond.
Diameter X Depth (in inch/feet or cm/meters) x 0.785 = Volume

4.) Round ponds where the top and bottom diameters are not the same but depth is the same over the entire pond. T = Top, B = Bottom
Diameter (T) X Diameter (B) X Depth (inch/feet or cm/meters) x 0.785 = Volume

5.) Round ponds where the top and bottom diameters may or may not vary but depth varies. Determine an average depth by taking at least 10 samples and use formula 3 or 4.

6.) Irregular ponds are complicated. The best and most accurate way is to measuring the amount of water needed to completely fill them.

Conversion numbers.

1.) Cubic feet from inches ....Total inches divided by 1728 = cubic feet
2.) Cubic feet to gallons............Cubic Feet x 7.48 = U.S. Gallons
3.) Cubic meters to liters...........Cubic meter x 1000 = total liters
4.) Cubic feet to cubic meter..... Cubic meter = 35.31 cubic feet

Oxygen Saturation ppm O_2

O_2 Saturation in Saltwater

°C	°F	20 ppt	30 ppt
10	50	9.9	9.0
15	59	8.9	8.1
20	68	8.1	7.4
22	72	7.8	7.1
24	75	7.5	6.9
26	79	7.2	6.6
28	83	7.0	6.4
30	86	6.8	6.2
35	95	6.6	5.8

O_2 Saturation in Freshwater

°C	°F	Sea Level	2,000 ft (600 M)
10	50	11.3	10.5
15	59	10.1	9.4
20	68	9.1	8.4
22	72	8.7	8.1
24	75	8.4	7.8
26	79	8.1	7.5
28	83	7.8	7.3
30	86	7.5	7.0

Amount of Chemical, at 1 ppm with 100% Active Ingredient

2.72 lbs/acre foot
1,233 grams/acre foot
0.0283 grams/cu foot
0.0000624 lbs/cu foot
0.0038 grams/gallon
0.00584 grains per gallon
1 milligram per liter
0.001 gram per liter
8.34 lbs/million gal of water

Estimation Of Fish Weight

(Number of fingerlings) X (Initial weight of fingerlings) = Initial total fingerling weight.

(Total pounds of feed fed) divided by 1.75 (approximate conversion ratio) = pounds of gain

(Initial total fingerling weight) + (pounds of gain) = Current weight of fish.

(Current weight of fish) divided by (Number of fish = Average current weight of fish.

Estimating Total Fish Biomass & Feed Levels

Fish Size	Avg. #/lb	Est Wt /K lbs	Protein	Feed Size	Est Lbs Feed/1K
1"	1,428.0	0.7 lbs	36%	1/16"	0.021/K
2"	322.6	3.1 lbs	36%	1/16"	0.093/K
3"	113.6	8.8 lbs	36%	1/8"	0.264/K
4"	52.3	19.1 lbs	36%	1/8"	0.573/K
5"	28.3	35.3 lbs	36%	3/16"	1.059/K
6"	17.0	58.8 lbs	36%	3/16"	1.764/K
7"	10.9	91.0 lbs	36%	3/16"	2.730/K
8"	7.5	133.3 lbs	32%	1/4"	3.999/K

Estimated quantity is based on water temperature of 75° F and a feeding rate of 3% of the biomass. Data based on catfish. (Gold Kist) Index: K = 1000

Conversion for ppm in Proporation & Percent

ppm	proportion	percent
0.1	1:10,000,000	0.000010
0.25	1:4,000,000	0.000025
0.5	1:2,000,000	0.00005
1.0	1:1,000,000	0.0001
2.0	1:500,000	0.0002
3.0	1:333,333	0.0003
4.0	1:250,000	0.0004
5.0	1:200,000	0.0005
8.4	1:119,047	0.0008
10.0	1:100,000	0.001
15.0	1:66,667	0.0015
20.0	1:50,000	0.002
25.0	1:40,000	0.0025
50.0	1:20,000	0.005
100.0	1:10,000	0.01
200.0	1:5,000	0.02
250.0	1:4,000	0.025
500.0	1:2,000	0.05

Calculating Amount of Chemical Needed

Amt. Chem = V x C.F. x ppm x 100% Active Ingredient

V = Determine the volume of the unit to be treated.
C.F. = Conversion factor is that weight of chemical used to give one ppm in one unit volume of water. (table left)
ppm = Concentration desired of chemical
% A.I. = 100 divided by the percent of active ingredient of the chemical.
Example: masoten 80% active needed 0.5 ppm active ingredient in a pond 100' x 100' x 4' deep.
V = L x W x Depth = cu ft of pond or tank
Divide by 43,560 cubic feet to convert to large numbers into acre-feet.

$$0.92 \times 2.72 \text{ lb} \times 0.5 \text{ ppm} \times \frac{100}{80\%} = 1.55 \text{ lbs}$$

If you use cu ft use .0000624 lb conversion number from tbe table on left etc.

FRICTIONAL & PRESSURE RELATIONSHIPS

Friction Loss of Water in Plastic Pipe Schedule 80

Loss are given in Velocity (Vel) measured in ft/sec and pressure (ft of head) per 100 ft of pipe

	1/2" pipe		3/4" pipe		1" pipe		1 1/4" pipe		1 1/2" pipe		2" pipe		3" pipe		4" pipe	
GPM	Vel	Loss	Vel	Loss	Vel	Loss	Vel	Loss	Vel	Loss	Vel	Loss	Vel	Loss	Vel	Loss
2	2.74	6.72	1.48	1.51												
4	5.48	24.2	2.97	5.45	1.79	1.54	1.00	0.39	0.73	0.18						
6	8.23	51.2	4.45	11.5	2.68	3.34	1.50	0.82	1.09	0.38	0.65	0.11				
8	11.0	86.9	5.94	19.6	3.57	5.69	2.00	1.39	1.45	0.64	0.87	0.18	0.61	0.08		
10	13.7	132.0	7.42	29.6	4.46	8.60	2.50	2.10	1.82	0.96	1.09	0.28	0.49	0.04		
12			8.91	41.5	5.36	12.0	3.00	2.94	2.18	1.35	1.30	0.34	0.57	0.06		
15			11.1	62.7	6.7	22.9	3.76	4.45	2.72	2.04	1.63	0.59	0.73	0.08		
18			13.4	87.9	8.03	25.5	4.50	6.25	3.27	2.86	1.96	0.82	0.87	0.12		
20			14.5	107	8.92	30.9	5.00	7.57	3.63	3.47	2.17	0.99	0.97	0.14	0.56	0.04
25					11.2	58.8	6.25	11.4	4.55	5.25	2.71	1.51	1.21	0.21	0.70	0.06
30					13.4	65.3	7.50	16.0	5.45	7.38	3.26	2.11	1.44	0.30	0.84	0.08
35					15.6	86.9	8.75	21.3	6.38	9.78	3.80	2.81	1.70	0.40	0.97	0.10
40					17.9	111	10.0	27.3	7.26	12.5	4.35	3.59	1.94	0.51	1.12	0.13
45							11.2	33.9	8.26	15.6	4.89	4.46	2.18	0.63	1.25	0.16
50							12.5	41.3	9.08	18.9	5.43	5.41	2.42	0.77	1.40	0.20
55							13.7	49.2	10.0	32.0	5.98	6.44	2.67	0.91	1.53	0.24
60							15.0	57.8	10.9	26.5	6.32	7.61	2.92	1.07	1.67	0.28
65							16.1	67.0	11.8	30.7	7.06	8.84	3.14	1.25	1.81	0.32
70							17.5	77.1	12.7	35.3	7.61	10.1	3.39	1.43	1.95	0.37
75							18.8	87.4	13.6	40.1	8.15	11.5	3.64	1.62	2.08	0.42
80							20.0	98.2	14.5	45.2	8.69	12.9	3.88	1.83	2.23	0.48
85							21.2	110	15.4	50.3	9.03	14.5	4.10	2.04	2.34	0.53
90							22.5	122	16.3	55.9	9.78	16.1	4.33	2.27	2.51	0.59
95									17.2	62.0	10.3	17.8	4.57	2.51	2.64	0.65
100									18.2	68.2	10.9	19.6	4.85	2.76	2.79	0.72
110									20.0	81.3	12.0	23.4	5.33	3.29	3.07	0.86
120									21.8	95.4	13.0	27.4	5.80	3.87	3.35	1.00
130									23.6	111	14.1	31.8	6.30	4.48	3.63	1.16
140									25.4	127	15.2	36.5	6.80	5.12	3.91	1.33
150											16.3	41.5	7.27	5.87	4.19	1.52

These values are typical of clean, used plastic pipe and seawater. Assumptions are: water temperature of 70°F (21°C), kinematic viscosity = 1.1 x $10^{-5}ft^2$/sec and equivalent sand roughness = 4 x 10-5 ft, SCH 80 PVC pipe. Since SCH 80 pipe has relatively smaller inside diameter a more conservative value for SCH 40 pipe or thinner walled pipe.

ENERGY RELATIONSHIPS

Conversion for Units of Energy

From	To BTU	To Joule	To Foot Pound
BTU	1	1055	778
Joule	0.0009478	1	0.7376
Foot Pound	0.001285	1.3558	1

BTU = British Thermal Unit. A unit of heat equal to 252 calories , or the quantity of heat required to raise the temperature of one pound of water from 62°F to 63° F.

JOULE = A unit of electrical energy or work equivalent to the work done to raise one coulomb of electricity one volt, or in maintaining for one second a current of one ampere against a resistance of one ohm.

FOOT POUND = a unit of work equal to the amount of energy required to raise a weight of one pound a distance of one foot.

PRESSURE RELATIONSHIPS

Converting Water Pressure (PSI) to Feet Head

Pounds/IN^2	Feet Head	Pounds/IN^2	Feet Head
1	2.31	100	230.90
2	4.62	110	253.91
3	6.93	120	277.07
4	9.24	130	300.16
5	11.54	140	323.25
6	13.85	150	346.34
7	16.16	160	369.43
8	18.47	170	392.52
9	20.78	180	415.61
10	23.09	200	461.78
15	34.63	250	577.24
20	46.18	300	692.69
25	57.72	350	808.13
30	69.27	400	922.58
40	92.36	500	1154.48
50	115.45	600	1385.39
60	138.54	700	1616.30
70	161.63	800	1847.20
80	184.72	900	2078.10
90	207.81	1000	2309.00

One Pound of pressure per square inch of water equals 2,309 feet of water at 62° Fahrenheit.

Converting Feet Head of Water to PSI

Feet Head	Pounds/IN^2	Feet Head	Pounds/IN^2
1	0.43	100	43.31
2	0.87	110	47.64
3	31.30	120	51.97
4	1.73	130	56.30
5	2.17	140	60.63
6	2.60	150	64.96
7	3.03	160	69.29
8	3.46	170	73.63
9	3.90	180	77.96
10	4.33	200	86.62
15	6.50	250	108.27
20	8.66	300	129.93
25	10.83	350	151.58
30	12.99	400	173.24
40	17.32	500	216.55
50	21.65	600	259.85
60	25.99	700	303.16
70	30.32	800	346.47
80	34.65	900	389.78
90	38.98	1000	433.00

One pound of water at 62° Fahrenheit equals 0.433 pounds pressure per square inch.

Inches of Water (IN/H_2O) x 0.07353 = IN./Hg
Inches of Mercury (IN/Hg) x 13.6 = IN/H_2O
Inches of Water (IN/H_2O) x 0.036 = PSI
Feet of Water (FT/H2O) x 0.433 = PSI
Inches of Mercury (IN/Hg) x 0.490 = PSI

Centimeters of Mercury (Cm/Hg) x 0.193 = PSI
Millimeters of Mercury (Mm/Hg) x 0.193 = PSI
Microns of Mercury (μ/Hg) x 0.000019337 = IN/Hg
Microns of Mercury (μ/Hg) x 0.00003937 = IN/Hg

Conversion For Pressure Equivalents (Item X Number)

FROM to	POUNDS/IN^2	FEET WATER	IN. MERCURY	ATMOSPHERES
Pounds/IN^2	1.0	2.309	2.036	0.068
Feet Water	0.433	1.0	0.882	0.0295
Inches Mercury	0.491	1.134	1.0	0.033
Atmospheres (atm)	14.7	33.9	29.92	1.0
Kilopascals (kPa)	0.145	0.335	0.295	0.010

Inches Mercury = a unit of pressure as measured by a manometer equal to the pressure balanced by the weight of a one inch column of mercury in the instrument.
Atmosphere = the weight of the atmosphere per square inch of surface; the pressure of 14.659 pounds per square inch exerted in all directions at sea level by the atmosphere.
Kilopascal = 1,000 pascals = a unit of force equal to one Newton per square meter (N/m^2); typically used as pascal second (Pa - s), or 10 poise designating absolute viscosity.
Bar = a metric unit of measure often used, equal to 100 kilopascal (kPa) or 14.5 lb/in.^{2}